Cross-sectoral Policy Developments in Forestry

Cross-sectoral Policy Developments in Forestry

Edited by

Yves C. Dubé

Forestry Department
Food and Agriculture Organization of the United Nations, Rome, Italy

and

Franz Schmithüsen

Department of Environmental Sciences
Swiss Federal Institute of Technology Zurich, Switzerland

Published by
Food and Agriculture Organization of the United Nations

www.cabi.org

CABI is a trading name of CAB International

CABI Head Office
Nosworthy Way
Wallingford
Oxfordshire OX10 8DE
UK
Tel: +44 (0)1491 832111
Fax: +44 (0)1491 833508
E-mail: cabi@cabi.org
Website: www.cabi.org

CABI North American Office
875 Massachusetts Avenue
7th Floor
Cambridge, MA 02139
USA
Tel: +1 617 395 4056
Fax: +1 617 354 6875
E-mail: cabi-nao@cabi.org

A catalogue record for this book is available from the British Library, London, UK.

Library of Congress Cataloging-in-Publication Data

Cross-sectoral policy developments in forestry/edited by Y.C. Dubé and F. Schimithüsen.
p. cm
Includes index.
ISBN 978-1-84593-250-3 (alk.paper)
1. Forests and forestry. 2. Forest policy. 3. Forest management.
I. Dubé, Yves C. II. Schimithüsen, Franz. III. Title.

SD373. C665 2007
634.9--dc22.

2006039238

ISBN-13: 978 1 84593 250 3

Food and Agriculture Organization of the United Nations (FAO),
Viale delle Terme di Caracalla, 00100 Rome, Italy
Tel: +39 06 57051, Fax +39 06 57053152, Email: webmaster@fao.org
Website: http://www.fao.org

Typeset by SPi, Pondicherry, India.
Printed and bound in the UK by Biddles, Kings Lynn.

Contents

Contributors ix
Preface xiii
Introduction xv
Franz Schmithüsen and Yves C. Dubé

I GLOBAL ASPECTS
1 **Cross-sectoral Tropical Forest Cover Impacts: What Matters?** 1
Sven Wunder and Ahmad Dermawan
2 **The Role of Decentralization in Integrating Cross-sectoral Demands on Forests** 14
Stefanie Engel and Charles Palmer
3 **Wood-product Trade and Policy Issues** 24
Osamu Hashiramoto
4 **Ecosystem Approaches to Sustainable Forest Management: Changing Realities** 36
Jeffrey A. Sayer, Stewart Maginnis and Simon Rietbergen
5 **Agriculture and Forestry: Enhancing Cross-sectoral Policy Planning** 45
Randy Stringer
6 **Policy Impact of the Kyoto Protocol on Sustainable Forest Management** 52
Dieter Schoene
7 **The United Nations Convention to Combat Desertification: a Global Framework for Cross-sectoral Policy Coordination Addressing Sustainable Forest Management** 58
Camilla Nordheim-Larsen, Eleonora Canigiani and Paule Hérodote
8 **Bringing Forest Ecosystems into National Income Accounts: an Environmental Accounting Approach** 64
Glenn-Marie Lange
9 **The Model Forest Experience in Using Broad-based Partnerships for Sustainable Forest Management** 71
Peter Besseau, Washington Alvarado, Ramon Calumbay, Purificación Daloos, Rosalie Imperial, Lisa Jones, Cyprain Jum, Christa Mooney, Johan Svensson, Genrikh P. Telitsyn and Lourdes Wagan

II REGIONAL AND COUNTRY ASPECTS

Africa

10 The World Agroforestry Centre's Experience in Cross-sectoral Policy Planning in Africa 82
Jan G. Laarman

11 Cross-sectoral Policy-planning Issues in Forest Management: Lessons Learnt from Two Case Studies in Ethiopia 89
Samuel Gebre-Selassie

12 Gender and Off-reserve Forest Management in the High Forest of Ghana 95
Elizabeth Ardayfio-Schandorf

13 Multiple-stakeholder Policy-planning Processes: Lessons from Community-based Projects in Mozambique 103
Boaventura M. Cau

14 Enhancing Cross-sectoral Linkages in Nigerian Forestry 110
Sylvester A. Okonofua

15 External Policy Impacts on Miombo Forest Development in Tanzania 117
Gerald C. Monela and Jumanne M. Abdallah

Asia

16 China Environmental and Economic Accounts for Forestry: Case Studies of Jiaohe County and Fengman District in Jilin Province 126
Guangcui Dai and Dong Huang

17 Non-timber Forest Product Linkages in the Tribal Economy of the Western Ghats Region of Karnataka, India 133
Mogenahalli R. Girish and Mamatha Girish

18 Road Construction and Population Effects on Forest Conversion in Rubber Villages in Sumatra, Indonesia 140
Motoe Miyamoto

19 Community Forest Management in the Terai Region of Nepal: Contribution to the Local and National Economy 147
Maheshwar Dhakal and Misa Masuda

20 Cross-sector Linkages in Northern Thailand Mountain Development 153
Mingsarn Kaosa-ard and Jeff Rutherford

Europe

21 Public Policy Impacts on European Forest Sector Development 163
Franz Schmithüsen, Volker Sasse and Carsten Thoroe

22 Comparative Analysis of Framing the 'Forest Sector': Case Studies from Austria and the Netherlands 174
Evelien Verbij, Esther Turnhout and Heiner Schanz

23 Cross-sectoral Linkages between Forestry and other Sectors in Romania 183
Ioan V. Abrudan

24 Importance of the Turkish Forestry Sector in the National Economy: an Input–Output Analysis 190
Mustafa F. Türker

Americas

25 **The Experience of the US Forest Sector in Cross-sectoral Impact Analysis** 195
Denise C. Ingram, Ralph J. Alig and Karen L. Abt

26 **Cross-sector Challenges for Public Land Management in the Western Canadian Sedimentary Basin** 204
Marian L. Weber

27 **Urbanization and Timberland-use Change in the Southern USA** 212
Rao V. Nagubadi and Daowei Zhang

28 **Towards Joint Transboundary Watershed Management in Patagonia: Lessons from Argentina and Chile** 226
Samuel J. Francke-Campaña

29 **Environmental Policy and Law Influencing Forest Management Practices in Chile** 231
Enrique Gallardo Gallardo

30 **Cross-sectoral Linkages in Mexican Forestry** 237
Víctor Sosa Cedillo

31 **Major Policy Changes in National Forest Management in the USA, 1975–2005** 248
Dennis C. Le Master and James B. Snow

Index 260

Contributors

Jumanne M. Abdallah, *Sokoine University of Agriculture, Morogoro, Tanzania, email: jumanne_mushi@yahoo.com*

Ioan V. Abrudan, *Dean, Transilvania University of Brasov – Faculty of Silviculture and Forest Engineering, 1, Sirul Beethoven, 2200-Brasov, Romania, email: abrudan@unitbv.ro*

Karen L. Abt, *US Forest Service, Department of Agriculture, 201 Fourteenth Street, SW, Washington DC 20250, USA, email: kabt@fs.fed.us*

Ralph J. Alig, *US Forest Service, Department of Agriculture, 201 Fourteenth Street, SW, Washington DC 20250, USA, email: ralig@fs.fed.us*

Washington Alvarado, *Gerente General, Bosque Modelo Araucarias del Alto Malleco, O'Higgins 990, Lonquimay, IX Region, Chile, email: wservet@hotmail.com*

Elizabeth Ardayfio-Schandorf, *Professor, Department of Geography and Resource Development, University of Ghana, Legon, Ghana, email: ardayfel@ug.edu.gh*

Peter Besseau, *Executive Director, International Model Forest Network Secretariat, PO Box 8500, Ottawa, Canada K1G 3H9, email: pbesseau@idrc.ca*

Ramon Calumbay, *Department of Environment and Natural Resources, Region 8, Sto. Niño Extension, Tacloban City, Leyte, Philippines 6500, email: rncalumbay1457@yahoo.com*

Eleonora Canigiani, *The Global Mechanism of the UNCCD, IFAD, Via del Serafico 107, 00142 Rome, Italy, email: e.canigiani@ifad.org*

Boaventura M. Cau, *Department of Geography, Eduardo Mondlane University, PO Box 257, Maputo, Mozambique, email: caubm@yahoo.com.br*

Víctor Sosa Cedillo, *Senior Forestry Consultant, Mexico, Nicolas San Juan, 968-501, Col. Del Valle, CP 03100, Mexico, email: sosacedillo@yahoo.com.mx*

Guangcui Dai, *China National Forestry Economics and Development Research Center, State Forestry Administration, Beijing, China, email: daigc@forestry.gov.cn*

Purificación Daloos, *Department of Environment and Natural Resources, Region 8, Sto. Niño Extension, Tacloban City, Leyte, Philippines 6500, email: psdaloos@yahoo.com*

Ahmad Dermawan, *CIFOR, Bogor, Indonesia, email: a.dermawan@cgiar.org*

Maheshwar Dhakal, *Graduate School of Life and Environmental Sciences, Laboratory of International Forest Resources Economy, University of Tsukuba, Japan, email: Maheshwar_dhakal@hotmail.com*

Yves C. Dubé, *Forestry Officer (Planning), FAO Forestry Department, Rome, Italy, email: yves.dube@fao.org*

Stefanie Engel, *Professor, Environmental Policy and Economics, Swiss Federal Institute of Technology (ETH), Zurich, Switzerland, email: stefanie.engel@env.ethz.ch*

Samuel J. Francke-Campaña, *Forest Engineer, Corporación Nacional Forestal, CONAF, Paseo Bulnes 259, Santiago de Chile, email: sfrancke@conaf.cl*

Enrique Gallardo Gallardo, *Abogado, Corporación Nacional Forestal, CONAF, Paseo Bulnes 259, Santiago de Chile, email: egallard@conaf.cl*

Samuel Gebre-Selassie, *Director, Ethiopian Economic Policy Research Institute, PO Box 34282, Addis Ababa, Ethiopia, email: eea@telecom.net.et*

Mamatha Girish, *Agricultural Economist, # 103 LIG, 6th Cross, KHB Colony, II Stage, Basaveshwaranagar, Bangalore 560 079, Karnataka State, India, email: mamathagirish@yahoo.com*

Mogenahalli R. Girish, *Assistant Professor, AICRP-CS-ECF Scheme, First floor, DE's Office Building, University of Agricultural Sciences (UAS), Hebbal, Bangalore 560 024, Karnataka State, India, email: mrgirish2000@yahoo.co.in*

Osamu Hashiramoto, *Forestry Officer (Wood Products Trade), FAO Forestry Department, Rome, Italy, email: osamu.hashiramoto@fao.org*

Paule Hérodote, *The Global Mechanism of the UNCCD, IFAD, Via del Serafico 107, 00142 Rome, Italy, email: p.herodote@ifad.org*

Dong Huang, *China National Forestry Economics and Development Research Center, State Forestry Administration, Beijing, China, email: fmhd@163.com*

Rosalie Imperial, *Department of Environment and Natural Resources, Region 8, Sto. Niño Extension, Tacloban City, Leyte, Philippines 6500, email: pbesseau@idrc.ca*

Denise C. Ingram, *Policy Analysis, US Forest Service, Department of Agriculture, 201 Fourteenth Street, SW, Washington DC 20250, USA, email: cdingram@fs.fed.us*

Lisa Jones, *Communications and Extension Officer, Foothills Model Forest, 1176 Switzer Drive, Hinton, Alberta T7V 1X6, USA, email: lisa.jones@gov.ab.ca*

Cyprain Jum, *CIFRO, PO Box 2008, Yaoundé, Cameroun, email: c.jum@cgiar.org*

Mingsarn Kaosa-ard, *Social Research Institute, Chiang Mai University, Thailand, email: mingsarn@chiangmai.ac.th*

Jan G. Laarman, *Deputy Director General (Programmes), International Centre for Research in Agroforestry, United Nations Avenue, PO Box 30677-00100, Nairobi, Kenya, email: J.Laarman@cgiar.org*

Glenn-Marie Lange, *Senior Research Scholar, The Earth Institute at Columbia University, Hogan Hall Room B-15, 2910 Broadway, NY 10027, USA, email: gl2134@columbia.edu*

Stewart Maginnis, *Head, Forest Conservation Programme, The World Conservation Union (IUCN), 28 Rue Mauverney, CH-1196 Gland, Switzerland, email: stewart.maginnis@iucn.org*

Dennis C. Le Master, *Professor Emeritus, Department of Forestry and Natural Resources, Purdue University, West Lafayette, IN 47907, USA, email: dclmstr@comcast.net*

Misa Masuda, *Graduate School of Life and Environmental Sciences, Laboratory of International Forest Resources Economy, University of Tsukuba, Japan, email: masuda@sakura.cc.tsukuba.ac.jp*

Motoe Miyamoto, *Forestry and Forest Products Research Institute, 1 Matsunosato, Tsukuba, Ibaraki, 305-8687, Japan, email: motoe@ffpri.affrc.go.jp*

Gerald C. Monela, *Sokoine University of Agriculture, Morogoro, Tanzania, email: monela@suanet.ac.tz*

Christa Mooney, *Communications Officer, International Model Forest Network Secretariat, PO Box 8500, Ottawa, Canada K1G 3H9, email: cmooney@idrc.ca*

Rao V. Nagubadi, *Post-doctoral Fellow, School of Forestry and Wildlife Sciences, Auburn University, AL 36849-5418, USA, email: nagubve@auburn.edu*

Camilla Nordheim-Larsen, *Financial Strategy Officer, The Global Mechanism of the UNCCD, IFAD, Via del Serafico 107, 00142 Rome, Italy, email: c.nordheim-larsen@ifad.org*

Sylvester A. Okonofua, *Assistant Director of Forestry, Federal Department of Forestry, Federal Ministry of Environment, P.M.B. 135, Area Eleven, Garki, Abuja, Nigeria, email: aroboiokonofua@yahoo.com*

Charles Palmer, *Senior Researcher, Environmental Policy and Economics, Swiss Federal Institute of Technology (ETH), Zurich, Switzerland, email: charles.palmer@env.ethz.ch*

Simon Rietbergen, *Acting Coordinator, Ecosystem Management Programme, IUCN – The World Conservation Union, 28 rue Mauverney, CH-1196 Gland, Switzerland, email: simon.rietbergen@iucn.org*

Jeff Rutherford, *Social Research Institute, Chiang Mai University, Thailand, email: tjeffrutherford@hotmail.com*

Volker Sasse, *Forestry Officer, FAO Regional Office, Budapest, Hungary, email: volker.sasse@fao.org*

Jeffrey A. Sayer, *Senior Associate, Forests for Life Programme, WWF-International, Avenue du Mont Blanc, CH-1196 Gland, Switzerland, email: jsayer@wwfint.org*

Heiner Schanz, *Professor, Forest and Environmental Policy, University of Freiburg, Germany, email: heiner.schanz@ifp.uni-freiburg.de*

Franz Schmithüsen, *Professor Emeritus, Forest Policy and Forest Economics, Swiss Federal Institute of Technology (ETH), Zurich, Switzerland, email: franz.schmithuesen@env.ethz.ch*

Dieter Schoene, *Senior Forestry Officer (Climate Change), FAO Forestry Department, Rome, Italy, email: dieter.schoene@fao.org*

James B. Snow, *JD, Special Council for Real Property, US Department of Agriculture, Washington DC, USA, email: james.snow@usda.gov*

Randy Stringer, *Professor, School of Agriculture, Food and Wine, University of Adelaide, Waite Road, Urrbrae, Australia 5064, email: randy.stringer@adelaide.edu.au*

Johan Svensson, *Swedish Forest Agency, Skedom 107, SE-881 92 Solleftea, Sweden, email: johan.svensson@skogsstyrelsen.se*

Genrikh P. Telitsyn, *Supervisor of Science Projects, Gassinski Model Forest Association, Kalinin Str. 106, Troitskoe Nanaiski Rayon, 682350 Khabarovski Krai, Russia, email: ikp2@forest.kht.ru*

Carsten Thoroe, *Professor, Director, Institute for Economics, Federal Research Centre for Forestry and Forest Products, Leuschnerstrasse 91, 21031 Hamburg, Germany, email: c.thoroe@holz.uni-hamburg.de*

Mustafa F. Türker, *Professor, Karadeniz Technical University, Turkey, email: mft@ktu.edu.tr*

Esther Turnhout, *Assistant Professor, Forest and Nature Conservation Policy Group, Wageningen University, Wageningen, The Netherlands, email: ester.turnhout@wur.nl*

Evelien Verbij, *Forest and Nature Conservation Policy Group, Wageningen University, PO Box 47, 6700 AA Wageningen, The Netherlands, email: evelien.verbij@wur.nl*

Lourdes Wagan, *Department of Environment and Natural Resources Building, Visaya Avenue, Diliman, Quezon City, Philippines 1100, email: ludycw@edsamail.com.ph; Ludycw@yahoo.com*

Marian L. Weber, *Alberta Research Council, 250 Karl Clark Road, Edmonton, Alberta T6N 1E4, email: Marian.Weber@arc.ab.ca*

Sven Wunder, *CIFOR, Belém, Brazil, email: s.wunder@cgiar.org*

Daowei Zhang, *Professor, School of Forestry and Wildlife Sciences, Auburn University, AL 36849–5418, USA, email: zhangdw@auburn.edu*

Preface

As an essential resource for economic and social development, forests call for consideration in several policy domains. Forest protection and forestry practices are closely linked, for example, to public policies that address climate change, biodiversity, water management and agriculture. Public policy linkages across sectors are therefore essential not only for advancement of sustainable forest management, but for sustainable development in general.

This publication has been prepared to improve awareness and understanding of potential positive and negative impacts of policies outside the forest sector on sustainable forest management, and to lead to more harmonized policies among forestry administrations and other sectoral agencies and stakeholders. It was inspired by a technical session on cross-sectoral linkages in forestry organized by the Food and Agriculture Organization of the United Nations (FAO) and the Swiss Federal Institute of Technology (ETH) at the 22nd World Congress of the International Union of Forest Research Organizations (IUFRO) in Brisbane, Australia, in 2005. We have selected a number of papers presented at the session for inclusion in this publication. Additional invited contributions have been requested to broaden and substantiate the theme.

Emerging from the overall picture is a need for improved policy planning and coordination across several political levels and economic sectors, as well as a call for regional and country networks with a view to creating and sharing information and knowledge on best practices.

The publication is intended to inform a large audience concerned about the fate of forests and the maintenance of the earth's natural resources: interested citizens, landowners and land users, policy-makers and representatives of non-governmental organizations, researchers, teaching staff and students. It should thus contribute to a better understanding of the need to promote policy and institutional changes in order to develop more integrated and effective development approaches at the country and local levels.

We hope that this book contributes to further enhancing cross-sectoral policy planning, which is indispensable if the United Nations Millennium Development Goals of reducing hunger and poverty and protecting the environment are to be reached.

Y.C. Dubé
F. Schmithüsen

Introduction

Franz Schmithüsen and Yves C. Dubé

The international dialogue on forests following the United Nations Conference on Environment and Development in 1992, and the World Summit on Sustainable Development in Johannesburg in 2002, highlights the importance of cross-sectoral public policy linkages in the advancement of sustainable forest management. The decisions taken at the World Summit on Sustainable Development place forests in the context of sustainable development as a substantial part of the natural resource base for economic and social advancement. They acknowledge the multiple and varying productive functions and environmental services of forests for poverty alleviation, as raw material and energy sources, and as natural habitats. The implementation plan highlights the role of forests in several policy domains and points to the fact that forest protection and forestry practices are closely linked to public policies that address measures on climate change, biodiversity and the institutional framework for sustainable development.

Sustainable forest management is identified as a critical means for significantly reducing deforestation, halting the loss of forest biodiversity, and avoiding land and resource degradation. Its role in improving food security and access to safe drinking water and affordable energy is specifically mentioned. Sustainable forest management provides multiple benefits of both natural and planted forests and trees, and contributes to the well-being of the planet and of humanity. Its achievement, nationally and globally, through partnerships between governments and a large number of stakeholders, including the private sector, indigenous and local communities and non-governmental organizations (NGOs), is an essential goal in maintaining the natural resource base.

An expression of the multilevel cross-sectoral policy context in which forests are placed is the emerging international forest regime, which is based on five pillars (Schmithüsen, 2000; FAO, 2003; Dubé *et al.*, 2006):

1. International legal instruments such as conventions, agreements and declarations directly or indirectly addressing forests and forestry.
2. Worldwide political processes involving governments, non-governmental organizations, the private sector and indigenous and local communities within the United Nations system.
3. Regional forest-related initiatives currently taking place in several continents and developing their own forest and forest sector political agenda.
4. Criteria and indicator processes that provide ecoregion-specific standards for sustainable forest management certification systems.
5. National forest programme processes, which are largely based on stakeholder agreement and make it possible to concretize international commitments by individual countries.

The goals of good governance strongly determine policy, legal and institutional reforms. Participation in decision-making, access to information, transparency and accountability are starting to influence policy and legislative planning in many sectors. Institutional frameworks are being changed to reflect policies promoting local decision-making and community-based initiatives in a wide variety of fields. Decentralization is becoming important as a strategy for reform of the public sector. It also becomes evident that maintaining the natural resource base and managing it in a sustainable manner cannot be tackled by one single public policy domain or one body of specific legislation alone. Effective solutions for most societal problems have to be found through the different agencies, policy actors and stakeholders concerned. Effective coordination among the goals and instruments set out in different public policy domains has to be achieved. In forestry, water management, fisheries, land use, wildlife and other areas, the emphasis is on moving away from exclusive state competencies to stronger management responsibilities and property rights vested in private individuals and companies, subnational government bodies and local cooperatives and municipalities.

This book provides a rich and multifaceted documentation of current progress being made in creating the political, economic and social conditions indispensable for sustainable and multifunctional use of forest resources, and notes the obstacles that need to be removed in order to reach this goal. The first part introduces general and global aspects that have to be considered in the context of cross-sectoral policy coordination. This includes discussions on the impact of external shocks such as a sudden oil price increase on forest management, the impact of energy or trade policies on global wood markets and the role of decentralization in integrating multiple demands on forests. Policy changes induced by international agreements such as the Kyoto Protocol and the Convention to Combat Desertification, and the implications for new approaches in ecosystem management versus sustainable forest production, are equally of importance. Other relevant subjects are common as well as divergent interests between agriculture and forestry in sustainable rural development, experiences with partnerships for sustainable forest management among different stakeholders, and the introduction of more comprehensive national accounting systems to demonstrate the links between decision-making in the forest sector and national development planning.

The second part of the book deals with regional, national and local issues of cross-sectoral policy linkages. The chapters on *Africa* focus largely on the improvement of land management practices such as agroforestry, land tenure and gender issues, more integrative policies in promoting reforestation and afforestation, multiple-stakeholder planning processes and external policy impacts in protecting and managing Miombo forests. In *Asia*, important subjects appearing in several chapters are the need to develop environmental and economic accounts for forestry, and to demonstrate more clearly the great importance of non-timber forest-product linkages, road construction and population effects on forest conversion, community forest management contributions to the local and national economy, and cross-sectoral policy links in the development of mountainous areas are other issues addressed. In *Europe*, both environmental problems as well as strong trends towards developing a competitive forest and wood-processing sector determine public policy development to a considerable extent. This can be seen from leading policy scenarios that have been identified, from the changes in perceptions with regard to the forest sector and from the policy issues at national level that are presented.

A somewhat similar pattern of competing policy demands between resource use, industrial expansion and strong environmental demands emerges from the contributions dealing with the *Americas*. This refers to the experiences of the USA in cross-sectoral impact analysis, the lessons to be learned from the long and conflictual history in managing the national forests, as well as from the resource conflicts described between forestry uses, oil and gas development and environmental protection in the boreal regions of Canada. Challenges in improving the prevailing public policy networks exist in the Mexican forestry sector, in Chile due to strong environmental policy demands on forest management practices, and with regard to establishing more effective transboundary watershed management systems undertaken in Patagonia.

The overall picture emerging from such diverse and complex presentations is a mosaic of global, regional and country situations determined very much by local socio-economic and ecological conditions. This calls for improved policy planning and coordination across several political levels and economic sectors. It also calls for the establishment of regional and country networks with a view to sharing and creating information and knowledge on best practices. Multiple efforts, both from the public and private sector, are required in order to maintain and develop forests as a local, national and worldwide resource that can be managed for sustainable production of wood as well as for a great variety of other goods and services, and for preserving a habitable environment on our planet.

References

Dubé, Y.C., Lange, G.-M. and Schmithüsen, F. (2006) Cross-sectoral policy linkages and environmental accounting in forestry. *Journal of Sustainable Forestry* 23(3), 47–66.

FAO (Food and Agriculture Organization of the United Nations) (2003) Cross-sectoral policy impacts between forestry and other sectors. Forestry Paper 142. Rome, Italy.

Schmithüsen, F. (2000) The expanding framework of law and public policies governing sustainable uses and management in European forests. In: Schmithüsen, F., Herbst, P. and Le Master, D.C. (eds) *Forging a new Framework for Sustainable Forestry – Recent Developments in European Forest Law. IUFRO World Series* 10, 1–27. IUFRO Secretariat, Vienna.

1 Cross-sectoral Tropical Forest Cover Impacts: What Matters?

Sven Wunder and Ahmad Dermawan

Introduction

Most research on the importance of cross-sectoral impacts focuses on deforestation, which constitutes the most important threat to tropical forests. While deforestation happens in all three tropical continents, its sources and the driving forces behind it differ across regions. The growth of cattle ranching in Latin America since the 1960s has been the key driving force of deforestation in that region. In South-east Asia, logging activities for the harvesting of rich timber resources have opened up forest land for conversion into cash or estate crops, particularly in Malaysia and Indonesia. In Central Africa and South Asia, the swidden production system for staple food crops has been the leading factor behind deforestation. What drives tropical deforestation is usually increasing demand for land rather than merely an increasing demand for timber. It is also becoming increasingly clear that deforestation happens less because of single biased policy incentives spearheading forest loss (Repetto and Gillis, 1988) than because, in production terms, it proves profitable for land users to convert the land to a variety of alternative uses (Kaimowitz, 2002).

The aim of this chapter is to present key issues on the impacts of macro-level factors and cross-sectoral influences on tropical forest cover. The term 'cross-sectoral' refers to factors outside of forests and forestry that have a significant effect on forests. This chapter does not look at the reverse effect, such as the impact of timber harvesting and rents on society-wide governance structures (Ross, 2001). Nor does it consider specific degradation factors – for example, pollution impacts from industries or the alleged impact of high energy consumption on global warming and forest biodiversity. Our main concern here is with forest cover change equivalent to the definition[1] of deforestation by the Food and Agriculture Organization of the United Nations (FAO). This directs particular attention to the spatial transition zone between tropical forests and converted land uses or forest margins. What are the economic and policy factors that accelerate frontier expansion at these land use margins, and which ones tend to slow it down?

The chapter draws on a decade of research at the Center for International Forestry Research (CIFOR) on the underlying causes of deforestation, supplemented with selected empirical results from other sources. It first discusses the impacts of a resource boom (oil bonanza) as an example of impacts from an extra-sectoral shock. The next section summarizes other cross-sectoral influences as described in the corresponding literature, while

[1]A 'forest' is defined as land spanning more than 0.5 ha with trees higher than 5 m and a canopy cover of more than 10% (or trees able to reach these thresholds *in situ*), but excluding all land that is predominantly under agricultural or urban use (FAO, 2000).

the last section presents conclusions and perspectives. A key result from the assessment is that what happens to tropical forests is in most cases much more determined by events outside the forest arena than by what happens inside the forest sector. In other words, cross-sectoral impacts will often be more important than the new forest law, the participatory tree-planting project or the environmental education programme that is implemented at the forest margins.

Resource Boom

The impact of a resource boom – oil or mineral export revenues accruing as a windfall to tropical commodity producers – provides a case study of macroeconomic impacts on deforestation, which is interesting for us for two reasons. First, using the data from the *State of the World's Forests* (FAO, 1995), an update of the *Global Forest Resources Assessment* (FAO, 1990, 2000), it was found that oil-rich and mineral-rich countries in the tropics on average retain a greater share of forest cover. For example, the size of forest cover in oil- and mineral-exporting countries was 38.5% in 1995, and increased to 47.7% in 2000. They also tend to lose these remaining forests at a slower pace, even after third factors in the respective multiple regression models are controlled for (Mainardi, 1998; Sunderlin and Wunder, 2000). Second, due to the high volatility of world market oil prices, these countries have faced extreme macroeconomic instability, characterized by boom and bust cycles, making them 'macroeconomic laboratories' that allow also a retrospective study of extra-sectoral impacts on forests.

With this pattern in mind, Wunder (2003) carried out national case studies of the impacts of oil and mineral booms in eight specialized oil and mineral exporters.[2] The core reason is that a large influx of foreign exchange from mineral exports comes to these countries, allowing higher government spending levels to attract urbanization. At the same time, an increased real exchange rate makes both agriculture and timber extraction less competitive than in non-mineral countries. This underdevelopment of agriculture and forestry has a protective impact on forests, especially if the accompanying policies have a potential impact on forest conservation. Hence, the development path of these countries becomes more urban-based. The urban population still consumes resources that leave an 'ecological footprint' on forests, but peri-urban cultivation systems tend to be more land-intensive than those practised by a rural-based population. In most cases, urbanization is thus unambiguously good for forest conservation.

However, a number of accompanying policy measures can significantly alter this core effect. Using oil money for ambitious road building into forested areas and for other rural development policies can significantly extend land use. Food crop import protection policies can significantly shelter the land-using agricultural sector from the Dutch disease. Partial analysis thus reveals that there are usually effects both to increase and to decrease pressures on forests, the net effect of which depends on the balance between them. An alternative to partial analysis is to use a computable general equilibrium (CGE) model, but this requires a huge amount of information that is normally unavailable in the tropics and there are difficulties involved in dealing with the spatial specificity of deforestation, and the outcome of the model depends to a considerable extent on the assumptions in the CGE parameters. This chapter presents a partial recursive approach instead, since we believe it exhibits greater transparency than the CGE approach. Figure 1.1 provides a stylized overview of the assumed linkages.

The analysis starts from the external level (1) of foreign exchange inflow and continues from the macroeconomic level (2) down to sectoral production (3), land use (4) and forest levels (5). At the external level, oil wealth is often correlated with patterns of foreign borrowing against future oil revenues, together determining the size of major external foreign exchange inflows into the economy. Some of the public spending effects have direct forest implications. When oil money is used to finance large increments of road budgets, resettlement projects in forested areas or subsidies for transport (making it cheaper to extract products from the agricultural frontier) and agriculture, this usually increases pressures on forests. The opposite

[2] The core study covers Gabon, Venezuela, Cameroon, Ecuador and Papua New Guinea. Additional, secondary studies cover Indonesia, Nigeria and Mexico.

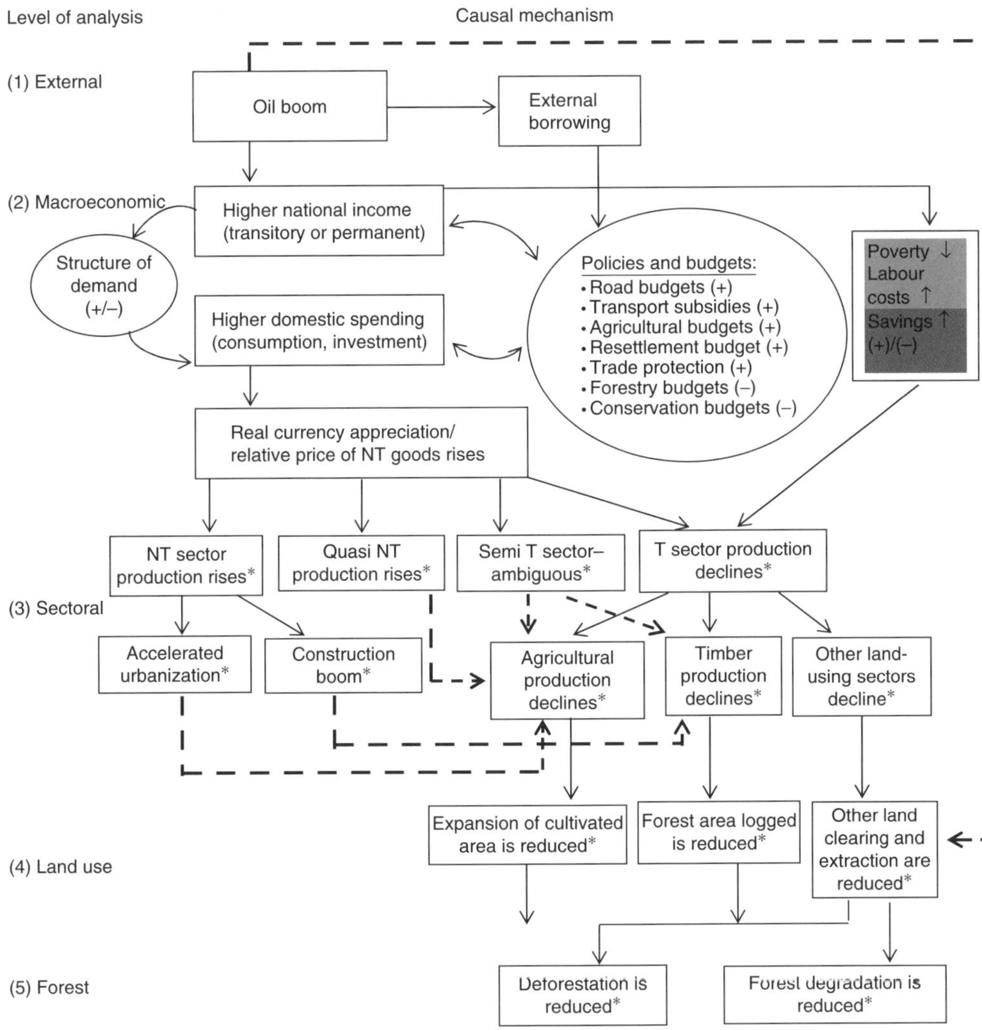

Notes:

T	Traded sector
NT	Non-traded sector
*	Relative to pre-existing trends (growth and structural change)
⟶	Core-hypothesis causality – 'Oil wealth protecting forests' (trend indicated in target box)
- - ->	Counter-hypothetical causality 'Oil wealth eliminates forests' (counteracting trend indicated in target box)
(+)	Factor expected to accelerate forest loss and degradation
(–)	Factor expected to decelerate forest loss and degradation
(+/–)	Factor/factors jointly ambiguous in their effect

Fig. 1.1. Linking resource booms to forest cover outcomes. T = traded sector; NT = non-traded sector; * = relative to pre-existing trends (growth and structural change); →1 = core-hypothesis causality – 'Oil wealth protecting forests' (trend indicated in target box); – →1 = counter-hypothetical causality 'Oil wealth eliminates forests' (counteracting trend indicated in target box); (+) = factor expected to accelerate forest loss and degradation; (–) = factor expected to decelerate forest loss and degradation; (+/–) = factor/factors jointly ambiguous in their effect.

could be the case to the extent that part of the oil wealth is used to finance conservation (e.g. creating new national parks) and strengthening forestry regulation (reducing illegal logging and encroachment).

Returning to the macroeconomic level, higher incomes cause the macroeconomic Dutch disease effect, with real exchange rate appreciation and changes in relative prices. But higher income also shifts the structure of consumption. In some cases, richer consumers buy more meat and dairy products, which tends to increase deforestation for cattle ranching. However, if the increase in income shifts the consumption of land-extensive inferior tubers to more land-intensive and higher-value rice farming, that leads in the opposite direction, as was the case in Central Africa. In addition, reduced poverty in a booming economy goes hand in hand with higher labour costs, which tends to reduce deforestation as a labour-intensive activity. However, there will also be more funds available for investment, which can promote more forest clearing in capital-scarce frontiers. The net effect can differ from case to case.

At the sectoral level (3), Dutch disease causes traded sectors to decline, while non-traded (construction, services) and quasi-traded sectors (industry) rise. If timber production and agriculture are fully traded sectors with perfect foreign substitutes, Dutch disease at the land use level (4) will unambiguously alleviate pressures to log new forest areas and convert forests to agriculture and other traded-sector land uses – at least compared with pre-existing sectoral trends. If these cross-sectoral effects are dominant in their impact on forests (5), deforestation and forest degradation would be reversed or at least reduced. Whether the forest protection effect is proportional to the land use effect depends on specific parameters, such as the number of species extracted per hectare or the extent to which abandoned agricultural areas convert back to forests.

Contrary effects may occur to the extent that logging and agriculture are semi-traded or quasi-traded sectors, for example, through import protection of part of the sector. In that case, greater urbanization may cause a construction boom with higher domestic timber demand or a beef-and-dairy boom accelerating deforestation for domestic ranching. The partial approach does not produce a bottom-line calculus of whether these contrary effects in net terms reverse the forest-protecting effect. Yet it is a flexible framework for pulling together the patchwork of land use factors, in spite of incomplete data coverage and variable spatial scenarios.

From the above, it is already clear that economy-wide extra-sectoral impacts on forests can be significant. However, what about the direct extra-sectoral impact of oil production on forest cover? How much forest is being felled to make room for oil-related infrastructure? In Gabon, most oil production is in the coastal production zone; it affects both forests (moist forests, mangroves) and savannahs. However, several studies have shown that the direct deforestation impacts in this country prove to be very limited, in total approximately 10,000 ha or 0.05% of forest area (Wunder, 2003, chapter 4). In Venezuela, most oil production is onshore, but it is concentrated in savannah areas, and only a minor share comes from deciduous forests and the mangroves of the Orinoco delta. Again, direct deforestation effects are thus negligible. In Cameroon, oil was produced offshore without impacts on forests. Forest impacts of oil were the largest in Ecuador. No less than 99% of Ecuador's oil production comes from the Amazon forest, and although the direct effects have been limited to 3000–6500 ha (0.04–0.09% of the forest in Ecuador's Amazon region), indirect deforestation effects have been much greater (Wunder, 2003, chapter 7). In Papua New Guinea, oil deforestation effects are also negligible. This overview shows that the direct forest cover effects of tropical oil production are clearly subordinate to the 'invisible' economy-wide effects that oil has, compared to the shifting cultivation of plantains, new oil palm plantations or the establishment of new pastures for cattle (Table 1.1).

Next, we will present comparative macroeconomic cycles and deforestation trends to test for the correlation between oil booms and reduced forest pressures (conversely, oil busts and deforestation hikes). There are two levels of potential confirmation of the core hypothesis that oil incomes help protect forests. Oil booms could coincide with a reversal of deforestation in absolute terms (absolute confirmation). Alternatively, deforestation rates could slow down under an oil boom, meaning that other driving factors (e.g. population growth) were still at play, but

Table 1.1. Direct forest effects of petroleum production in selected countries. (From Wunder, 2003.)

Country	On/offshore	In forests	Main effects	Extent/location
Gabon	Mostly onshore	Rainforest, mangroves, savannah	Deforestation, direct	National: up to 10,000 0.05% of forest area over time
			Deforestation, indirect	10,000–15,000 ha
Venezuela	Mostly onshore	Savannah (mostly) some deciduous forest some mangroves	Deforestation, direct	Negligible
			Deforestation, indirect	Historical – Llanos and Maracaibo regions
Cameroon	Offshore	No	None	None
Ecuador	Onshore	99% in rainforest	Deforestation, direct	3000–6500 ha (0.04–0.09% of Amazon forest)
			Deforestation, indirect	Strong road-related timber extraction and colonization
			Degradation	Strong pollution
Papua New Guinea	Mostly onshore	Mostly forest	Deforestation, direct	1200–1300 ha (Kutubu)
			Deforestation, indirect	Negative (clearing reduced)

were being dampened significantly by the extra-sectoral effect (relative confirmation). Table 1.2 provides an overview for the five primary study countries.

Gabon provides an absolute confirmation of the core hypothesis. With its transformation into an oil exporter in the early 1970s, this low-population country became the peak rent economy in our sample. With the total neglect of rural areas and road building, and the massive stimulation of urban employment and rent-seeking, there was an exodus of young people from the countryside, and abandoned agricultural fields grew back into forest. Although we do not have national forest inventories for sub-periods, the two point estimates plus a number of case studies give us quite a solid basis for saying that high oil wealth led to a small absolute net expansion of forest area, in spite of high population growth during the same period.

Venezuela's historical transformation from a specialized agrarian exporter prior to the Second World War showed many similarities with the Gabon story and was an even stronger case confirming our core hypothesis in absolute terms: there was a marked net forest regrowth in abandoned agricultural areas from 1920 to 1950. While Venezuela combined with its neighbours still had quite low forest loss rates during 1950–1980, the trend was an accelerating one. Since the Second World War, Venezuela has thus been a case that confirms the core hypothesis in relative terms, although roads and cattle increasingly emerged as confounding factors.

Cameroon clearly had the most pronounced economic cycles among our country sample. The boom (1979–1985) had a clear urban bias and accelerated rural–urban migration, which led to a slowdown in deforestation. In the crisis period (1986–1994), per capita gross domestic product

Table 1.2. Country macroeconomic cycles versus land use trends. (From Wunder, 2003.)

Country	Macroeconomic cycles	Deforestation cycles	Core hypothesis
Gabon	1960–1973: Pre-boom	1960–1990s: Forest regrowth (short and long run)	Absolute confirmation
	1974–1985: Boom	Recent: Probably peri-urban clearing under mini-crisis years	
	1986–1989: Mini-bust		
	1990s: Fluctuations		
Venezuela	1920/1930s: Oil transformation	1920–1950: Forest regrowth	Pre-Second World War: absolute confirmation
	1956–1958: Mini-boom	1950–1980s: Slow loss	Post-Second World War: relative confirmation
	1974–1983: Boom	1980s/1990s: More rapid loss	
	1984–: Crisis and mini-booms		
Cameroon	1960–1978: Pre-boom	1973–1985: Slow loss	Relative confirmation
	1979–1985: Boom	1986–1994: High loss	
	1986–1994: Bust, fixed CFA	After 1994: Probably high loss	
	1995–: Devaluation, recovery		
Ecuador	1960–1973: Pre-boom	1975–1990: High loss	Rejection
	1974–1981: Boom (rising)	1990s: Probably lower loss	– absolute and relative
	1982–1985: Boom (declining)		
	1986–1995: Bust		
	1996–: Mini-booms		
Papua New Guinea	1972–1994: Mineral boom (rising), fixed overvalued kina	Whole period: Probably stable, low loss	Relative confirmation
	1995–: Oil boom, financial capital outflows, devalued kina	After 1994: Perhaps loss acceleration	

almost halved, especially declining in urban sectors. A significant return migration to rural areas and an upsurge in slash-and-burn food production (notably plantain) was sufficient to cause a strong rise in deforestation in the forest zone. Since the devaluation of 1995, cash crops have experienced a slow revival, and deforestation continues to be high. In contrast to neighbouring Gabon, oil wealth was not high enough to reverse forest loss and cause forest area to expand; rents were simply not high enough to do more than cancel out the effect of other factors, notably rural population growth. However, there was an abrupt reduction in the pace of deforestation. Cameroon thus provides a strong relative confirmation of the core hypothesis.

The case of Ecuador was rather different. The size of the boom was greater than in Cameroon, but much smaller than in Venezuela and Gabon. Due to rising oil production, oil revenues remained quite high until the mid-1980s. A bust followed, combined with a strong political crisis in the 1990s, so although there have been mini-oil bonanzas, the entire post-1985 period is best characterized as an economic downturn cycle. Nevertheless, deforestation actually accelerated during the oil boom, stayed high for a decade after and then probably decelerated. This is because the government deliberately spent a large share of oil revenues on issues that promoted extensive land uses, notably new roads (especially connecting the highlands with the lowlands), but also transport subsidies and land colonization programmes and support. As in Venezuela, a second factor was cattle: a relatively large urban middle class used part of their new purchasing power to buy cattle-derived proteins, thus creating the level of demand to match the supply-side factors that enabled the large-scale extension of pastures. In Ecuador, these contrary factors were stronger than the Dutch disease forest protection ones and policy was decisive in creating that balance. Although there was an upsurge in agriculture after the devaluation, as the core hypothesis would suggest, on the whole this country case performed contrary to the hypothesis, in both absolute and relative terms.

Papua New Guinea became a substantial exporter of minerals from 1972 onwards. Copper and gold predominated until significant petroleum production began in 1992. Strong macroeconomic management succeeded in stabilizing the impact of export revenues until the 1990s, but then fiscal control was increasingly lost. The hard fixed kina exchange rate introduced in 1975 greatly hampered non-mineral exports until 1994, but then a floating kina implied a considerable real devaluation. In other words, the whole period is characterized by fairly high mineral revenues, but with two different economic management cycles. In terms of land use, cash crops never became a large source of deforestation: even after the kina devaluation, rural violence, the impediments of insecure land-tenure arrangements and infrastructure problems were too great. This country is thus an example in which price competitiveness remained one among several key obstacles to cash crop development; only the oil palm sector managed to cause some deforestation. As in Gabon, government policies ignored rural infrastructure. On the national scale, deforestation was driven by the shifting cultivation of food crops. However, the rigid land tenure structure favoured land intensification, so even this source of deforestation remained restricted. Forest loss rates were thus stable and low. On the whole, mineral wealth did not stop deforestation in absolute terms. It does seem to have slowed down forest loss, in particular by curtailing the cash crop sector. However, factors such as land tenure, crime and political instability were at least as important as mineral wealth in shaping land use.

Hence, Table 1.2 exhibits patterns that one would expect from the core hypothesis about oil wealth and forests. Agriculture is generally the prime victim of the Dutch disease, and declines in agricultural land demand tended to be mirrored by decreased forest pressures, and vice versa: generally, forests were the default land cover in most countries. Besides deforestation, many of the effects described in this section are also valid as determinants of logging, being a main source of forest degradation in the five primary study countries. Oil booms tend to slow down the (expansion of) logging, since the timber industry tends to be a traded sector that is hit by the Dutch disease. The findings confirm a strong impact on the competitiveness of logging exports. However, in middle-income countries with strongly expanding domestic timber markets, as in the case of Ecuador, Venezuela and Cameroon, home market demand also rises with urban income and population, because domestic production is fully or partially import-protected. The relative weight of the forest protection Dutch disease effect and the forest-unprotecting urban boom effect varies widely, so that policies to influence extraction levels need to be tailored accordingly (Wunder, 2005).

Other Macroeconomic Effects

In the previous section, we used oil as an extreme example of an extra-sectoral factor working its way from the macroeconomic sphere to the forest. Obviously, there are many other effects, both inside and outside the macroeconomic sphere. In this section, we will take a look at some of these, drawing mainly on a review of some 150 theoretical and empirical models of deforestation (Kaimowitz and Angelsen, 1998). Table 1.3 gives an overview of specifically some other macroeconomic factors affecting forests. All the models that analyse devaluation find that it increases deforestation, i.e. the opposite scenario of the exchange rate appreciation happening under an oil boom. General trade liberalization leads to higher deforestation in all but one case. For foreign debt, the effect is much more dubious, in some cases increasing deforestation, in others having mixed or no significant effect. This is because the impact on forests will ultimately

Table 1.3. Effect of selected open-economy variables on deforestation. (From Kaimowitz and Angelsen, 1998, pp. 64, 85.)

			Independent variables		
Type	Study	Country	Devaluation	Trade liberalization	Foreign debt
CGE	Aune *et al.* (1996)	Tanzania	Increase		
	Barbier and Burgess (1996)	Mexico		Increase	
	Cruz and Repetto (1992)	Philippines	Increase	Increase	
	Lopez (1993)	Ghana		Increase	
	Mwanawina and Sankhayan (1996)	Zambia	Increase		
	Thiele and Wiebelt (1994)	Cameroon		Reduction	
	Wiebelt (1994)	Brazil	Increase		
MCR	Bawa and Dayanandan (1997)				Increase
	Burgess (1991)				Increase
	Capistrano (1990)			Increase	Mixed
	Gullison and Losos (1993)				No effect
	Inman (1993)				Mixed
	Inman (1990)				Mixed
	Kahn and McDonald (1995)				Increase
	Kahn and McDonald (1994)				Increase
	Kant and Redantz (1997)				Increase
	Kimsey (1991)				No effect
	Mainardi (1998)		Increase		Increase
	Rudel and Roper (1997a)				No effect
	Rudel and Roper (1997b)				Mixed
	Rudel and Roper (1996)				Increase
	Shafik (1994a,b)		Increase		No effect

Blank space = variable not analysed in the particular study; MCR = multi-country regression model; CGE = computable general equilibrium model.

depend on the macroeconomic management to any increase in foreign debt.

Effects of price and profitability changes

In most places, deforestation is fairly well explained by expansions in agriculture, which is the largest tropical land use competitor for forests. Hence, a lot of changes happening to forests can be traced to changes in agricultural profitability caused by alterations in prices of inputs, outputs or technology adoption. Higher agricultural prices in most cases stimulate more forest clearing. An increase in agricultural output price will stimulate most farmers to produce more of the given product – unless they are maximizing leisure time and are satisfied with a certain target income. Hence, they tend to increase their income by cultivating more land themselves, or newcomers will be attracted, thus increasing deforestation. On the other hand, an increase in agricultural input prices (e.g. fertilizers, insecticides) will normally reduce the use of such inputs, which may be compensated for by increasing areas under cultivation. Changes in relative prices between agricultural products can also alter the balance between land uses, which affects deforestation. In particular, if farmers produce both land-extensive food crops and land-intensive cash crops, and choose mainly between these two livelihood options in their land use, a rising relative price of food crops over cash crops will tend to cause higher deforestation. In Cameroon, for instance, this dynamic was observed in relation to cocoa production (cash crop) versus tubers and plantains (food crop) (Ndoye and Kaimowitz, 2000). Higher timber prices can also stimulate deforestation, although the evidence is weaker than for agriculture commodities. This happens because higher prices tend to stimulate a more rapid harvesting rate, which indirectly opens up forested areas for conversion, mainly through road building. Higher rural wages, or higher labour opportunity costs in terms of new employment options, will tend to reduce deforestation. This is because forest clearing is a particularly labour-intensive activity, which will suffer from higher labour costs.

How will improved agricultural technologies affect forest cover? This is one of the most heated debates. The so-called Borlaug hypothesis, named after the Nobel Prize laureate Norman Borlaug, states that improved technologies can save the conversion of natural forests and wildlands, since the quantities of agricultural production needed will come from smaller areas. This line of reasoning probably still holds at the global level, although the needs are not fixed and increased welfare also tends to increase land demand. However, at the sub-global scale, the hypothesis has been severely challenged. While the impacts are conflictive, complex and dependent on multiple context factors, one can generally say that most local progress in agricultural technology increases deforestation: just like a price increase, it simply makes production locally more profitable and expandable. It often takes special scenarios – for example, extreme labour scarcity or improved crops in prime agricultural areas out-competing crops in marginal lands – to achieve the opposite result. A main factor is whether the technological change is widely enough adopted to raise market output and drive down the prices of the commodity. If impacts at this scale are achieved, the Borlaug hypothesis may hold.

Transport costs and access to forests

Road building near or in forest areas is the single most important factor causing deforestation. It lowers transport costs for both timber (extracted from the forest) and agricultural products (replacing forest land), so that these commodities can 'pay their way out' to the marketplace. By making viable a series of economic activities and enabling intensive human settlement, roads are thus often the first but decisive step towards forest conversion. Road improvement (e.g. paving of existing roads) has a similar effect of spreading agricultural activities to new spatial zones that were previously forested. However, it also provides higher value to already cleared land, so that intensified land uses here will become more attractive (Andersen *et al.*, 2002). On aggregate, one would expect it to cause higher deforestation, as for instance predicted for the road improvements planned by the Brazilian government in the Amazon. Large energy subsidies (e.g. cheap gasoline) have a similar, though non-spatially specific effect, of making transport cheaper, thus helping make activities in the agricultural frontier cheaper and increase deforestation. Many people think that energy subsidies have the opposite effect, because they promote the shift from wood fuels to cheaper substitutes. While this effect can reduce forest degradation in some places, fuel wood harvesting does not normally cause deforestation, making it an irrelevant argument for most of the land use change debate.

Land tenure

Land tenure security is an ambiguous factor relative to the determination of forest loss. People who have insecure tenure and access rights can only plan for limited time periods and focus more on short-term benefits. If tenure becomes long-term, the benefits are more likely to be captured by the current land user, so that he or she will be inclined to invest in the land. In general (more) secure tenure will thus help the land user adopt solutions that are more profitable in the longer term. In some circumstances, that will favour forest management, but in many cases it will not. It depends on whether forestry is actually the most profitable productive option in the long run – or whether that is cattle ranching, oil palm estates, soybean fields. Depending on the socio-economic context, secure tenure appears to have a more positive effect on tree planting and agroforestry than on natural forest management. Trees take time to grow, so the decision to allocate land almost per definition requires control over the land until harvest. But there is nothing in and of itself to ensure that more secure tenure will lead to more forests in the landscape; examples from Latin America in particular show the opposite, because pastures for cattle ranching often are the most rewarding and convenient land use option in the long run.

One factor to consider is thus what land uses are favoured by secure tenure, but another is the process by which this secure tenure is established in the first place. Forest clearing is often seen as a sign of active occupation – 'the land is being worked' – that discourages others from taking possession. On the other hand, forest land is often seen as 'idle' territory inviting invasion. This means that deforestation often helps establish property rights – homesteading – whether by informal tenure recognition among a group of land-colonizing settlers, or by the process of obtaining formal land tenure through a state agency controlling land that has actively been cleared. Homesteading thus promotes excessive, speculative deforestation – beyond what can be explained by a purely economic rationale. People may clear forests simply to obtain control over the land, regardless of what the most profitable land use option is.

Income and population growth

Long-term trends in income and population growth characterize most tropical countries – how do these trends affect deforestation? Many environmental problems have been found to worsen during the initial stages of economic development before improving again later on, following a U-curve that has been called the Environmental Kuznets Curve. For deforestation, many tests in cross-country regressions (Kaimowitz and Angelsen, 1998) and a few long-term time series assessments have been carried out on this issue. There is mixed evidence, but little sign that tropical developing countries are capable of growing their way out of deforestation. The most stable relationship is that rising income also goes hand in hand with higher forest loss. Even the rate of deforestation does not seem to go down systematically with higher incomes (Culas and Dutta, 2002) – except for special cases such as the oil countries, where the service economy comes to play a dominant role, in the same way as in high-income developed economies. By the same token, poverty and its reduction over time have an ambiguous effect on forest cover. On the one hand, poverty alleviation typically goes along with higher wages, which tends to reduce deforestation. On the other hand, when people become less poor, they start to consume more protein-rich foods such as meat and dairies, which require larger land areas for the necessary calorie production, often promoting deforestation. In this and other ways, they increase their 'ecological footprint' on forests. Often, these effects will dominate the outcome on forest cover.

In ideological terms, it is a matter of heated debate whether population growth has a greater or lesser impact on deforestation than increasing wealth. In the sample presented by Kaimowitz and Angelsen (1998), 19 out of 28 cross-country regression models and 11 out of 14 regional regression models confirm a significant effect of high population growth on accelerating forest loss. Higher population both increases the 'mouths to fill' (food demand) and the 'hands to work the land' (labour supply), both of which will tend to increase pressures on forests. However, the linkage is clearest at more aggregated levels (Palo, 1994). At more local levels, it may well be that, for instance, food is bought from other regions and the work force is specialized in non-agricultural activities that do not demand land conversion. However, through trade their activities may well cause deforestation pressures elsewhere, which will appear when the relationship is compared at higher aggregation levels.

Conclusions and Perspectives

There are few reliable quantitative assessments of how different causal factors comparatively contribute to tropical deforestation. Nevertheless, from many studies at different levels of aggregation, it seems clear that extra-sectoral factors are overwhelmingly dominant in determining what happens to forest cover. The emergence of a new agricultural export opportunity, a crisis in the urban economy or a new road building programme tend to have a much larger impact on forest cover than a new forest code, a log export ban or a new log-extraction technology. Probably the only significant exception to this pattern of extra-sectoral dominance over forest-specific interventions is the creation of protected areas, which have been shown to slow down deforestation significantly (Bruner *et al.*, 2001) – although the question remains of the extent to which this protection is additional or spatially moves clearing pressures to different forest areas.

Macroeconomic effects with cycles of booms and crisis can have significant effects on forest cover. It is not, as is sometimes claimed (Wood *et al.*, 2000), that all these macroeconomic changes concurrently increase the loss of forests and biodiversity – some changes are bad news for forests, others are good news, and the net outcome depends on the balance between them. When these effects are being analysed, it is important to look at the whole economy, and not just at the production pattern of one particular sector, as was demonstrated for the case of petroleum. Agriculture is the great land use competitor for tropical forests, and by far the most deforestation occurs in order to increase farmlands. A great deal of extra-sectoral analysis will thus involve agricultural sector analysis, including livestock activities. Most policies and interventions that favour the expansion of agricultural production also come to decrease forest areas – this is at least true for forest-abundant agricultural frontier areas.

Population and income growth are slowly creeping root factors that expand pressures on forests, although mediated by other factors. Spatially specific interventions, such as road building in particular, can also have a decisive impact on forest loss. Often, different causal factors work together inseparably in what has been called deforestation tandems (Geist and Lambin, 2001) – it is hard or impossible to single out one factor alone. Also, there are different regimes of which driving forces dominate. In some cases, agricultural expansion is driven by an increasing poor population growing food crops with land-extensive swidden cultivation. In others, it is the emergence of new market opportunities that drives the process. In a third set of cases, it is the clashes between these two processes that provide the main impetus.

Notably, even agricultural intensification that increases per-hectare yields can accelerate forest loss. Intensification is still often seen as an area- and forest-saving factor, but that effect is highly context-dependent and scale-dependent. The assumptions certainly hold for widespread intensifying innovations that reduce the total market price through their general supply-boosting effect. Yet, where adoption is limited, innovators increase production but prices remain high, so they will in most cases scale up their now more profitable production. Hence, despite reduced poverty, they will tend to deforest more, rather than less. Almost any agricultural investment in frontier areas with flexible labour supply promotes deforestation. So it is hard to design agricultural programmes in these regions without a negative effect on remaining forests. In providing policy recommendations as to what could be done strictly with the aim of stabilizing the forest margins, we could distinguish between those factors that directly affect land extension through a spatial effect and those that work through the macro-level context.

What does all this mean for policy-making? A politically problematic implication from the case studies on oil-producing countries was that many of the policies there that came to protect forests were at the same time dubious or directly harmful development policies, especially for rural areas. These de facto conservation successes are the result of blind strategies and unintentional side effects from macro-level policies. Conversely, many good development policies (for economic growth and poverty reduction) are bad for forest conservation. The ten-point list below – with nine of the ten items representing extra-sectoral factors – were the de facto forest protection policies that appeared to be particularly influential in this country sample:

1. Neglect rural road building.
2. Sell gasoline at its normal, unsubsidized price.
3. Ignore smallholder agriculture at the expense of agroindustrial white elephants.
4. Stop moving or directing people into forests.
5. Promote urban labour absorption and rural–urban migration through urban spending.
6. Avoid large currency devaluation.
7. Liberalize imports of food crops, meat, dairy and timber products.
8. Increase logging taxes to effectively capture stumpage values.
9. Reduce population growth rates through family-planning programmes.
10. Create a rent-seeking environment where few people find it worthwhile to produce.

For instance, to neglect the rural road network was possibly the most important policy decision for maintaining forest cover, with Papua New

Guinea and Gabon being the main successful examples of maintaining large forest areas by not providing access to them. In contrast, Ecuador and Indonesia were examples of massive investments in rural road building, which also went along with high deforestation scenarios. Of the ten policies listed here, only 2, 7, 8 and 9 would possibly be accepted as good development policies; some (4, 6) are debatable in character, while the others (1, 3, 5, 10) would clearly be seen as perverse development policies that in the long run decrease the welfare of the country's citizens. This illustrates that the trade-offs outweigh the synergies with respect to the extra-sectoral policies promoting forest conservation. On the other hand, in this country sample we still observed some strategies that were probable 'win–win' options for both forests and people: (2) elimination of energy subsidies; (7) selective import liberalization; (8) taxing away above-normal timber rents; and (9) family planning efforts as reducing a driving force behind long-term forest loss. Only one of these four factors was related to the forestry sector itself, the rest being extra-sectoral in nature.

References

Andersen, L.E., Granger, C.W.J., Reis, E.J., Weinhold, D. and Wunder, S. (2002) *The Dynamics of Deforestation in the Brazilian Amazon*. Cambridge University Press, Cambridge.

Aune, J.B., Glomsrod, S., Iversen, V. and Wiig, H. (1996) Structural adjustment and land degradation: a CGE analysis for Tanzania. In: *Structural Adjustment Policies and Environmental Degradation in Tanzania, Zambia, and Ethiopia*. Preliminary Seminar Report, 9–10 January. Agricultural University of Norway, Ås, Norway.

Barbier, E.J. and Burgess, J.C. (1996) Economic analysis of deforestation in Mexico. *Environment and Development Economics* 1, 203–239.

Bawa, K.S. and Dayanandan, S. (1997) Socio-economic factors and tropical deforestation. *Nature* 386, 562–563.

Bruner, A., Gullison, R.E., Rice, R.E. and da Fonseca, G.A.B. (2001) Effectiveness of parks in protecting tropical biodiversity. *Science* 291, 125–128.

Burgess, J.C. (1991) *Economic Analysis of Frontier Agricultural Expansion and Tropical Deforestation*. MS Thesis, Environmental and Resource Economics, University College London.

Capistrano, A.D. (1990). Macroeconomic Influences on Tropical Forest Depletion, a Cross Country Analysis. Ph.D. Dissertation, University of Florida, Gainesville, Florida.

Cruz, W. and Repetto, R. (1992) *The Environmental Effects of Stabilization and Structural Adjustment Programs: The Philippine Case*. World Resource Institute, Washington, DC.

Culas, R. and Dutta, D. (2002) *The Underlying Causes of Deforestation and the Environmental Kuznets Curve: A Cross-Country Analysis*. University of Sydney, Sydney.

FAO (Food and Agriculture Organization of the United Nations) (1990) *Global Forest Resources Assessment 1990*. Food and Agriculture Organization of the United Nations, Rome.

FAO (1995) *State of the World's Forests*. Food and Agriculture Organization of the United Nations, Rome.

FAO (2000) *Global Forest Resources Assessment 2000*. Food and Agriculture Organization of the United Nations, Rome.

Geist, H.J. and Lambin, E.F. (2001) *What Drives Tropical Deforestation? A Meta-Analysis of Proximate and Underlying Causes of Deforestation Based on Subnational Case Study Evidence*. University of Louvain, Louvain-la-Neuve, Belgium.

Gullison, R.E. and Losos, E.C. (1993) The role of foreign debt in deforestation in Latin America. *Conservation Biology* 7, 140–147.

Inman, K. (1990) Social Causes of Deforestation in Thrid World Countries. Master's Thesis, University of Georgia, Athens, Georgia.

Inman, K. (1993) Fueling expansion in the Third World: population, development, debt, and the global decline of forests. *Society and Natural Resources* 6, 17–39.

Kahn, J.R. and McDonald, J.A. (1994) International debt and deforestation. In: Brown, K. and Pearce, D. (eds) *Causes of Tropical Deforestation*. UCL Press, London, pp. 57–67.

Kahn, J.R. and McDonald, J.A. (1995) Third-world debt and tropical deforestation. *Ecological Economics* 12, 107–123.

Kaimowitz, D. (2002) Amazon deforestation revisited. *Latin American Research Review* 37(2), 221–235.

Kaimowitz, D. and Angelsen, A. (1998) *Economic Models of Tropical Deforestation: A Review*. Center for International Forestry Research, Bogor, Indonesia.

Kant, S. and Redantz, A. (1997) An econometric model of tropical deforestation. *Journal of Forestry Economics* 3, 51–86.

Kimsey, M. (1991) An Analysis of Spatial and Temporal Variation in Tropical Deforestation. Ph.D. Dissertation, University of Georgia, Athens, Georgia.

Lopez, R. (1993) *Economy-Wide Policies, Agricultural Productivity, and Environmental Factors: The Case of Ghana*. Workshop on Economy-Wide Policies and the Environment, World Bank, Washington, DC.

Mainardi, S. (1998) An econometric analysis of factors affecting tropical and subtropical deforestation. *Agrekon* 37(1), 23–45.

Mwanawina, I. and Sankhayan, P.L. (1996). A computable general equilibrium model to measure the effects of structural adjustment programmes on ecology and development in Zambia. In: *Structural Adjustment Policies and Environmental Degradation in Tanzania, Zambia, and Ethiopia*. Preliminary Seminar Report, 9–10 January. Agricultural University of Norway, Ås, Norway.

Ndoye, O. and Kaimowitz, D. (2000) Macro-economics, markets, and the humid forests of Cameroon, 1967–1997. *Journal of Modern African Studies* 38(2), 225–253.

Palo, M. (1994) Population and deforestation. In: Brown, K. and Pearce, D.W. (eds) *Causes of Tropical Deforestation*. UCL Press, London, pp. 42–56.

Repetto, R. and Gillis, M. (eds) (1988) *Public Policy and the Misuse of Forest Resources*. Cambridge University Press, Cambridge.

Ross, M.L. (2001) *Timber Booms and Institutional Breakdown in Southeast Asia*. Cambridge University Press, Cambridge.

Rudel, T.K. and Roper, J. (1996) Regional patterns and historical trends in tropical deforestation, 1976–1990: a qualitative comparative analysis. *Ambio* 25, 160–166.

Rudel, T.K. and Roper, J. (1997a) Forest fragmentation in the humid tropics: a cross national analysis. *Singapore Journal of Tropical Geography* 18(1), 99–109.

Rudel, T.K. and Roper, J. (1997b) The paths to rain forest destruction: cross national patterns of tropical deforestation, 1975–1990. *World Development* 25, 53–65.

Shafik, N. (1994a) Macroeconomic causes of deforestation: barking up the wrong tree? In: Brown, K. and Pearce, D. (eds) *Causes of Tropical Deforestation*. UCL Press, London, pp. 86–95.

Shafik, N. (1994b) Economic development and environmental quality: an econometric analysis. *Oxford Economic Papers* 46, 757–773.

Sunderlin, W.D. and Wunder, S. (2000) The influence of mineral exports on the variability of tropical deforestation. *Environmental and Development Economics* 5, 309–332.

Thiele, R. and Wiebelt, M. (1994) Policies to reduce tropical deforestation and degradation: a computable general equilibrium analysis for Cameroon. *Quarterly Journal of Internatioanl Agriculture* 33, 162–178.

Wiebelt, M. (1994) Stopping deforestation in the Amazon: trade-off between ecological and economic targets? *Weltwirtschaftliches Archiv–Review of World Economics* 131, 542–568.

Wood, A., Stedman-Edwards, P. and Mang, J. (eds) (2000) *The Root Causes of Biodiversity Loss*. Earthscan Publication, London.

Wunder, S. (2003) *Oil Wealth and the Fate of the Forest: A Comparative Study of Eight Tropical Countries*. Routledge, London.

Wunder, S. (2005) Macroeconomic change, competitiveness and timber production: a five country comparison. *World Development* 33(1), 65–86.

2 The Role of Decentralization in Integrating Cross-sectoral Demands on Forests

Stefanie Engel and Charles Palmer

Introduction

The integration of cross-sectoral demands on forestry has for some time been a considerable policy challenge (de Montalembert, 1995; World Bank, 2002; FAO, 2003). Non-forestry sector policies frequently affect forest degradation and poverty more than forest policies (Kaimowitz and Angelsen, 1999). Schmithüsen (2003) defines cross-sectoral policy linkages between the forest sector and other sectors as those resulting from public policies that have a direct or indirect influence on the behaviour of landowners, forest users, governmental agencies and non-governmental organizations (NGOs).

This chapter considers three categories of direct demands on forests. First, economic development and commercial demands for timber have increasingly placed considerable pressure on forest resources, for example, through logging, mining and construction of hydroelectric dams (Engel and López, 2004; WRI, 2005). Second, there are demands to utilize forests for the alleviation of poverty. In developing countries, forests play a particularly important and direct role in rural livelihood through, for example, the collection of non-timber forest products (NTFPs) and firewood, and as hunting and grazing grounds. Hundreds of millions of people in the tropics derive a significant proportion of their livelihood from forest products (Sayer, 1998). Third, forests carry out numerous ecological functions at various levels. At the global level, these include carbon sequestration and storage, and biodiversity conservation. Water retention services, erosion prevention, recreational services and scenic beauty are examples of ecological services at a more local or regional level.

These demands sometimes overlap, but frequently also stand in conflict with one another (Kaimowitz and Angelsen, 1999). For example, the livelihood needs of local populations are interlinked with local ecological functions. Similarly, local people can potentially share in the benefits from economic development, for example, by receiving a share of logging profits directly through logging agreements, or indirectly through employment in logging activities. Larger-scale ecological functions could also be potentially translated into local livelihood benefits and economic development through ecotourism or carbon payments. In reality, however, the three types of demand often conflict. For example, the World Resources Institute (WRI) highlights how the livelihood of the poor has increasingly been affected by being in 'direct conflict with extractive industries such as large-scale fishing, logging and mining' (WRI, 2005, p. 4). Commercial logging and mining, as well as the overexploitation of non-timber forest products by local people, often result in forest degradation, and consequently may have a negative impact on forest ecological functions.[1]

This chapter explores the ways in which recent decentralization and empowerment trends[2] have affected the drive towards the integration of different demands on forests. Have these trends made it easier to integrate demands? Cross-sectoral demands are typically studied at an aggregate level, with relatively little attention being given to the ways in which local people might manage such demands and the ways in which incentives might be used to influence local behaviour (Broadhead and Dubé, 2002). This chapter is a contribution to this gap in the literature. The discussion will focus largely on the context of a developing country, but several of the arguments are equally applicable to developed countries. Suggestions are presented on how some of the current problems with decentralization could be mitigated and how the benefits of decentralization for cross-sectoral integration could be enhanced.

Impact of Decentralization on Cross-sectoral Demands on Forests

Potential for integration

The decentralization of forest sectors in many developing countries, coupled with the empowerment of local people, has resulted in local communities being granted some rights – typically use and management rights – over the forest. This has led to people taking more control over the use of the local resource base. Local participation in decision-making may thus result in a greater sense of ownership over these decisions, and thereby in the local establishment of rules regulating resource use (Ostrom, 1990). This should lead to an increased willingness to abide by locally made rules, hence increasing their effectiveness (Carney, 1995). Moreover, local communities can draw on informal sanctioning mechanisms, which can represent a cost-effective alternative to government control (McKean, 1995). By contrast, distant state authorities are often less effective in monitoring and enforcing forest regulations, due to the vast size of the forests under their control and a lack of financial and institutional capacity (Arnold, 1998; Carney and Farrington, 1998).

Other actors, such as companies interested in the commercial exploitation of natural resources – for example, logging and mining companies – are present at the local level. The interactions between these actors and local communities may generally take the form of conflict or negotiation (Engel *et al.*, 2006). Recently, there has been quite a high incidence of negotiated solutions, in which local communities bargain for financial and in-kind benefits in exchange for allowing the commercial actors access to forest resources (Mayers and Vermeulen, 2002; Bray *et al.*, 2003). This phenomenon has arguably enabled economic development that is more sensitive to local needs and has led to some improvement in the livelihood of local people (Palmer and Engel, 2006). Moreover, there is evidence that, to some degree, some local ecological service demands were also being met. As shown in Box 2.1, many Indonesian communities negotiated for environmental controls and took responsibility for the enforcement of these.

Limitations and problems

There are a number of institutional constraints that undermine the benefits of decentralization in integrating cross-sectoral demands. First, decentralization as a means of transferring management responsibilities to the local level has only rarely been truly implemented (Larson, 2004). Substantial decision-making powers are still highly centralized in many countries. Also, institutional frameworks in numerous developing countries have failed to clarify precisely where management responsibilities actually lie. In various Asian countries, for example, it is not clear which institutions are responsible for managing

[1]The market failure in which users do not capture the full value of preserving tropical forests is the dominant, underlying cause of deforestation (Pearce, 1996).

[2]Over the last decade, the governments of over 60 countries worldwide have decentralized at least some management responsibilities and powers from the centre to local entities, be they local governments, communities or user groups (Agrawal, 2001; Ribot, 2002).

Box 2.1. Negotiated demands on forests in Indonesia. (From Palmer and Engel, 2006.)

The recent decentralization of Indonesia's forest sector resulted in forest-dependent communities exerting property rights over forests. Consequently, many communities made legalized agreements with logging firms in exchange for financial and in-kind benefits. In a study of 55 agreements made in East Kalimantan, the potential trade-offs made between financial and environmental provisions were analysed. Thirty-five of the sampled agreements (64%) contained some kind of environmental provision, such as reforestation and limits on the diameter of trees to be cut.

A comparison was made of contracts with and without environmental provisions. With respect to the contractual provisions negotiated, the results indicated no significant differences between the contracts containing any environmental provisions and those containing none at all. Thus, there was no statistical evidence of any negotiation trade-offs between the environment and other contractual provisions negotiated.

With regard to actual fee payments made to the communities (as opposed to those negotiated), the statistical results indicated a significant difference between the two group means. Specifically, actual payments are slightly higher for contracts containing environmental provisions.

With regard to the proportions of communities experiencing compliance with in-kind payments, job provision and agricultural developments, the tests indicate no significant differences between the two groups. Thus, it appears overall that there is no statistical evidence of trade-offs in actual contract outcomes as a result of environmental provisions in community–company agreements. Rather, communities that negotiated contracts containing environmental provisions appear to have had a greater capacity to self-enforce agreements than those who did not. This result supports the findings of Engel and López (2004), and Engel *et al.* (2006) that the ability of communities to self-enforce property rights over the forest is a prerequisite for obtaining any kind of benefit at all.

Interestingly, however, there is little evidence of significant differences between the perceptions of selected environmental impacts resulting from contracts that contained environmental provisions and those that did not. While this confirms that communities that negotiated environmental provisions in their contracts were definitely not worse off than those that did not negotiate these, the results also shed some doubt on the effectiveness of the environmental provisions.

forest watersheds that cross administrative boundaries (Asia Forest Network, 1995). Moreover, as Edmunds *et al.* (2003) put it, decentralization has frequently been used by central governments to 'outsource costs while maintaining control'. The case of India (Box 2.2) highlights the fact that devolution of rights has often not kept pace with the devolution of responsibilities (Engel, 2004). Governments thus maintain control through extensive bureaucratic procedures, by withholding information, or through a lack of capacity building (Behera and Engel, 2006b).

An uneven transfer of rights and responsibilities reduces communities' incentives for sustainable resource management, reinforces tenure insecurity and thus encourages overexploitation for quick economic benefits. Even where management responsibilities have been properly devolved and are relatively clear, many developing countries suffer from weak state local institutions, leading to poor forest governance. Rent-seeking and corruption undermine the effectiveness of the state in implementing its own rules.[3] Consequently, the state is unable to enforce a common set of property rights, sometimes resulting in conflicts among forest stakeholders (Palmer, 2006).

A second problem with decentralization approaches for integrating cross-sectoral demands is that some demands are simply not represented at the local level. In particular, many of the benefits from forest ecological functions occur at higher levels. This tends to lead to socially suboptimal decisions by local actors even when the rights over the forest resource are clearly defined, complete and secure. In general, ecological demands may be regional, national or international. These

[3]Forest-rich countries such as the Democratic Republic of the Congo, Cameroon and Indonesia may be more prone to rent-seeking behaviour in the administration of their forests due to the easy availability and high levels of rents from the sale of timber (Palmer, 2005).

Box 2.2. Incomplete rights in India's Joint Forest Management Programme. (From Behera and Engel, 2006b)

India's Joint Forest Management (JFM) programme began in 1990, when the central government issued guidelines to state governments giving the latter the authority to devolve everyday forest management responsibilities to village-level institutions (Forest Protection Committees, or FPCs). Moreover, the guidelines prescribed benefit-sharing arrangements following regeneration. Thus, JFM is based on administrative orders and, as such, its rules only become formally binding once a user organization such as an FPC signs up to the programme. The programme effectively constitutes a contract between individual FPCs and the corresponding Forest Department (FD), with differences in individual rules and their enforcement. The degree to which specific rights over the forest were actually transferred from the FD to local communities differs significantly across states (Damodaran and Engel, 2003). In 12 out of 23 states that have implemented JFM, the FD has unilateral power to cancel the JFM agreement and in most cases even to dissolve the FPC. Most state programme guidelines do not provide long-term rights to communities. The legal status of FPCs also differs among states.

Under JFM, members of FPCs obtained withdrawal rights – rights to use several non-timber forest products and to keep a share of the proceeds from the sale of timber. However, for some lucrative forest products, withdrawal rights have not been fully devolved to communities, and timber harvesting rights are restricted. The share of forest benefits obtained by communities varies between states and among types of products. Moreover, in the event that FPCs fail to protect the forests, the government has retained the right to withdraw the usufruct rights from the communities unilaterally.

Management rights over the forests have been partially transferred to the FPCs. *Exclusion rights* – rights to assign access rights under JFM – have been formally granted to the FPCs, although restrictions remain. Moreover, *alienation rights* – rights to sell or transform a resource – have been transferred to communities only to a very limited degree. No ownership or lease rights over forest land can be given to communities. However, other rights (withdrawal, management and exclusion) over forests can be transferred by the community from one generation to another. The right to transform the resources is also limited to afforestation activities.

To conclude, community property rights under JFM are insecure because they are only administrative rather than legal rights, and as such can be withdrawn at any time. The transfer of property rights to communities has been only partial, and its degree differs widely across states. This incomplete transfer of rights is likely to result in reduced community incentives for sustainable resource management and provides incentives to over exploit forests for short-term economic benefits.

externality effects are not considered at the local level unless incentives for their consideration are provided by higher-level policies, or actors are present at the local level who represent these effects. While one could argue that NGOs and international donors do to some degree represent higher-level ecological demands, it is well known that the private provision of public goods leads to underprovision (Tietenberg, 2006). Thus, one would expect that NGO and donor activities correct for externalities only to a very limited extent. In many cases, their activities are limited to education and awareness building. Such activities can help to ensure a full internalization of local environmental values in the decision-making of local communities (with the caveat of unequal representation, explained below); but again, there is no reason why these communities should internalize higher-level environmental values unless mechanisms are in place that translate external benefits into real financial incentives at the local level. In this regard, it is not surprising that local communities in so many countries opt for selling their newly established rights over resources to commercial actors for short-term economic benefits at high environmental cost. Similarly, effective forest management frequently requires coordination over administrative boundaries, for example, for watershed management. This coordination over policy can sometimes only be dealt with at a higher level of governance.

A third limitation lies in the fact that local actors often differ considerably in their ability to defend their interests in their interactions with other actors. For example, the Indonesian case (Box 2.3) illustrates the relatively strong position of logging companies vis-à-vis local communities, as even the most favourable logging agreements

Box 2.3. Uneven power relations between communities and firms in Indonesia. (From Engel and Palmer, 2006a)

Further results from the study described in Box 2.1 indicated that there was wide variation in the actual pay-offs received by communities from community–company logging agreements. These ranged from IDR 2500 (US$0.28) to IDR 106,322 (US$11.81) per cubic metre. By contrast, domestic timber prices during this study period ranged from US$30 to US$70 per cubic metre, while timber from East Kalimantan was sold for US$80–125 per cubic metre in its most important international market (Palmer and Obidzinski, 2002). Thus, relative to timber prices, the actual share of rents claimed by communities was low, while firms claimed profit margins of approximately 20% or more. Overall, these figures suggest that firms hold more power over negotiations relative to the communities controlling access to commercially valuable forests.

yield only a small share of logging profits to local communities. Power imbalances also exist within local communities. It is naive to think of local communities as homogeneous units automatically acting in the interest of the group. Several studies show that decentralization and related participatory approaches to natural resource management may enhance existing power imbalances between local-level interest groups (Platteau and Gaspart, 2003). Individual households within the community often differ widely in their actual use of the resources as well as their participation in collective management activities and related decision-making processes (Sarin, 1996). In particular, devolution carries the risk that wealthier and more powerful segments of the community may benefit disproportionately and manipulate devolution outcomes in their own interests (Shackleton *et al.*, 2002). For example, a comparative study between Joint Forest Management (JFM) and non-JFM villages in the Indian state of Jharkhand found that richer sections of the communities benefited under JFM at the expense of poorer ones because community rules favour long-term timber benefits through forest closure and plantation of high-value species (Kumar, 2002). As a consequence, poor, forest-dependent households are marginalized as they suffer from the reduced availability of, and access to, NTFPs. Agarwal (2001) discusses how seemingly participatory institutions often exclude significant sections of the community, such as women. Behera and Engel (2006b) point out that high transaction costs may discourage the participation of poorer groups in the decision-making of forest protection committees, thereby freeing the way for richer segments to adopt rules that are biased towards their own interests. By transferring more power to local elites, decentralization may thus prevent the state from exercising an important role in ensuring the inclusion of marginalized groups.

Decentralization programmes and participatory approaches increasingly aim to overcome this problem by ensuring that marginalized groups have membership in the executive committees of local user organizations and by encouraging these groups' attendance of important meetings at which decisions over the local resource base are taken. Evidence from India's JFM programme, however, highlights the difficulties of overcoming traditional hierarchies (Box 2.4). While the programme was found to have been successful in encouraging the participation of marginalized groups in executive committees and meetings, actual decision-making processes continued to be dominated by traditional elites. This may lead to the continued marginalization of the poor, who may then encroach further into the forests.

Balancing Cross-sectoral Demands

There are 'top-down' conditions and 'bottom-up' processes that could complement one another in order to deal with the limitations and problems described in the previous section. First, the institutional framework for decentralization should be clarified and the institutional conditions improved. This may need to be instigated by central governments, but dialogue among institutions on settling responsibilities is crucial. Second, there is a strong need to consider

Box 2.4. Participatory resource management: evidence from Andhra Pradesh, India. (From Behera and Engel, 2006a.)

In a study in the Indian state of Andhra Pradesh, 660 households within 55 JFM communities were interviewed regarding levels of programme participation. The study distinguishes three different levels of participation: attendance at decision meetings, membership in executive committees and influence over decisions taken. While 82% of the interviewed households indicated that they had attended recent forest management plan meetings, only 41% of these had expressed opinions at the meeting, and only 18% felt that they had influenced outcomes.

In a two-stage econometric analysis of the determinants of different levels of participation, it was found that, *ceteris paribus*, poorer and female-headed households are more likely to attend JFM meetings. Policies to include poor and marginalized groups in executive forest management committees (MCs) under JFM also show some success, as it was found that being land-poor and belonging to a lower caste increase the likelihood of a household being selected into the MC. Female-headed households were also somewhat more likely to be MC members. Less-educated people are, on the other hand, less likely to become MC members.

Most importantly, however, the results indicate that despite these apparent successes, richer and more educated people in the community are still significantly more likely to influence the decisions taken. Older people, male-headed households and those belonging to larger caste groups are also more likely to influence decisions. The dominance of richer, male and more-educated segments of the community in decision-making processes implies that marginalized groups, although formally given some authority in the JFM process, are still de facto excluded from decision-making processes.

The results highlight the difficulties in establishing democratic structures in a context of deeply ingrained traditional hierarchies. In particular, the study shows that education is a crucial determinant of a household's ability to influence decisions. This points to strong synergies between general policies aimed at improving access to education for poor and marginalized groups and the success of participatory approaches.

policies that can effectively internalize externalities, which could be either a top-down or bottom-up process. The efficacy of such policies may, however, be contingent on the quality of the institutional arrangements. Third, the uneven representation of local demands by local actors differing in power implies a need for policies to strengthen weaker groups and mitigate conflicts. All this suggests that for decentralization to be effective in integrating cross-sectoral demands on forests, it needs to be complemented by other policies.

Institutional arrangements

In any decentralized system of forest governance, legislation and guidelines that clearly define property rights and management responsibilities are crucial for effectively integrating cross-sectoral demands on forests. Moreover, even where central government has a diminished role in decentralized natural resource management, it still has to ensure that all stakeholders comply with the rules. If decentralization is to fulfil its potential of achieving improvements in forest management by providing a sense of ownership, the first step required is a more complete transfer of rights to local communities. Furthermore, the central government has an important role in providing legal backing for community-established rules and community complaints about company non-compliance in community–firm agreements.

Given weak local state institutions in forest areas and the need to reduce their dependence on raising revenues through forest exploitation, central government could provide resources to local governments for securing property rights. Where state institutions are desperately weak or non-existent, a second-best solution would be to build up the self-enforcement capacities of local (community) institutions. For example, in Indonesia, Engel and Palmer (in press) show that the ability of communities to self-enforce property rights over the forest increases with the community's valuation of the standing forest and with the community's capacity for collective action, and decreases with the community's discount rate and opportunity costs of time. Thus, poverty alleviation, awareness building on local environmental values, enhancing community ability for collective action, etc. could help to improve the ability of the community to achieve self-enforcement. Government and the international community can also strengthen communities by providing information on community rights and on lessons learnt and cases of success from other communities. These kinds of policy in a decentralized system of forest governance would also help to counteract the imbalance of powers (see below). Box 2.5 summarizes some important roles for the state and the international community in a decentralized system.

Externalities

Forest policies made at the local level to meet the demands of local interests alone inevitably create externalities, which are not internalized by local actors – be they communities, companies or local governments. There is a strong need to consider policies that can effectively internalize externalities and promote policy coordination. As explained earlier, traditional approaches to regulate natural resource use in

Box 2.5. Important roles for the state and international community. (From Bulte and Engel, 2006, Baland and Platteau, 1996; Larson and Ribot, 2002; Ribot, 2002.)

State

- Setting national environmental priorities and standards (boundaries to community management, zoning)
- Dealing with national and regional externalities
- Providing a legal framework which enables communities 'to obtain legally enforceable recognition of their rights and to call upon the state as an enforcer of last resort'
- Providing legal backing for community-established use and access rules
- Providing conflict resolution mechanisms
- Providing technical assistance, monitoring equipment, scientific information, awareness building
- Providing economic incentives for conservation (e.g. poverty alleviation, payments for environmental services)
- Providing information on best practices, experiences of other communities
- Ensuring inclusion of marginalized groups, promoting pro-poor rules
- Promoting grass-roots participation and downward accountability (e.g. through popular elections, forums for discussions and negotiations, mandated financial reports, social audits or vigilance committees)

International community/donor agencies/NGOs

- Implementing mechanisms to translate international externalities into economic incentives for sustainable resource management and conservation in developing countries
- Supporting the state in its functions
- Supporting horizontal coordination between community organizations (e.g. regional forest user associations) to address broader-scale problems and increase bargaining power versus the state
- Supporting good governance (rule of law, secure land tenure, democratic elections, government accountability)
- Helping overcome 'culture of distrust' between communities and the state, build community confidence in its collective action ability, building up or adapting local institutional arrangements for resource management
- Capacity building at local level (financial and administrative management, technical skills, problem-solving)
- Awareness building at local level (providing information on rights, success stories)
- Promoting participation by marginalized groups (e.g. by helping them to raise their voice in defence of their interests and demand transparency and accountability)

developing countries have not only failed, but have also created social externalities vis-à-vis the rural poor who depend on these resources for their livelihood. A complete transfer of rights, however, can only help to achieve a representation of local environmental demands on the forest. For correcting externalities beyond the local level, new approaches such as regional resource management committees that encourage policy coordination among local communities and the establishment of markets for environmental services appear promising. Payments for environmental services, for example, can provide an alternative income source for local people by putting a value on non-extractive (or more sustainable) uses of the forest. In addition, they can potentially provide alternative economic opportunities and strategies for economic development (e.g. conservation concessions rather than logging concessions). Yet, their design can be complex (Engel and Palmer, 2006b).

Power imbalances

The uneven representation of local demands by local actors differing in power points to the need for policies to strengthen weaker groups and mitigate conflicts. On one hand, governments can help to strengthen local communities' bargaining position and capacity for self-enforcement vis-à-vis commercial actors. On the other hand, even in a decentralized system, the government has a crucial role in preventing the marginalization of the poor. Results from India, for instance, highlight the potential for education policies in promoting genuine participation by marginalized segments in community decision-making. Grassroots participation and downward accountability can also be promoted through popular elections, forums for discussions and negotiations, mandated financial reports, social audits or vigilance committees.

Conclusions

We have shown that a balance among competing demands is in principle possible, although some trade-offs are inevitable and a number of limitations and problems remain.

With regard to economic demands, commercial actors and local governments in developing countries usually, but not always, aim for short-term benefits, while society as a whole might want to ensure long-term sustainable income from forests. The quality and quantity of the ecological functions of tropical forests are both high, thus affecting many actors beyond the local level. Furthermore, with the rapid growth of populations in tropical regions and the continued degradation of forest ecosystems, it seems that demands on forests, at the aggregate level, can only increase. One way of slowing or even reversing this trend is to assign, and then try to capture, forest non-market values.

After the failure of traditional 'command and control' centrally regulated forest policies, decentralization has allowed for more local-level negotiations with respect to forest management among forest stakeholders, and without interference from distant and bureaucratic central governments. This trend may provide opportunities to support rural communities in managing forests beyond local-level interests. However, globalization has brought new challenges to tropical forests – in particular, increasing the commercial and large-scale exploitation of resources contained in and under forests. With strong economic growth continuing in countries such as India and China, it seems likely that this demand will continue to rise, despite the growth in new technologies that may increase the efficiency of resource use.

Decentralization has enabled local communities to benefit from the influx of commercial actors into forest areas as well as attempt to counter the local ecological impacts of commercial exploitation. Nevertheless, it should be made clear that communities have, by and large, acted in their self-interest due to forest dependence. However, any move away from forest dependence, for example, through increased off-farm employment opportunities in the course of economic development – could lead to a fall in livelihood demands on the forest. This could potentially remove local-level constraints on firms' behaviour, thus allowing companies to focus completely on generating short-term profits from the commercial exploitation of forests. All of this suggests that there is still an important role for government in regulating firms' behaviour – in particular, higher-level state institutions representing non-local level forest demands.

References

Agrawal, A. (2001) The decentralizing state: nature and origins of changing environmental policies in Africa and Latin America 1980–2000. Paper presented at the 97th Annual Meeting of the American Political Science Association. San Francisco, California, 30 August–2 September 2001.

Agarwal, B. (2001) Participatory exclusions, community forestry, and gender: an analysis for South Asia and a conceptual framework. *World Development* 29(10), 1623–1648.

Arnold, J.E.M. (1998) *Managing Forests as Common Property*. Forestry Paper 136. Food and Agriculture Organization of the United Nations, Rome.

Asia Forest Network (1995) Transitions in forest management: shifting community forestry from project to process. *Research Network Report* 6, Asia Forest Network, Manila, Philippines.

Baland, J.M. and Platteau, J.P. (1996) *Halting Degradation of Natural Resources: Is There a Role for Rural Communities?* Food and Agriculture Organization of the United Nations, Rome, and Clarendon Press, Oxford, UK.

Behera, B. and Engel, S. (2006a) Who forms local institutions? Levels of household participation in India's Joint Forest Management program. *ZEF-Discussion Papers on Development Policy* 103. Center for Development Research, Bonn, Germany.

Behera, B. and Engel, S. (2006b) The four levels of institutional analysis of the evolution of Joint Forest Management in India: a new institutional economics approach. *Forest Policy and Economics* 8(4), 350–362.

Bray, D., Merino, L., Negreros, P., Segura, G., Torres, J.M. and Vester, H. (2003) Mexico's community-managed forests as a global model for sustainable landscapes. *Conservation Biology* 17(3), 672–677.

Broadhead, J. and Dubé, Y.C. (2002) Cross-sectoral policy impacts in forestry. Paper submitted to the Secretariat of the 12th World Forestry Congress, Food and Agriculture Organization of the United Nations, Rome.

Bulte, E. and Engel, S. (2006) Conservation of tropical forests: addressing market failure. In: López, R., Stiglitz, J. and Toman, M. (eds) *Sustainable Development: New Options and Policies*. Oxford University Press, New York. Available at: http://www0.gsb.columbia.edu/ipd/pub/ConservationofTropicalForests11_29_04.pdf

Carney, D. (1995) Management and supply in agriculture and natural resources: is decentralization the answer? *ODI Natural Resource Perspectives* 4. Overseas Development Institute (ODI), London.

Carney, D. and Farrington, J. (1998) *Natural Resource Management and Institutional Change*. Routledge, London.

Damodaran, A. and Engel, S. (2003) Joint Forest Management in India: assessment of performance and evaluation of impacts. ZEF Discussion Papers on Development Policy 77. Center for Development Research, Bonn, Germany.

De Montalembert, M.R. (1995) Cross-sectoral linkages and the influence of external policies on forest development. *Unasylva* 46, 25–37.

Edmunds, D., Wollenberg, E., Contreras, A., Dachang, L., Kelkar, G., Nathan, D., Sarin, M. and Singh, N. (2003) Introduction. In: Edmunds, D. and Wollenberg, E. (eds) *Local Forest Management: The Impacts of Devolution Policies*. Earthscan, London, pp. 1–19.

Engel, S. (2004) Designing institutions for sustainable resource management and environmental protection. Habilitation thesis. Faculty of Agriculture, University of Bonn, Germany.

Engel, S. and Palmer, C. (2006a) Who owns the right? The determinants of community benefits from logging in Indonesia. *Forest Policy and Economics* 8(4), 434–446.

Engel, S. and Palmer, C. (2006b) Designing payments for environmental services with weak property rights and external interests. Working paper, Chair in Environmental Policy and Economics, ETH Zurich, Switzerland.

Engel, S. and López, R. (2004) Exploiting common resources with capital-intensive technologies: the role of external forces. ZEF Discussion Papers on Development Policy 90. Center for Development Research, Bonn, Germany.

Engel, S., López, R. and Palmer, C. (2006) Community–industry contracting over natural resource use in a context of weak property rights: the case of Indonesia. *Environmental and Resource Economics* 33(1), 73–93.

FAO (Food and Agriculture Organization of the United Nations) (2003) *Cross-Sectoral Policy Impacts between Forestry and Other Sectors*. Forestry Paper 142. Food and Agriculture Organization of the United Nations, Rome.

Kaimowitz, D. and Angelsen, A. (1999) *The World Bank and Non-Forest Sector Policies that Affect Forests* [mimeo]. Center for International Forestry Research (CIFOR), Bogor, Indonesia.

Kumar, S. (2002) Does 'participation' in common pool resource management help the poor? A social cost–benefit analysis of Joint Forest Management in Jharkhand, India. *World Development* 30(5), 763–782.

Larson, A.M. (2004) Democratic decentralization in the forestry sector: lessons learnt from Africa, Asia and Latin America. Paper presented at Interlaken Workshop on Decentralization in Forestry, 27–30 April 2004. United Nations Forum on Forests.

Larson, A.M. and Ribot, J.C. (2002) Democratic decentralization through a natural resource lens: an introduction. *European Journal of Development Research* 16(1), 1–25.

McKean, M.A. (1995) Common property: what is it, what is it good for, and what makes it work? Paper presented at the International Conference on Chinese Rural Collectives and Voluntary Organizations: between State Organization and Private Interest, 9–13 January. Leiden University, The Netherlands.

Mayers, J. and Vermeulen, S. (2002) *Community–Company Forestry Partnerships: From Raw Deals to Mutual Gains?* Instruments for Sustainable Private Sector Forestry Series. International Institute for Environment and Development (IIED), London.

Ostrom, E. (1990) *Governing the Commons: The Evolution of Institutions for Collective Action*. Cambridge University Press, Cambridge.

Palmer, C. (2005) The nature of corruption in forest management. *World Economics* 6(2), 1–10.

Palmer, C. (2006) *The Outcomes and their Determinants from Community–Company Contracting over Forest Use in Post-Decentralization Indonesia*. Peter Lang, Frankfurt am Main, Germany.

Palmer, C. and Engel, S. (2006) For better or for worse? Local impacts of the decentralization of Indonesia's forest sector. Working paper, Department of Environmental Policy and Economics, ETH Zurich, Switzerland.

Palmer, C. and Obidzinski, K. (2002) Indonesia case study prepared for the FAO report: higher international standards or rent-seeking race to the bottom? The impacts of forest trade liberalization of forest governance [mimeo]. Center for Development Research, Bonn, Germany.

Pearce, D.W. (1996) Global environmental value and the tropical forests: demonstration and capture. In: Adamowicz, L. and others (eds) *Forestry, Economics and the Environment*. CAB International, Wallingford, UK, pp. 11–48.

Platteau, J.P. and Gaspart, F. (2003) The risk of resource misappropriation in community-driven development. *World Development* 31(10), 1687–1703.

Ribot, J.C. (2002) *Democratic Decentralization of Natural Resources: Research Report*. World Resources Institute, Washington, DC.

Sarin, M. (1996) From conflict to collaboration: institutional issues in community management. In: Poffenberger, M. and McGean, B. (eds) *Village Voices, Forest Choices: Joint Forest Management*. Oxford University Press, Calcutta, India.

Sayer, J. (1998) Foreword. In: Wollenberg, E. and Ingles, A. (eds) *Incomes from the Forest: Methods for the Development and Conservation of Forest Products for Local Communities*. Center for International Forestry Research, Bogor, Indonesia, pp. ix–xi.

Schmithüsen, F. (2003) *Understanding Cross-Sectoral Policy Impacts: Policy and Legal Aspects*. Forestry Paper 142. Food and Agriculture Organization of the United Nations, Rome, pp. 5–44.

Shackleton, S., Campbell, B., Wollenberg, E. and Edmunds, D. (2002) *Devolution and Community-Based Natural Resource Management: Creating Space for Local People to Participate and Benefit?* Natural Resource Perspectives, 76. Overseas Development Institute, London.

Tietenberg, T.H. (2006) *Environmental and Resource Economics*, 7th edn. Chapter 4. Pearson/Addison-Wesley, Boston, Massachusetts, pp. 62–87.

World Bank (2002) *A Revised Forest Strategy for the World Bank Group*. World Bank, Washington, DC.

WRI (World Resources Institute) (2005) *World Resources 2005: The Wealth of the Poor – Managing Ecosystems to Fight Poverty*. United Nations Development Programme, United Nations Environment Programme, World Bank, World Resources Institute, Washington, DC.

3 Wood-product Trade and Policy Issues

Osamu Hashiramoto

Introduction

Production and trade in major wood products have been increasing during the last decade. One of the most significant changes in the global forest products market is the growth of processing industries in rapidly developing regions in the world. The first part of this chapter provides an overview of markets and trade in wood products. It focuses on the emerging producers and consumers of major wood products, and on competition among producers. The second part discusses the relevance of recent trends in wood-product markets and trade to cross-sectoral policy issues such as the competitiveness of wood-producing countries, wood energy use, and the procurement of sustainably and legally produced products in public and private sectors.

Wood-product Trade

Industrial roundwood

Global production and trade in industrial roundwood have increased steadily, primarily due to production increases in temperate and boreal forest regions. The production of industrial roundwood surpassed 1.6 billion cubic metres in 2004. Trade in industrial roundwood reached 120 million cubic metres – about 7% of its production during the same year. The USA is the largest producer, accounting for more than 400 million cubic metres or 25% of global production in 2004, followed by Canada. Production and trade by the Russian Federation and Eastern Europe have expanded rapidly after a severe decrease in production in the 1990s during the economic transition. In 2004, Russia exported 42 million cubic metres of industrial roundwood, accounting for 35% of global trade in the product. The production of industrial roundwood also increased in forest plantations in Oceania and South America. On the other hand, production has decreased in major producing countries in Asia. China has grown quickly to become the largest importer, with 28 million cubic metres or 23% of the world trade in industrial roundwood in 2004 (Fig. 3.1).

Sawnwood

Production of sawnwood has been increasing. However, global production in 2004, which was about 400 million cubic metres, was lower than at the peak in the late 1980s when production in the former USSR was highest. The USA and Canada accounted for 36% of global production in 2004. Trade has been continuously increasing, particularly due to exports from Canada to the USA, from Europe to the USA and Japan, and from Eastern Europe and Russia to Western

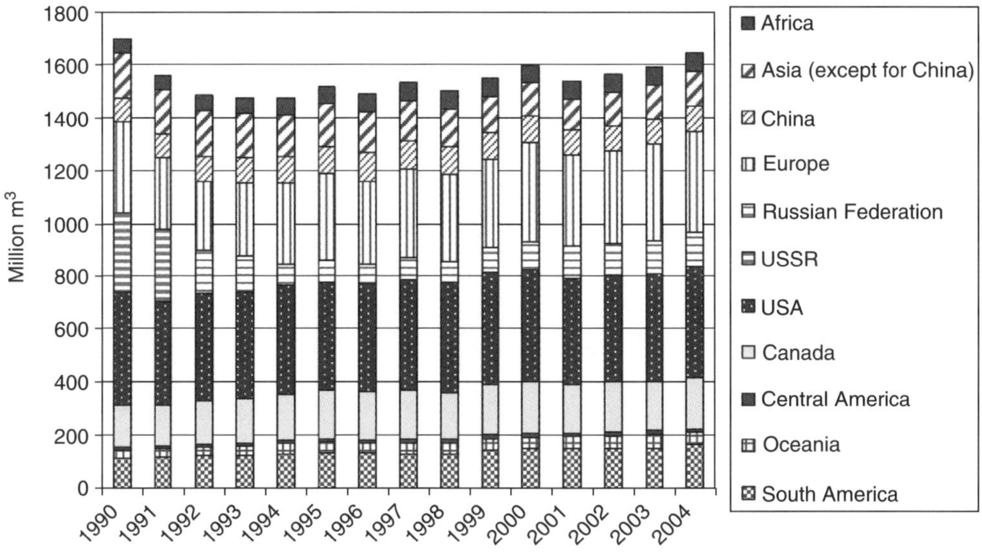

Fig. 3.1. Production of industrial roundwood, 1990–2004. (From FAO, 2005.)

Europe (Fig. 3.2). Canada is the largest exporter, accounting for 32% of global trade in 2004, with the country exporting 68% of its production. The USA has an approximately 30% share of global consumption, followed by Japan at 5% of global consumption.

Wood-based panels

Wood-based panels include plywood, particle board including oriented-strand board (OSB), and fibreboard. In 2004, the production of wood-based panels reached 225 million cubic metres, of which particle board accounted for 43%, plywood 30% and fibreboard 23%. A total of 78 million cubic metres of wood-based panels, or 32% of production in 2004, was traded. Production and trade in particle board have been growing rapidly in Western Europe, and production and trade in OSB have accelerated with rising wood-frame housing construction in North America. Production and trade have grown faster for fibreboard than for any other category of wood-based panels. Western Europe and North America have been major producers. China has increased its fibreboard production threefold since 2000 and became the largest producer, surpassing the USA. China also became the largest producer of plywood. The production of wood-based panels has also increased in Eastern Europe, Malaysia and Brazil. With regard to consumption, the USA used 63 million cubic metres and China 45 million cubic metres, accounting for 28% and 20% of world production in 2004, respectively. Wood-based panels have replaced sawnwood on construction sites in many countries, mainly because of new technology for using small trees and wood particles, and changes in construction methods (Fig. 3.2).

Pulp and paper

Production and trade of paper and paperboard have increased steadily in accordance with the development of a global economy. Europe reached record production and trade levels in 2004, and became the largest producing region, surpassing North America. China has become the largest producer of paper and paperboard in Asia and the second largest importer in the world next to the USA. Production has also been increasing in Brazil and Indonesia. Consumption of paper and paperboard has increased in Asia reflecting its rapid economic growth. China's economy is growing at a rate of 9.5% a year and India's at about 6.5% a year (World Bank, 2005). In 2004, the USA consumed 92 million tonnes of paper and paperboard,

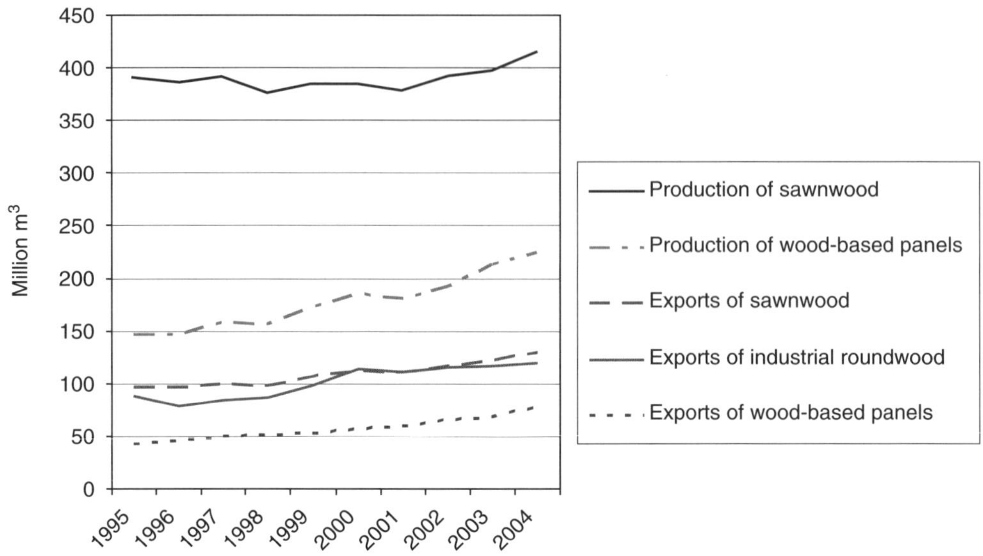

Fig. 3.2. Production and trade in solid wood products, 1995–2004. (From FAO, 2005.)

followed by China, consuming 60 million tonnes. These two countries consumed 26% and 17%, respectively, of global paper and paperboard production in 2004 (FAO, 2005). With regard to pulp for paper, North America, Europe and Asia are major producers. Canada is the largest exporter in the world. Recently, production has increased in Western Europe, Russia, Indonesia and Brazil. On the other hand, production has decreased in the USA. The proportion of recovered paper used in paper and paperboard production rose from 38% in 1995 to 42% in 2004. Much of the growth in the consumption of paper has involved the use of recovered paper (Fig. 3.3).

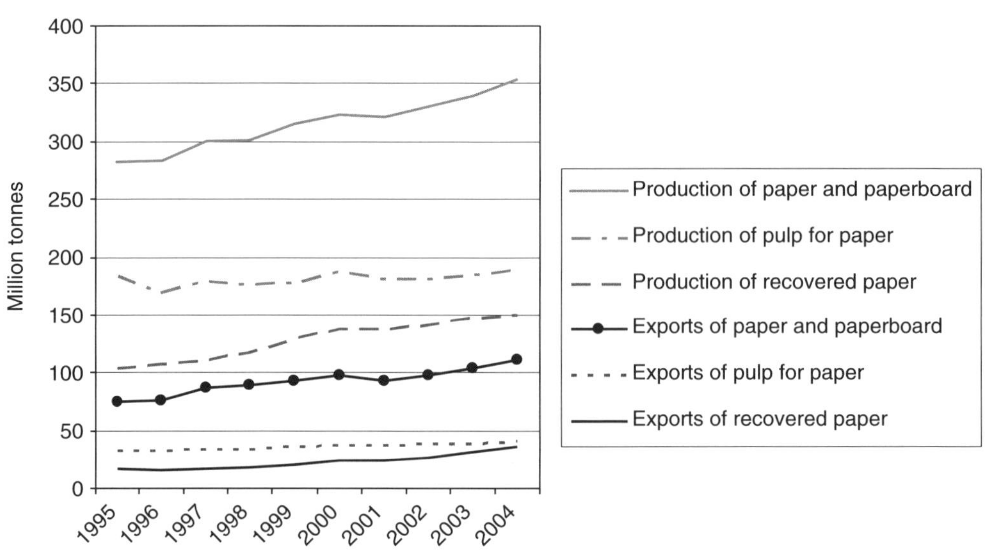

Fig. 3.3. Production and trade in pulp and paper products, 1995–2004. (From FAO, 2005.)

Trade value and secondary processed wood products

Recent strong economic conditions in global markets have driven the value of wood-product trade to US$240 billion in 2004, including the value of secondary processed wood products such as wooden furniture and parts (Fig. 3.4). This amount equals about 3% of the global trade value of all commodity products. The export value of paper and paperboard reached US$86 billion, accounting for 35% of the total trade value of wood products, in 2004. The value of exported secondary processed wood products has increased rapidly during the last decade and reached US$63 billion, or 26% of the total export value of wood products, in 2004. Primary processed wood products such as industrial roundwood, sawnwood and wood-based panels account for modest proportions of global wood trade values, in spite of the large volume traded (Fig. 3.4).

Europe has been the largest market for wood products as well as the largest exporter. Europe imported about US$119 billion and exported US$132 billion of wood products, including secondary processed wood products, in 2004, accounting for 47% and 55%, respectively, of the global import and export value. These large shares are mainly due to the high proportion of paper products in trade. North and Central America and Asia accounted for 26% and 22% of global imports and exports. Asia's share amounted to 23% of global imports and 15% of exports in 2004 (Figs 3.5 and 3.6).

Most international trade in wooden furniture used to take place in Western Europe and North America. However, since the 1990s, the production and export of secondary processed wood products have expanded in South-east Asia, Eastern Europe, and Central and South America, and recently in China on a larger scale. China became the world's largest exporter of secondary processed wood products, with exports valued at US$9.5 billion in 2004 (UN, 2005). With the development of processing capacity in these regions, the international trade in intermediate products has increased. For example, manufacturers in the USA have taken a contract-manufacturing approach by outsourcing to countries such as China (Xu *et al.*, 2003).

The USA is the largest market, importing US$22 billion of secondary processed wood products in 2004. The world's five largest importers – the USA, Germany, France, the UK and Japan – imported US$41 billion or 65% of the world import value of secondary processed wood products in 2004.

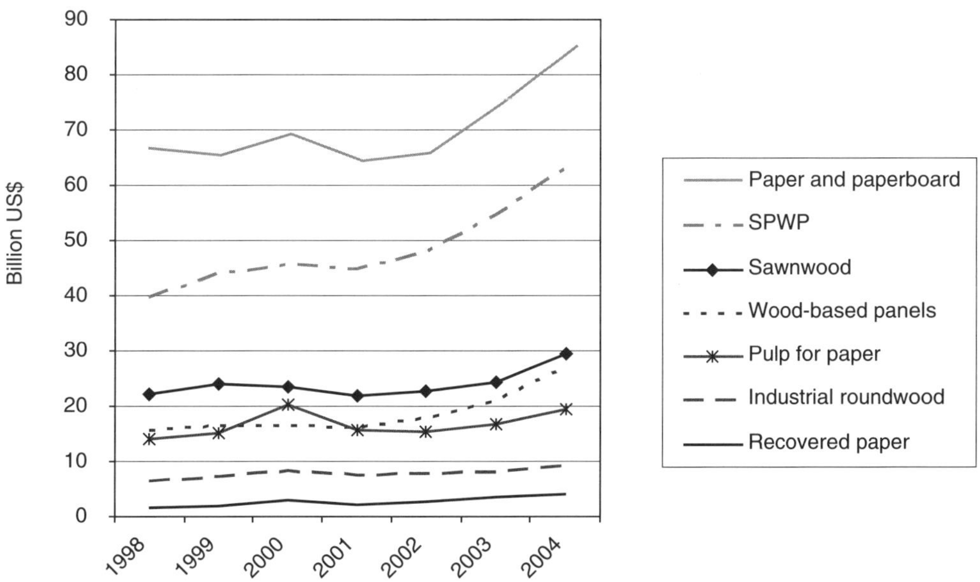

Fig. 3.4. Trade value of wood products, 1998–2004. (From FAO, 2005; United Nations, 2005.)

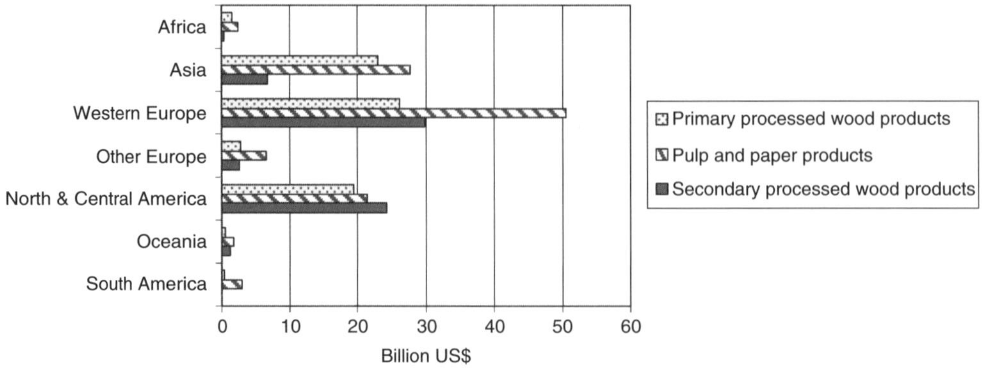

Fig. 3.5. Import values by region. (From FAO, 2005.)

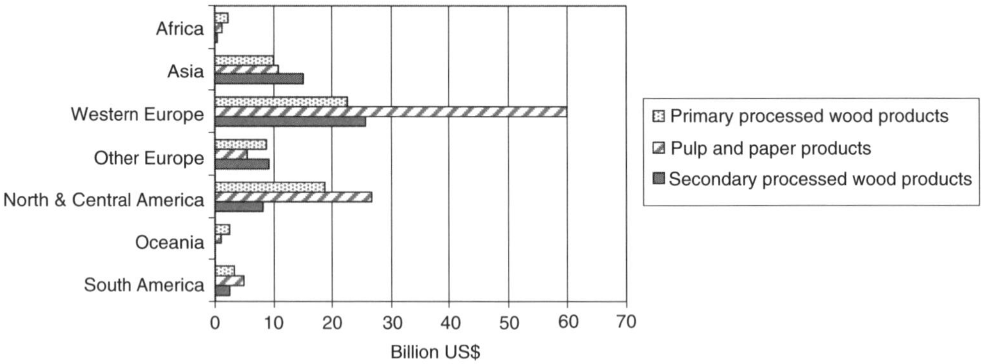

Fig. 3.6. Export values by region. (From FAO, 2005.)

More than half of US imports of wooden furniture were from Asia, mainly China. Germany, France and the UK imported mainly from other European countries. However, the market share of Asian products in these countries has been increasing.

Price trends and competition

The expansion in wood-processing capacity on a global scale and the subsequent increase in the supply of products have intensified competition among industries. This change has been reflected in downward trends in the trade prices of major products since the middle of the 1990s. These downward trends in the nominal export price of major wood products do not take into account the effects of inflation. Taking inflation in other commodities into account in this period, the fall in the real prices of wood products has been significant (Figs 3.7 and 3.8).

The most conspicuous fall was observed in the price of plywood, which has been facing competition with particle board and fibreboard, the prices of which also fell in this period. From 1995 to 2004, the production of wood-based panels increased by 50%. Exports of wood-based panels increased by 80% in the same period. The reduction in the production costs of processed products is explained by the sharp fall in the prices for these products in comparison with the modest decline in industrial roundwood prices up to 2002. Like solid wood products, the production and export of paper and paperboard also increased by 25% and 50%, respectively, from 1995 to 2004, with downward trends in prices.

Lower processing costs among emerging producers have influenced the trade pattern for certain types of industrial roundwood, reflecting resource scarcity in natural forests. In the case of tropical timber industrial roundwood, the relatively higher price level has been maintained since 1990. Chinese

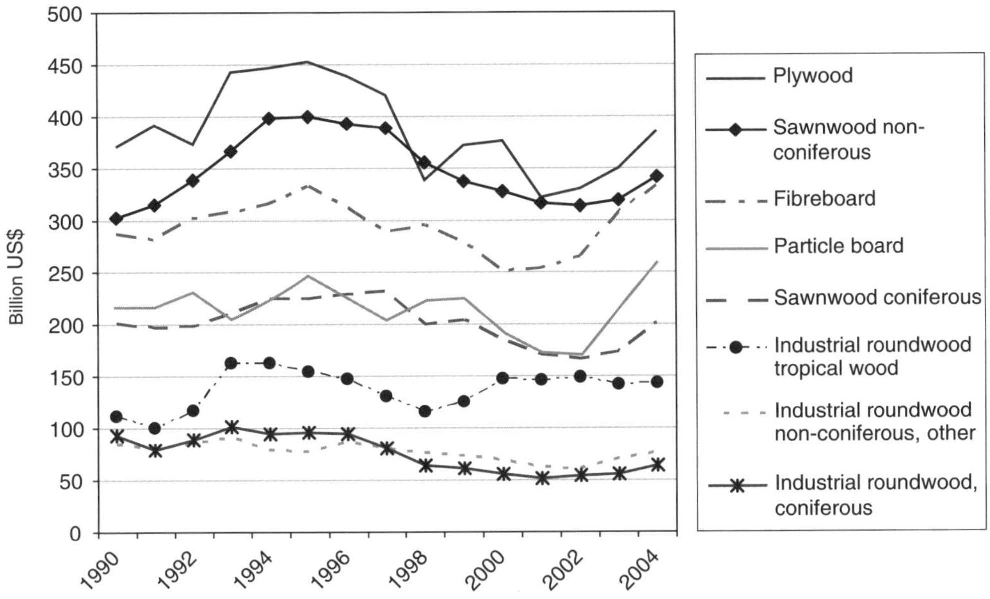

Fig. 3.7. Export prices for solid wood products, 1990–2004. (From FAO, 2005.)

and Indian traders have often offered higher prices than traditional Japanese and European importers in international trade. Other important evidence of price competition is the increase in China's imports of hardwood timber from the USA and Western Europe, due to China's rapidly expanding furniture production. Taking advantage of low production costs, there is some evidence that Chinese traders have been able to offer higher prices for certain wood raw materials for processing than traditional producers in the USA (Schuler *et al.*, 2003). On the other hand, declining prices for

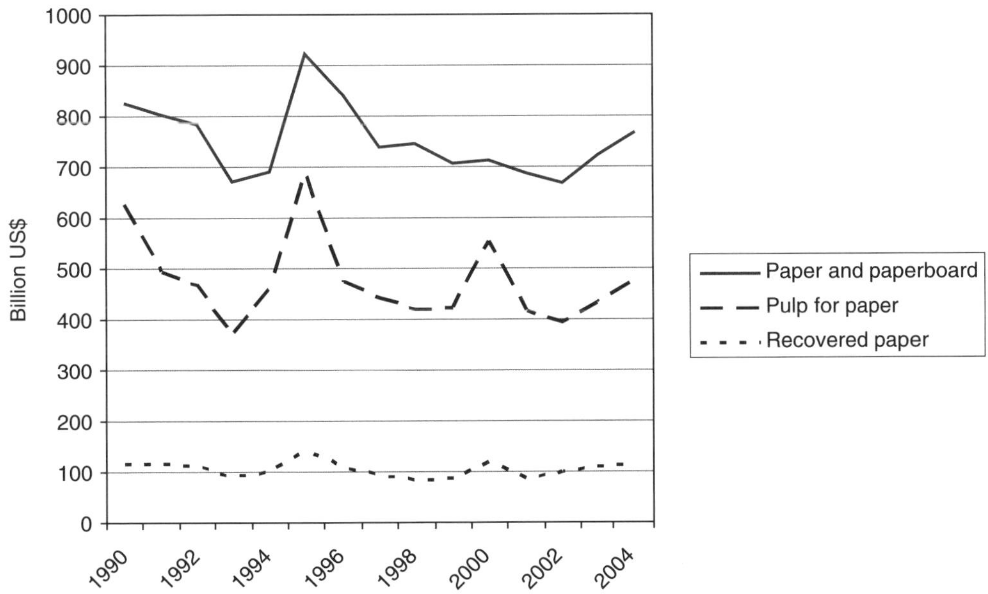

Fig. 3.8. Export prices for pulp and paper products, 1990–2004. (From FAO, 2005.)

coniferous industrial roundwood have discouraged timber production by forest owners in some parts of Western Europe (UNECE, 2005). The price decrease for coniferous wood in Europe is mainly due to increased production in the region, including Russia, where more forest resources grow than are harvested. In Europe, an average of about 45% of the annual increment in growing stock was harvested in 2000 (UNECE/FAO, 2005a).

Export prices for major wood products appeared to recover in 2003 and 2004 (Figs 3.7 and 3.8). However, this was partly due to the recent high appreciation of the euro against the US dollar. In 2004, the euro countries accounted for 30% and 34% of global import and export value, respectively, of wood products, excluding secondary processed wood products. One of the important factors in the price recovery is the strong housing market in the USA. The USA accounted for 26% and 28% of the global import value in wood-based panels and sawnwood, respectively, in 2004. Other factors include economic growth in Asia, particularly for pulp and paper products, and Eastern Europe. China has become a large importer of industrial roundwood, pulp for paper, and paper and paperboard, accounting for 27%, 17% and 8%, respectively, of the global import value in 2004. However, import prices there for wood products are still lower than in other major markets in developed countries, except for industrial roundwood in some cases.

In the wooden furniture sector, with more and more producers entering into the global market, prices have been declining. For example, a free-on-board price for a type of wooden furniture in major Asian exporting countries has decreased more than 30% from 1995 to 2002. There is evidence that the recent increase in exports of wood furniture was simply the result of incremental inputs in items such as labour, raw materials and equipment. On the other hand, value added per employee during production has not corresponded to the increase in export growth (Ratnasingam and Ioras, 2005). A case study in South Africa has also indicated the problem seen in Asia of decreasing prices for products (Kaplinsky *et al.*, 2001). In the face of competition and shrinking profits, the capacity to develop good designs and a marketing strategy targeting major international markets are becoming important in the furniture industry in developing regions.

Policy Issues

Private investment

The rapid development in the processing industries has often been initiated or supported by increasing investment from other regions. Enterprises in Western Europe have invested in establishing sawnwood, wood-based panels and furniture mills in some countries in Eastern Europe. In particular, large capital inflows have been observed in the countries that joined the European Union (EU) in 2004, such as the Baltic States, the Czech Republic and Poland. Enterprises in Taiwan, Singapore and other Asian countries have invested in China, establishing plywood, furniture and paper mills. In the pulp and paper sector, large-scale investments from northern Europe have been made in China and Brazil. Foreign capital was also invested in the management of forest plantations in these regions in order to secure the supply of wood raw materials for pulp and paper production.

Industrial investment strategies largely depend on the relative competitiveness of producing countries. There are several factors that have an influence on the country's competitiveness, such as low labour and production costs, sufficient capacity to absorb advanced technologies, government incentives to invite foreign investment, support for research and development, stable social and political conditions, and a growing domestic economy. Many of these factors have cross-sectoral implications. Proximity to forest resources and major markets is another fundamental factor. However, as has been noted in exports from China, low production costs to a certain extent offset higher transportation costs for reaching forest resources and global markets.

Policies that affect these factors and consequently the level of competitiveness are often cross-sectoral in nature. In some countries, governments have enhanced competitiveness through incentive policies to attract foreign investment. For example, in some provinces in China, preferential tax treatment has been granted to foreign investors. This includes reduced income tax, particularly for enterprises producing for export, and reductions of, or exemptions from, land use charges. In addition, in the case of China, the low exchange rate of the country's currency against the US dollar has had a direct impact on the increase in exports of wood products.

In some countries in Eastern Europe, measures such as a reduction in, or exemption from, corporate tax, reduced import tariffs on equipment, and other forms of subsidy have been introduced since the 1990s to attract foreign investment (Kaneko, 2005). Although policies preferential to foreign investment have created unequal conditions for domestic industries, they have contributed to increased national income through the development of export industries with access to global markets. Foreign investment has also played a role in technology transfer and in the development of infrastructure. On accession to the EU in 2004, some Eastern European countries had to reduce these incentives in order to meet EU criteria. None the less, accessions to the EU will provide advantages for the new members, including the elimination of import tariffs, improved management, reductions in transaction costs through simplified trading procedures and financial support from Western Europe. On the whole, it can be concluded that there is a widening gap among countries that are able to produce competitive products, making use of foreign investment, and those that are not able to keep pace with developments. In Africa, for example, there is still only minor development in the value-added processing industry and much of the tropical industrial roundwood is being exported. While investment has increased in Eastern Europe as a whole, investors are reluctant to invest in some countries of the region because of the associated high risks.

Trade measures

In some South-east Asian countries, governments have implemented trade measures to promote domestic processing such as curbing exports of primary commodities through the imposition of levies, taxes or even total export bans, sometimes accompanied by high tariffs on imports of value-added products. Fiscal incentives for the value-added producing sectors have supported the growth in these sectors (ITTO and ITC, 2002). Competition for raw materials has intensified as a result of the expanding production capacity. In 2005, Malaysia announced a ban on the export of sawn rubberwood. It is estimated that 80% of sawn rubberwood is consumed by the furniture industry. Indonesia has banned the export of rough-sawn timber and unprocessed rattan in order to be able to supply these materials to the domestic furniture industry. The South-east Asian countries, which are major sources of industrial roundwood and at the same time have a large processing capacity, are caught between supplying their own industries and meeting China's growing demand (ITTO, 2005).

Intensive competition in low-cost production has affected trade measures in some countries. In the wood-product sector, tariff rates in markets in the developed countries have been rather low since the 1990s. A generalized system of preference rates (often zero) has been applied to many products from developing countries, although its benefits have been diluted due to low 'most favoured nation' rates. In comparison, tariff rates have been higher in developing country markets. However, in the face of rapidly increasing imports from certain developing countries, import tariffs have been raised in some markets in the developed countries. In Europe, anti-dumping duties of up to 66.7% have been imposed on imports of okoume plywood from China since 2004. The USA imposed an anti-dumping duty on imports of a certain type of wooden furniture from China. These policy changes have increased investors' interest in other countries such as Vietnam. In the USA, the rapid growth in plywood imports from Brazil has resulted in the rescinding of Brazil's duty-free status (UNECE/FAO, 2005b).

Energy from wood resources

Energy policies may have a considerable influence on wood-product markets. Wood biomass is a major renewable energy source, representing a significant proportion of the rural energy supply, particularly in developing regions. In recent years, there has been increasing awareness of promoting the use of wood energy in Europe. Concerns about sustainable energy supplies and commitments to the Kyoto Protocol have been major influences on the promotion of wood energy policies. For EU countries, the emission cut was set at 8%. In contrast to the consumption of fossil fuel, the use of sustainably produced biofuels does not result in a net release of carbon dioxide into the atmosphere (UNECE/FAO, 2003). Increasing prices for fossil fuels and the availability of wood raw material stocks – growing faster than removals in European forests – have attracted increased attention to the potential of wood biomass as an energy source.

In 1997, the European Commission set a target of doubling the share of renewable energy to 12% by 2010 (EC, 1997). Recently, policies that support the promotion of the use of wood for energy production have been initiated in several member countries, including Austria, Finland, Germany, Hungary, Norway and Sweden. Policy instruments used for this purpose include investment in biomass plants, taxation of fossil fuels, tax relief on bioenergy, support for the development and commercialization of new technology, and other subsidies and low-interest loans (UNECE, 2005). Recently, the wood-based panel industry, as well as the pulp and paper industry, has expressed concern regarding public incentives for promoting the use of wood for energy production. The industry's major concern is a possible shortage of wood fibre from small-diameter logs, wood chips and particles. The increased competition for, and anticipated price increases in, wood raw materials could make these industries less competitive on the world market (UNECE/FAO, 2003).

In Western Europe, the annual consumption of wood fuel increased from 30 million cubic metres in 1998 to 37 million cubic metres in 2004. Net imports increased from 0.1 million cubic metres to 1.2 million cubic metres during the same period. On the other hand, in Eastern Europe, net exports doubled and reached about 1 million cubic metres in 2003 (FAO, 2005). Trade in wood fuel has also been growing in Europe. In some Eastern European countries, an increase in wood fuel prices was recently observed due to competition with other uses of wood (UNECE, 2005).

Energy policy may also have an influence on wood markets, as the new demand for wood may become an incentive for forest owners to produce more timber and use residues in the forests more intensively. Sawmills may make a profit from selling wood residues and sawdust produced during processing. However, as the total volume consumed for wood fuel is relatively small in comparison with other wood products in markets in the developed countries, there is as yet no clear evidence of the potential impact of the new energy policy on the wood-product market.

Public procurement policies

The problem of illegal logging and related trade has become a priority agenda item at international conferences. At the United Nations Forum on Forests, as well as at the World Summit on Sustainable Development in 2002, a consensus was reached to take immediate action with regard to domestic forest law enforcement and illegal international trade in forest products. The Forest Law Enforcement and Governance process has promoted national initiatives on tackling the problems of illegal logging and trade.

In response to growing concerns about such problems in relation to sustainable forest management, schemes have been developing in several countries to take action through public procurement policies. In the UK, voluntary guidelines advising government departments to purchase timber and timber products from sustainable and legal sources were issued in 1997. In 2000, the regulation became a binding commitment. Similar public procurement policies have been developed or are under development in Belgium, Denmark, France, Germany, Japan and the Netherlands. These policies have often been developed as part of, or with relevance to, increasing 'green' purchasing policies that deal with products across industrial sectors. These attempts have implications for the private sector, and consequently on markets for wood products.

These emerging policies share targets and many common elements. However, differences are also seen with regard to the criteria used, the sources and coverage of products, and methods of verification. Some schemes only apply to tropical timber, while others cover timber from all regions, including domestic products. Some policies are mandatory for central government and others are voluntary. Several schemes include roundwood, sawnwood, plywood, but do not include processed products such as furniture and paper. Some schemes include specific categories of timber from sustainable sources. Denmark's schemes, for instance, require proof of legality as a minimum standard, with a stepwise approach towards sustainable sources being used (Brack, 2005). Third-party certification systems are often mentioned as an appropriate method of verification for sustainable forest management. Although it is too early to evaluate the influence of these differences in implementation, communication among the relevant parties will become important. Using simple criteria and procedures will be a prerequisite for practical and transparent implementation. With regard to harmonization with multilateral trade agreements, the most relevant principle would be non-discriminatory application among regions and countries.

Private sector initiatives

Initiatives to promote the use of sustainably produced forest products have been progressing in the private sector. The US Green Building Council, a coalition of leaders from across the building industry, has developed what is known as the 'Leadership in Energy and Environmental Design Green Building Rating System'. This system is a voluntary national standard for developing high-performance sustainable buildings. It provides a comprehensive framework for assessing building performance and meeting sustainability goals. Verification is carried out through scoring, by adding up the total number of points achieved for the agreed measures. Points are awarded, as a part of many categories across industrial sectors, for the use of wood certified by the Forest Stewardship Council and for the use of locally harvested wood, in the context of reducing negative environmental impacts resulting from transportation. Registered projects have also been started in Brazil, Canada, China, India, Italy, Japan, Mexico and Spain (US Green Building Council, 2005). Similar initiatives to promote the use of wood from sustainable forest management include the Green Globes, the Model Green Home Building Guidelines from the National Association of Home Builders, and the Buildings Research Establishment's Environmental Assessment Method.

Other changes have been seen in investments by large European paper companies in forest plantation management in developing countries. Some companies have started to incorporate a chain of custody verification and certification of forest management into their investment projects in developing countries. The Chief Executive Officers of the International Council of Forest and Paper Association issued a leadership statement in June 2006 in which they committed themselves to support forest certification systems to promote sustainable forest management, combat illegal logging, support recovery of paper and wood products, meet the challenge of climate change and contribute to the well-being of workers and communities (ICFPA, 2006).

Phytosanitary measures

Trade measures aimed at conserving forests have implications for trade in other commodities. Global concerns relating to the movement of potential pests through trade in wood commodities and wood packaging materials have resulted in the adoption of a global standard for treating wood packaging material – the International Standard for Phytosanitary Measures No. 15, *Guidelines for Regulating Wood Packaging Material in International Trade* in the International Plant Protection Convention – in 2002 (FAO, 2002). This convention is a major international treaty that aims at securing action to prevent the cross-border spread of plant and plant-product pests. The international standard includes a basic framework for risk analysis and development of phytosanitary measures to minimize such cross-border movement.

As wood packaging materials are used for trade in many commodities, the implementation of phytosanitary measures may have an impact on trade in other commodities. Since January 2006, the EU and more than 20 countries, including major exporters and importers of industrial commodities, have introduced or are developing national standards in accordance with the International Standard for Phytosanitary Measures No. 15. This process will promote the control of pests and will contribute to international harmonization of phytosanitary measures, avoiding the use of unjustifiable measures as barriers to trade.

Conclusions

As can be seen from the above, cross-sectoral policies have an influence on wood-product markets. Production factors, such as low production costs and the capacity to produce quality products, have been influential in the competitiveness of wood-producing countries. Policies relating to these factors often involve cross-sectoral linkages. Growth in the processing industries in some developing regions has brought about shifts in the production base and changes in the trade patterns of wood products. Changes in countries' competitiveness will bring about new flows in trade in wood products.

Energy policies in Europe can potentially affect prices for certain wood raw materials and can create markets for such materials as well. The fluctuation in the oil price will influence wood-product markets, in which the final prices are decreasing while transportation costs are increasing, due to the shifts in the processing bases.

Wood procurement policies that have been developing in the public sector in response to concerns about sustainable forest management have often been incorporated into broader public procurement policies that affect several industrial sectors. Initiatives aimed at promoting the use of sustainably produced materials in the private sector also have often included components involving forest management in their broad category of construction materials.

As globalization progresses, cross-sectoral policies relevant to the competitiveness of wood-producing countries will continue to affect wood-product markets through investment and trade. As the global demand for final products is still largely dependent on the markets in the developed countries, initiatives to promote the use of sustainably produced materials will have a certain impact on markets and forest management. This applies particularly in cases in which the providers of wood raw materials are part of a global network that supplies finished products to environmentally sensitive markets.

References

Brack, D. (2005) *Public Procurement of Timber. EU Member State Initiatives for Sourcing Legal and Sustainable Timber*. Chatham House, London.

EC (European Commission) (1997) *Communication from the Commission: Energy for the Future: Renewable Sources of Energy*. White Paper for a Community Strategy and Action Plan. European Commission, Brussels.

FAO (Food and Agriculture Organization of the United Nations) (2002) *International Standards for Phytosanitary Measures: Guidelines for Regulating Wood Packaging Material in International Trade*. FAO, Rome.

FAO (2005) FAOSTAT forestry data. Available at: http://faostat.external.fao.org

ICFPA (International Council of Forest and Paper Association) (2006) *Forest Products Industry: A Commitment to Global Sustainability*. International Council of Forest and Paper Association, Rome.

ITTO (International Tropical Timber Organization) (2005) *Tropical Timber Market Report* 10(12), 1–15 August 2005. International Tropical Timber Organization, Yokohama, Japan.

ITTO (International Tropical Timber Organization) and ITC (International Trade Center (UNCTAD/WTO) (2002) *Tropical Timber Products. Development of Further Processing in ITTO Producer Countries*. International Tropical Timber Organization, Yokohama, Japan.

Kaneko, Y. (2005) Special economic zones in Poland: evaluation of foreign investment promotion policy. *Bulletin of the Japan Association for Comparative Economic Studies* 42(2), 13–25.

Kaplinsky, R., Morris, M. and Readman, J. (2001) The Globalisation of Product Markets and Immiserising Growth: Lessons from the South African Furniture Industry [mimeo]. Institute of Development Studies, Brighton, UK. Available at: http://www.ids.ac.uk/ids/global/pdfs/rkmmjr.pdf

Ratnasingam, J. and Ioras, F. (2005) The Asian furniture industry: the reality behind the statistics. *Holz als Roh- und Werkstoff* 61(63), 64–67.

Schuler, A., Taylor, R. and Araman, R. (2003) Competitiveness of U.S. wood furniture manufactures: lessons learned from the softwood moulding industry. *Forest Products Journal* July/August, 14–20.

UN (United Nations) (2005) UN Commodity Trade Statistics Database (COMTRADE). Department of Economic Affairs, United Nations, New York.

UNECE (United Nations Economic Commission for Europe) (2005) *UNECE Timber Committee, Country Market Statement*. United Nations, Geneva. Available at: www.unece.org/trade/timber/

UNECE/FAO (2003) *Forests, Wood and Energy: Policy Interactions*. Geneva Timber and Forest Discussion Paper, 42; ECE/TIM/DP/42. United Nations, Geneva. Available at: www.unece.org/trade/timber

UNECE/FAO (2005a) *European Forest Sector Outlook Study: Main Report*. Geneva Timber and Forest Study Paper, 20; ECE/TIM/SP/20. United Nations, Geneva. Available at: http://www.unece.org/trade/timber/docs/sp/sp-20.pdf

UNECE/FAO (2005b) *Forest Products Annual Market Review, 2004–2005*. ECE/TIM/BULL/2005/3; Timber Bulletin, 58. United Nations, Geneva. Available at: http://www.unece.org/trade/timber/docs/fpama/2005/2005_fpamr.pdf

US Green Building Council (2005) Available at: www.usgbc.org. Washington, DC.

World Bank (2005) WDI data query. Available at: http://devdata.worldbank.org/data-query

Xu, M., Cao, X. and Hansen, E. (2003) China's wood furniture industry. *Asian Timber* September/October, 35–37 (AsiaTimber.net).

4 Ecosystem Approaches to Sustainable Forest Management: Changing Realities

Jeffrey A. Sayer, Stewart Maginnis and Simon Rietbergen

Introduction

This chapter summarizes the result of a study on parallel concepts for forest conservation that have emerged during recent times in order to cope with new trends and challenges in sustainable forest management (Sayer and Maginnis, 2005). The findings reveal an unexpected level of progress in the use of ecosystem approaches in many of the countries studied.[1] Major shifts in forest management policy and practices have seen conventional, commodity production–oriented approaches replaced by more holistic, people-centred ecosystem-level approaches. In many ways, ecosystem approach practice has moved ahead of the theoretical discussions going on within international forest dialogues. However, there are real practical and institutional obstacles to its widespread application in the world's forests.

[1]In this chapter, we distinguish between the Ecosystem Approach (capitalized), a set of general principles that can be applied in a wide range of circumstances which was endorsed by the 5th Conference of the Parties to the Convention on Biological Diversity; and ecosystem approaches to forest management (not capitalized), a diverse range of operational forest management approaches developed in response to particular forest management challenges from the late 1980s onwards.

Two Concepts in Forest Conservation and Management

Most forest laws and institutions evolved over time with relatively simple agendas. They were mainly driven by the need to protect timber and hunting 'rights' of royalty and other elites from the subsistence needs of peasants. Today's forestry institutions exist in a very different world. Forest management now needs to integrate broader societal concerns and tackle conservation and sustainable use issues on a larger scale, using, for instance, multifunctional landscapes or ecoregions as the units of analysis and management. Similar trends towards integration and scaling up have also occurred in other sectors, including agriculture, grassland, coastal zone and marine management. These and other recent trends in forest management (Box 4.1) have created the need for management approaches that can take into account the greater complexities and trade-offs involved in today's world.

There is a common perception within forestry circles that the Rio Summit spawned two parallel concepts for forest conservation and management. One has been that of sustainable forest management, developed from classical forestry and pursued through the United Nations Forum on Forests (UNFF) process, and organizations such as the Food and Agriculture Organization of the United Nations (FAO) and the International Tropical Timber Organization (ITTO). The

Box 4.1. Recent trends in forest management. (From Sayer and Maginnis, 2005.)

Broadened objectives. At all scales, from the community to the global enterprise, foresters are being urged to deal with a much broader range of social and environmental issues than in the past. Forest management is moving from production objectives to multiple function objectives. Further, a patch of forest can now be claimed to have 'global values' that often do not correspond to the values perceived by local people. Society is making more explicit demands for longer temporal scales and broader spatial scales to be addressed in forest management.

Codifying good practice. Regulators, certifiers and civil society are developing criteria and indicators against which they can assess the 'quality' of forest management or the 'health' of forests. Governments want to apply norms and capture rents, local people want to defend rights and assets and environmental groups want to foster best practice.

Recognition of pluralism. There is increasing recognition that different forests support different stakeholders and require different management systems. There is no single solution to fit all conditions. Also, it is becoming clear that many different systems of ownership and use of forests can qualify as sustainable.

Decentralization – devolution. As the locus of decision-making on some forest issues moves from the national to the global level, many governments are decentralizing control of forests and divesting themselves of forest assets. Responsibility for forests is being placed in the hands of regional, municipal and local communities.

Globalization. Multinational corporations, banks and trade regulations all have a strong impact on forest management and usually take it out of local control. Forest issues are also firmly in the global arena, as they are being included in a growing number of international forums.

Climate change. The uncertainties created by the potential impacts of different climate change scenarios have major implications for forestry laws and institutions. Eco-climatic zones are shifting by hundreds of kilometres, new pest and disease problems are emerging, and invasive weed species pose threats. Climate change adaptation will be the major challenge for all forest managers in the future.

other has been that of the Ecosystem Approach, pursued primarily within the framework of the Convention on Biological Diversity (CBD). International forums on forests have discussed the two concepts and governments have committed to implementing them. Yet there is still a good deal of confusion about how the concepts relate to each other. To help clarify these two concepts, the CBD and the UNFF invited member governments and relevant organizations to provide clarification on these concepts and to develop proposals for integrating them.

Origin and Philosophy of the Two Concepts

While ecosystem approaches and sustainable forest management can be thought of as similar responses to the same set of underlying driving forces, there are important differences in the origins and philosophies of the two concepts. Sustainable forest management has been developed and debated by forestry professionals, with their primary focus on producing goods and services from land under their control (De Montalembert and Schmithüsen, 1993; FAO, 1994, 1999). It has been the object of extensive on-the-ground testing using criteria and indicators, including those developed for certification. There has been a short feedback loop and lots of opportunities for testing and learning. It is now firmly embedded in practical forest operations.

On the other hand, the debate around ecosystem approaches has been led by a more heterogeneous group of proponents more concerned with conservation. Thus, sustainable forest management emerged from a production agenda, while ecosystem approaches developed from a conservation agenda. As it was endorsed by the

CBD, the Ecosystem Approach represents a compromise between a rich country 'precautionary' agenda and a developing country 'development' agenda, where poverty reduction and economic growth are predominant concerns. Perhaps its main significance is as a negotiated statement of the middle ground between conservation and development.

Sustainable forest management is the latest in a line of forest management concepts that have sought to capture the notion of sustained flows of different forest goods and services and, recently, to expand the range of these 'sustainable' goods and services. Thus, the progression from sustained yield forestry to sustainable forestry to sustainable forest management has seen an increased emphasis on a broader set of social and environmental goals. Many forestry institutions now practise various forms of it, and a wide range of methods and tools are available that have been tested over time. The definition of the term sustainable forest management developed by the Ministerial Conference on the Protection of Forests in Europe, which has also been adopted by the FAO, is:

> The stewardship and use of forests and forest lands in a way, and at a rate, that maintains their biodiversity, productivity, regeneration capacity, vitality and their potential to fulfil, now and in the future, relevant ecological, economic and social functions, at local, national, and global levels, and that does not cause damage to other ecosystems.

An early milestone in the development of ecosystem approaches was the public challenge to conventional forest management approaches that took place in the US Pacific Northwest during the late 1980s, which resulted in the adoption by the US Forest Service of ecosystem management (Haynes *et al.*, 2005). This was basically a logical progression from sustainable forest management, with the incorporation of a broader set of management and participation objectives. As such, ecosystem management was firmly anchored in a set of sustainable forest management tools and methodologies, as a practical, managerial approach. Later, the CBD endorsed the Ecosystem Approach as a set of general principles that can be applied in a wide range of circumstances, and defined it as:

> The ecosystem approach is a strategy for the integrated management of land, water and living resources that promotes conservation and sustainable use in an equitable way. An ecosystem approach is based on the application of appropriate scientific methodologies focused on levels of biological organization, which encompasses the essential structures, processes, functions and interactions among organisms and their environment. It recognizes that humans, with their cultural diversity, are an integral component of many ecosystems.
>
> (CBD, 2000)

The CBD also developed a set of 12 principles for the Ecosystem Approach (Box 4.2). It can be seen that this approach, as used within the CBD context, is not linked to any particular operational procedures and does not include clear targets or guidance for practical application. The lack of an operational framework has been one of the key criticisms levelled at the Ecosystem Approach.[2]

Lessons Learnt from Country Case Studies

Australia

The debate on forest management in Australia has been shaped by simplistic media treatment that has led to two polarizations: state governments versus the Commonwealth (central) government, and rural forest users versus largely urban conservationists (Ferguson, 2005). The heat of the forest dialogues has also been turned up by strong national stakeholder groups (including the forest industry, unions, landholder and environmental NGOs) and by Commonwealth and State governments intent on making political capital from the issues. Typically, the Commonwealth government's position has been that a change in tenure from State Forest to National Park is the sole route to achieving effective conservation, while the State governments have argued that retaining State Forests is the only way to maintain wood production and dependent industries and employment.

[2] Nevertheless, Smith and Maltby (2003) have made a first attempt at examining how the CBD understanding of the Ecosystem Approach might be translated into operational terms. Shepherd (2004), working with the IUCN Commission on Ecosystem Management, has regrouped the 12 Ecosystem Approach principles into five steps, in order to make them more operational.

Box 4.2. The convention on biological diversity's 12 principles of the ecosystem approach. (From CBD, 2000.)

- The objectives of management of land, water and living resources are a matter of societal choice.
- Management should be decentralized to the lowest appropriate level.
- Ecosystem managers should consider the effects (actual or potential) of their activities on adjacent and other ecosystems.
- Recognizing potential gains from management, there is usually a need to understand and manage the ecosystem in an economic context. Any such ecosystem-management programme should:
  - reduce those market distortions that adversely affect biological diversity
  - align incentives to promote biodiversity conservation and sustainable use
  - internalize costs and benefits in the given ecosystem to the extent feasible
- Conservation of ecosystem structure and functioning, in order to maintain ecosystem services, should be a priority target of the ecosystem approach.
- Ecosystems must be managed within the limits of their functioning.
- The ecosystem approach should be undertaken at the appropriate spatial and temporal scales.
- Recognizing the varying temporal scales and lag-effects that characterize ecosystem processes, objectives for ecosystem management should be set for the long term.
- Management must recognize that change is inevitable.
- The ecosystem approach should seek the appropriate balance between, and integration of, conservation and use of biological diversity.
- The ecosystem approach should consider all forms of relevant information, including scientific and indigenous and local knowledge, innovations and practices.
- The ecosystem approach should involve all relevant sectors of society and scientific disciplines.

This polarized debate has been termed one of 'single tenure–single use', with any gains for one side translating into losses for the other. Things came to a head in 1992 after a massive public protest over a politically motivated renewal of a wood chip export license. Both the Commonwealth and State governments recognized that this type of political posturing was counterproductive to all stakeholder interests. In response, the government introduced the concept of negotiated regional forest agreements. This process introduced many elements of the Ecosystem Approach concept, seeking a balance between conservation and forest production and attempting to ensure broad public participation in the decision-making. Regional forest agreements were developed across the country, reflecting Ecosystem Approach principles in their codes of forest practice, management plans, sustainable yield calculations and environmental management systems. The same process also resulted in a national conservation reserve system that includes a wider range of forest types and a larger and more consistent system than had existed previously. While the regional forest agreements have also had their weaknesses, particularly in achieving resource security for wood-using industries and in taking account of indigenous heritage concerns, they represent a major step towards a more rational, balanced approach to forest management.

USA

The rapidly urbanizing society of the Pacific Northwest, whose ranks have been swelled in the past few decades by immigrants attracted by the area's environment and economy, has quite different expectations of forest functions to those of the resource-dependent rural communities (Haynes *et al.*, 2005). Landscape beauty and recreational opportunities are top priorities for these urban populations, and forest managers have had to respond to these demands by providing an acceptable mix of commodity production, amenity use, and environmental and biodiversity protection. Balance has had to be sought between addressing conservation concerns (including the endangered northern spotted owl, dependent on the area's old-growth forests) and employment and other economic concerns.

The previous regime used in managing publicly owned forests until the 1980s was dominated by clear-cutting, burning and replanting, with timber extraction as a primary objective (although within a multiple-use context). Strong public reactions to the visual and environmental impacts of this system, and the ensuing conflicts between the different stakeholder groups in the 'jobs versus owls' debate, resulted in the Pacific Northwest becoming a test bed for the development of operational ecosystem management, as the Forest Service sought ways of building consensus and defusing conflict. Similar ecosystem approaches have evolved elsewhere in North America, but nowhere was the process as controversial and contentious as in that region.

The ecosystem approach concept is reflected in the area's natural-resource management plans developed during the 1990s, which now focus on 'old-growth' and multi-resource ecosystem management to provide habitat for threatened and endangered species (notably the northern spotted owl and wild salmon), protect riparian zones and promote biodiversity. State regulations for forest management also include many of the principles behind the ecosystem approach. It is particularly interesting that the Pacific Northwest Forest Plan, developed in 1993 as a long-term policy for managing northern spotted owl habitat, includes the designation of adaptive management areas to allow for the testing and modification of conservation management assumptions and approaches.

India

In 1988, the government policy was changed to subordinate commercial timber interests to conservation and local communities' needs as the primary objectives of forest management (Saigal *et al.*, 2005). Conflicts between local communities and the forest bureaucracy, and public protests against the earlier policy played an important role in the government's reorientation of its forest policy. On the basis of this new policy and the encouraging results from some pioneering experiments in community-based forest management, the government started the ambitious joint forest management programme, which shares many of the same principles as ecosystem approaches.

Over the last two decades, the joint forest management programme has emerged as a major strategy in the country, and by September 2003 there were officially 84,632 joint management groups protecting and managing over 17 million ha of state forest lands. The positive impacts of this programme have included an improvement in the relationship between the Forest Department staff and local communities, increased income for participating communities and an improvement in the condition of forests. On the other hand, it has had several shortcomings, including the lack of a firm legal basis, domination of joint forest management groups by the village elite, and inequitable sharing of benefits within communities; in some cases, the programme has led to intercommunity conflicts. A key challenge has also been the limited empowerment of the groups in real terms and the de facto control that the Forest Department still retains over them. Despite these problems, the new joint forest management approach still represents a significant improvement.

Other initiatives have promoted ecosystem approaches, including the ecodevelopment programme and people's protected areas, both of which seek to address the conflicts between conservation and communities in and around protected areas. The sustainable forest management work of the Indian Institute of Forest Management, which began in 1998, and the current preparation of the national biodiversity strategy and action plan, which strongly advocates the use of ecosystem approaches, are further landmarks on the road to acceptance of this approach in India. The challenge is now to effectively coordinate these innovations and put into practice the new guidelines and recommendations on ecosystem management.

Central America

Historically, the relationship between people and forests in Central America has been a difficult one. Local people have seen little benefit in forest conservation, as land and property rights have been unequally distributed, forest landscapes have been highly fragmented and forest management sizes have been small (Campos Arce *et al.*, 2005). The region's high levels of poverty and population growth, small and stagnating economies, and weak public institutions present additional barriers to achieving sustainable forestry. Yet, in the face of these huge challenges, there are encouraging signs of progress. At the end of the 1980s, the ITTO reported that there were no good examples of sound forest management in the region, while today, the Forest Stewardship Council records show 691,346 ha of certified forests in 42 units of natural and planted forests, including the community concessions of the Maya Biosphere Reserve (Poore *et al.*, 1989; FSC, 2004).

This move towards sustainability has been driven not by the region's forest industries but by its research institutes. In particular, the Tropical Agricultural Research and Higher Education Center (CATIE) has spearheaded research and development for sustainable forest management based on an ecosystem approach. The academic attention on sustainable forest management and ecosystem approach principles was then taken up by governments and other actors throughout the region. The Ecosystem Approach found strong support from international donors and began to emerge in new policies and projects. The 1990s saw many integrated conservation and development projects, which proved the viability of small-scale sustainable forest management and showed the need for local participation and benefits. At the same time, there has been a gradual shift in how the region's peoples and governments view forests, following several natural disaster crises. Hurricane Mitch in particular revealed the link between climate, natural resources and people, and forced Central American governments and international cooperation agencies to reorient their development strategies to address social and ecological vulnerability, transparency, participation and local development. These natural disasters also emphasized the important role of forests in mitigating the impacts of such events and helped promote the development of integrated watershed management policies. An important example of government reform, favouring an ecosystem approach, is the recent decentralization and reorientation of natural-resource management in Costa Rica, which has set ecoregions as the basis for the country's national conservation area network.

The growing awareness among Central American states and societies that ecosystems produce significant goods and services has helped reduce the historical incentives for forest conversion, and this has been the key to creating an enabling environment for ecosystem approaches. Important reforms and innovations in the region that have further promoted ecosystem approaches have included (Campos Arce *et al.*, 2005):

- Modernization of central government institutions (as in Costa Rica);
- Regional integration among the countries;
- Strengthening of municipal governments, particularly in Honduras and Nicaragua;
- Establishment of forestry producer organizations (particularly in Costa Rica and Guatemala), and the mobilization of civil society at large;
- Establishment of community forestry concessions in Guatemala and the development of financial mechanisms that value forest ecosystem services, particularly in Costa Rica;
- Ecoregional approaches for the sustainable management of natural resources, such as the Mesoamerican Biological Corridor.

For ecosystem approaches to become widely established in the region, there is a need for further institutional and policy changes. In particular, forest producers (especially small-scale and medium-scale ones) will need access to technical and financial resources and clarity over property and use rights for forests before they can participate fully in sustainable forest management and conservation.

Congo Basin

The history of forest management in the Congo Basin did not provide a promising start for the development of ecosystem approaches (Sayer *et al.*, 2005). In the 1970s and early 1980s, most forest management and conservation efforts went into either regulating logging concessions or establishing and protecting parks and reserves. This situation was in many ways the antithesis of ecosystem management, as local communities were largely excluded from forest governance, forest use was sharply segregated into protection and production zones, and the central government controlled everything.

Even in the late 1980s, international donor support for forest management in the region paid very little attention to social or environmental objectives. The Tropical Forestry Action Plan (TFAP) for Cameroon, prepared with donor support, was rooted in a vision of sophisticated large-scale concession management for international markets with silvicultural treatments that would greatly increase future yields of commercial timber. The TFAP contained the implicit assumption that environmental benefits would be inevitable by-products of good forestry practice. The Cameroon TFAP was vigorously attacked by environmental NGOs for its pro-logging stance and its failure to address the needs of conservation and forest-dependent communities. The debate that followed influenced the development of a new and progressive forestry law in 1994, which contains many innovations that favour the interests of forest people and biodiversity.

Cameroon remains the main focus of innovation in forestry in the region, as intermittent periods of civil conflict in the Democratic Republic of Congo, Congo-Brazzaville and the Central African Republic have inhibited similar innovations in these countries. However, despite the civil unrest and declining or stagnant economies in the region, a raft of recent policy decisions and international commitments by the governments is pointing towards a more inclusive approach to forests. The preamble to the Congolese forest law of 2002, for example, refers to 'forest ecosystems', and the legal frameworks in all countries of the Congo Basin now show considerable progress towards the integration of ecosystem approach objectives. Progress on the ground is very patchy, though it is still early days since these legal commitments have been made. Economic difficulties across the region mean that their forest management capacities are stretched to the limit and innovation is difficult.

Still, there are a few success stories for forest ecosystem management in the region. One example is the work of the Wildlife Conservation Society (WCS) in and around the concession of the Congolaise Industrielle des Bois in northern Congo. Significant steps have been taken by the concessionaires, under pressure from environmental NGOs, towards sustainable forest management, including the provision of livelihood support for the resident BaAka pygmy communities and the control of hunting of threatened large mammal species in the area – recently rewarded with a Forest Stewardship Council certificate. Other successful efforts include the work of the World Wide Fund for Nature (WWF) in the Dzangha-Sangha region of the Central African Republic and the work of WCS in the Ituri Forest of the Democratic Republic of Congo. However, these examples all come from remote, sparsely populated and relatively inaccessible areas of the Congo Basin. The real challenge for ecosystem approaches here lies in achieving success in the more densely settled and accessible forest areas, especially near the coast and the main communication axes. Maintaining large mammal populations in such densely settled areas will remain a particular challenge (Sayer *et al.*, 2005).

Russia

The development of ecosystem approaches in Russia has seen some drastic alterations over the last two decades (Angelstam *et al.*, 2005). The 'command and control' approach to forest management, firmly in place up until the early 1990s, was a classic version of sustainable forest management that focused on timber production and conservation, while largely ignoring any social concerns. Following the country's transition to a market economy, the collapse of this state-run system and the privatization of most industrial forest enterprises created a situation of great uncertainty. Harvesting volumes dropped sharply, social tensions erupted and illegal forestry activities soared. Recent years have seen some serious attempts at reorganizing the forest sector to achieve broader-based and more sustainable approaches. This has been most evident in the north-western federal district and the republic of Karelia in particular.

Karelia's forests cover more than half of the Republic's territory, and timber exports are an important contributor to the area's economy. Currently, 12% of Karelia's forests are set aside as nature reserves, national parks and wilderness areas. Management policies for the remaining forest areas are now attempting to curb excessive felling while maximizing value added by promoting the exportation of processed timber. The citizens of Karelia are becoming more aware of the need for sustainably managed forests, as the basis for better employment opportunities and better livelihood.

The potential impact of public debate, seen in the studies of Australia and the USA as a driver of ecosystem approaches, is also evident here. One such debate started in the 1990s and centred on the fate of the virgin taiga forests at the border of Karelia (western Russia) and Finland. Local authorities and environmental groups took strongly opposed positions, national and international stakeholders became involved, and a boycott was taken up against timber shipments from these forests. A productive dialogue was finally established and, with funds from the EU and the Russian Federation budget, national parks were established to protect the forests. Model forests have been established in several parts of Russia, and these espouse ecosystem approaches. However, these successes are the exception to the norm, and examples of integration of environmental and social concerns within Russian forest management regimes are still hard to find. The ongoing Northern Eurasian forest law enforcement and governance (FLEG) process presents Russia with an opportunity to consider how it can scale up some of the lessons learned from the successful applications of ecosystem approaches.

Conclusions

The most important conclusion from our case studies is that many of the issues that the Ecosystem Approach principles highlight are being addressed on the ground in the countries and the regions examined. The sort of thinking that led to the Ecosystem Approach principles is alive and well in the real world and has inspired much of the reform of policies and practices of forest management that has occurred in the past decade. And, while there are clear differences between ecosystem approaches and the more traditional applications of sustainable forest management, many of the principles of the Ecosystem Approach have already been incorporated into the more holistic sustainable forest management experiences. The real value of the Ecosystem Approach, therefore, is not as a competing concept to sustainable forest management, but rather as a set of general guidelines that help enrich the debate and provide a broad conceptual framework for resource management.

Some surprising similarities emerge from the case studies. It is clear, for example, that the spark for developing ecosystem approaches often comes from adverse public reactions to inappropriate forest management policies. The impact of public pressure on governments and especially on large-scale forestry corporations cannot be underestimated. Another common thread running through the case studies is the need for fundamental shifts within forest agencies. Forest management objectives, forest managers' skills and attitudes all need to change as the role of forest departments changes from an expert-driven, enforcement-oriented one to a collaborative, consensus-building one. Forest agencies of the future will have to provide the following services:

- Facilitating a dialogue among all forest stakeholders to establish a vision for their forests and to determine the limits within which forest owners and managers may operate;
- Establishing and maintaining multiple-resource databases on forests to detect emerging trends, threats and issues and to allow for adaptive management;
- Providing a problem-solving research capacity to deal with emerging problems of pests and diseases and to determine management requirements for specific targets (such as conservation of endangered species);
- Providing the overview, analysis and verification needed to make environmental service payments effective in supporting the production of the public good values of forests;
- Developing, reviewing and enforcing regulations, and recommending any necessary adaptation of these regulations.

References

Angelstam, P., Kopylova, E., Korn, H., Lazdinis, M., Sayer, J.A., Teplyakov, V. and Tornblom, J. (2005) Changing forest values in Europe. In: Sayer, J.A. and Maginnis, S. (eds) *Forests in Landscapes: Ecosystem Approaches to Sustainability*. Earthscan, London, pp. 59–74.

Campos Arce, J.J., Villalobos, R. and Louman, B. (2005) Poor farmers and fragmented forests in Central America. In: Sayer, J.A. and Maginnis, S. (eds) *Forests in Landscapes: Ecosystem Approaches to Sustainability*. Earthscan, London, pp. 129–146.

CBD (2000) Decision V/6 of the Conference of the Parties to the Convention on Biological Diversity. Available at: www.biodiv.org/decisions/

FAO (Food and Agriculture Organization of the United Nations) (1994) *Readings in Sustainable Forest Management*. Forestry Paper 122. FAO, Rome.

FAO (1999) *Status and Progress in the Implementation of National Forest Programs: Basic Principles and Operational Guidelines*. FAO, Rome.

Ferguson, I. (2005) Australian forestry: 'beyond one tenure–one use'. In: Sayer, J.A. and Maginnis, S. (eds) *Forests in Landscapes: Ecosystem Approaches to Sustainability*. Earthscan, London, pp. 147–164.

FSC (2004) Forests certified by Forest Stewardship Council–accredited certification bodies. Consulted 23 June 2004. International Center, Bonn, Germany. Available at: www.fscoax.org

Haynes, R.W., Szaro, R.C. and Dykstra, D.P. (2005) Balancing conflicting values: ecosystem solutions in the Pacific Northwest of the United States and Canada. In: Sayer, J.A. and Maginnis, S. (eds) *Forests in Landscapes: Ecosystem Approaches to Sustainability*. Earthscan, London, pp. 101–114.

De Montalembert, M.R. and Schmithüsen, F. (1993) Policy and legal aspects of sustainable forest management. *Unasylva* 44(4), 3–9.

Poore, D., Burgess, P., Palmer J., Rietbergen, S. and Synott, T. (eds) (1989) *No Timber Without Trees: Sustainability in the Tropical Forest*. Earthscan, London.

Saigal, S., Mitra, K. and Lal, P. (2005) Empowering the forest-dependent poor in India. In: Sayer, J.A. and Maginnis, S. (eds) *Forests in Landscapes: Ecosystem Approaches to Sustainability*. Earthscan, London, pp. 75–100.

Sayer, J.A. and Maginnis, S. (eds) (2005) *Forests in Landscapes: Ecosystem Approaches to Sustainability*. Earthscan, London.

Sayer, J.A., Ndikumagenge, C., Campbell, B. and Usongo, L. (2005) Wildlife, loggers and livelihoods in the Congo Basin. In: Sayer, J.A. and Maginnis, S. (eds) *Forests in Landscapes: Ecosystem Approaches to Sustainability*. Earthscan, London, pp. 115–128.

Shepherd, G. (2004) *The Ecosystem Approach: Five Steps to Implementation*. IUCN, Gland, Switzerland.

Smith, R.D. and Maltby, E. (2003) *Using the Ecosystem Approach to Implement the Convention on Biological Diversity: Key Issues and Case Studies*. IUCN, Gland, Switzerland.

5 Agriculture and Forestry: Enhancing Cross-sectoral Policy Planning

Randy Stringer

Introduction

Twentieth-century experience shows that, ironically, many government policies frequently exacerbate forest conversion, producing intense and lasting impacts on forest resources. Examples include tax policies, terms of forest concessions, administered prices, controlled transportation of forest goods, land and tree tenure insecurity, tariff and non-tariff barriers to international trade, investment incentives, agricultural sector strategies and macroeconomic policies all affect economic motivations and management of forests. In many cases, these policies directly encourage or unintentionally subsidize deforestation and degradation.

Attempts at establishing effective strategies to protect forests involve an array of difficult choices. Forest land produces income when converted to agriculture or logged for timber, but markets do not exist for many other important forest services and benefits. The result is that many land use decisions are biased in favour of agriculture and forest production decisions in favour of timber. Some policies result in inefficient resource use because many essential forest and agroecosystem goods and services such as watershed protection, soil conservation, biological diversity, wildlife habitat and carbon sequestration are not priced. Today, countries are seeking to find appropriate economic policies, regulatory mechanisms, financial incentives, organizational structures and tenurial arrangements to promote balanced and sustainable development of their forestry and agricultural practices. In many countries, the search for policies takes place alongside a wider examination of the role of governments as land and forest managers, as regulators of the market place and as landowners. This examination is prompted partly by governments' own need to optimize resource efficiency, and partly by the perceived ineffective performances.

This most recent view of what forests and agroecosystems are and what they contribute requires national strategies and policies to integrate forests in rural development efforts and balance economic and environmental needs among local, national and international interests. Governments are searching for pragmatic policy frameworks that deal coherently with both the contributions of environmental services to development and the organizational structures required to make better use of these contributions. In addressing the wide spectrum of priorities between local and global perspectives and responding to interest groups that may have competing objectives, trade-offs are inevitable. Critically important issues of equity and morality arise when the interests and welfare of local communities, with limited options and capacity to find alternatives for their subsistence, differ with national or international priorities. Consulting and compensating those poorly served by the priorities selected are essential; the public must be involved in setting priorities.

For all these reasons, national governments are challenged to mesh people's needs with national and global interests; to use policies that determine agricultural and forest conditions in ways that help improve opportunities for people and communities; and to better understand how interactions across sectoral policies influence how people use forests and agriculture land and the consequences of that use on national development.

This chapter presents lessons emerging from recent research efforts at the Food and Agriculture Organization of the United Nations (FAO) aimed at extending current thinking about the environmental, social and economic roles of agriculture including forestry.[1] It also discusses the effectiveness of cross-sectoral policy strategies. A key motivation for the research is to provide policy guidance for improved development strategies, especially for understanding the ways in which policies aimed at one goal impacted on other sectors, both with positive and negative consequences. Often the consequences are unintentional and unrewarded when they are positive, and unpunished when they are negative. Much of this recent research and the country case study results[2] call attention to a diverse set of indirect environmental, social and economic contributions. The evidence suggests that these indirect contributions are not well understood, seldom analysed in the context of development, and rarely reflected in national and rural development policy strategies.

Research Project

Objectives

The FAO socio-economic and policy implications of the roles of agriculture projects aimed at extending current thinking about the economic and social roles of agriculture to shed new light on one of the oldest issues addressed by development economists: how does agriculture contribute to economic and social development (Stringer and Pingali, 2004)? While policy attention has long focused on agriculture's direct roles – to provide food, create jobs, earn export income and produce primary commodities for expanding industries – the research project recognizes these direct, market-mediated contributions, but goes on to incorporate agriculture's indirect contributions, examining the importance of the public and private benefits they provide. Many of these indirect non-commodity contributions are public goods, social service benefits and environmental services not captured by markets.

The research project focuses on the non-market and non-production roles of agriculture. What motivates the research is that these diverse indirect roles of agriculture have not been well identified or analysed in the context of development, and are rarely reflected in national and rural development policy formulation. The study results argue that the public policy implications of these roles must be taken into account. The central objective of the research project is to identify and evaluate the positive contributions of the agriculture sector that might currently be undervalued. The research seeks answers to a set of policy-relevant questions to better understand whether agriculture is underperforming because it is under-rewarded (FAO, 2004):

- When a rapidly growing agricultural sector is crucial for reducing hunger, poverty and inequality, do policy-makers recognize the total social value of those contributions?
- Do governments take into consideration the cross-sectoral policy consequences during their planning and analysis activities?
- Are governments providing the right policy signals to capture the net social benefits from agricultural activities?
- Are governments investing adequate levels of public resources in their agriculture to take advantage of these contributions?

Emerging lessons

An important concept for the research project is that agriculture's contributions to society and the economy evolve over time. In agrarian societies with few trading opportunities, most resources are devoted to the provision of

[1]The research project Socio-economic and Policy Implications of the Roles of Agriculture Project is funded by Japan's Ministry of Agriculture, Forestry and Fisheries.

[2]The case study countries are China, India and Indonesia in Asia; Ethiopia, Ghana, Mali, Morocco and South Africa in Africa; and Chile, the Dominican Republic and Mexico in Latin America.

food. Agriculture's shares of national output and employment are therefore high. As economic development proceeds, agriculture's shares of the gross domestic product (GDP) and employment fall. This long-term shift of jobs, production and populations from agriculture to industry and services is a widespread economic characteristic of development.

Policies influencing agriculture shift in ways that reflect the sector's importance and its evolving contributions to society and the economy. As per capita income rises, society is not only willing to pay more for environmental quality, but it also means that more income is available to protect environmental services. This does not imply that lower-income countries desire environmental quality less, but it is consistent with the fact that people in relatively wealthy countries have greater capacity to pay for more of everything, including higher environmental quality. Higher willingness to pay for environmental services at higher income levels helps explain why the multiple functions of agriculture attract a great deal of attention in the Organization for Economic and Cooperation Development (OECD) countries. Many industrial countries look to agriculture to provide rural amenities, attractive agricultural landscapes and cultural heritage. These particular non-commodity benefits reflected in OECD country examples are often relatively unimportant in developing countries with high poverty levels and large numbers of subsistence producers.

The country cases provide lessons on the policy combinations countries used to address ecosystems in general, and agroecosystems in particular. The lessons are analogous for the forestry sector too. For example, as development proceeds, population pressure on land increases and then declines; urban demand and prices for wood products and energy increase; urban income and savings rise; non-farm employment opportunities grow; road systems and water resource developments expand; governmental capacities to protect forests, to subsidize forest growth in agricultural areas and to cooperate with local populations in forest management increase; the strength of environmental relative to extractive interests and the extent of their international integration increase.

Thus, national development policy is constantly creating incentives and capacities both to utilize and to enhance forest and agricultural resources. These dynamics determine the motives to cut and to grow trees in different places and at different times. Forest conditions reflect the consequences of laws, programmes and policies that create and modify the opportunities for people to grow and cut trees. For example, relative market prices between agricultural and wood products, and between fossil and forest fuels, influence the growth of plantations and rates of natural forest depletion. Likewise, the development of roads, ports and other public infrastructure provides new incentives and influences price structures and relationships.

How agricultural land and forests are used in any one location depends on the degree of competition for non-tree uses of land, the degree of access to markets and resources for tree growth, the rights provided to local forest-dependent communities, the cost of alternative sources of forest values and the social basis for the negotiation and enforcement of rules. Such factors reflect the nation's economy and society as a whole, as well as the policies a nation chooses to affect them. Forests and agroecosystems are living systems that evolve over time with or without human intervention. Forests and the resources devoted to growing, maintaining and protecting them depend on combinations of many different policies: environmental, energy, land, commodity, trade, industrial and agricultural policies; price, wage, income and investment policies; and the terms of international agreements. The task is to relate policy combinations to forest consequences in diverse conditions and to identify those that are likely to serve local, national and international interests in the best way.

Other sectoral contributions are likely to follow different paths, with each development stage corresponding to a different type of policy environment and economic links with the rest of the economy. The country cases reveal that the forestry sector, together with agriculture and rural development strategies overall, took important steps towards understanding how to use policies to affect resource use, as national development strategies shifted from project-based to policy-oriented programmes.

Effectiveness of Cross-sectoral Policy Strategies

During the mid-1980s, policy analysts turned their attention to the impacts of intersectoral policy linkages on the forestry sector, recognizing the inability of traditional forestry strategies to slow the accelerating pace of deforestation and forest degradation. Policy studies highlighted that the roots of forest degradation and depletion often lay outside the forestry sector. In the industrial countries, the effects of pollution (acid rain) on temperate forests highlighted this problem. In the developing countries, population growth, land tenure systems and agricultural sector policies were seen as underlying causes of deforestation.

One common and useful way of conceptualizing how these intersectoral policy linkages affect forests is to visualize a set of concentric circles moving outwards from the forest (FAO, 1994). At the hub are policies that directly affect forest management: forest revenue structures; tenurial institutions governing the privatization of forest land and enforcement of traditional use rights; reforestation incentives; and administration of timber harvesting concessions. In the next circle are policies directly influencing the demand for forest products: trade and investment incentives to promote wood-using industries and energy pricing to encourage fuelwood substitutes. In the third circle are policies directly affecting extensions of the agricultural frontier and the rate of conversion of forest land: agricultural credit, tax and pricing incentives for frontier agriculture, including policies affecting the relative price of new forest land; incentives for cultivation at the intensive as opposed to the extensive margin; and the concentration of landholdings as well as public investments that indirectly spur frontier expansion in the form of road building and public services, such as agricultural research and extension. In the outer circle are macroeconomic policies that indirectly affect deforestation: exchange rate policies affecting tropical forest product exports; policies affecting capital markets which influence investors' time horizons; demographic policies; trade and investment policies affecting labour absorption; and rural–urban migration.

It is important to note that the macro scale could just as easily be placed at the hub or in one of the intermediate circles, depending on a country's particular circumstances and where the primary interest resides. The primary interest may not be in the forest unit. Forest sector–specific policies have proved to be remarkably ineffective without a macro policy context allowing them to work properly. How these policy linkages are defined and interpreted depends on whether forest issues are viewed from a national (macro) or a forest unit (micro) perspective; evaluated using development-oriented or resource-oriented concepts of capital, space and location; and analysed with macroeconomic or microeconomic methods. The empirical evidence suggests that overall effectiveness of cross-sectoral policy strategies depends on how well governments address two key issues: market failures and incentive structures; and policy failures (FAO, 1994).

Market failure

Market failure occurs when incentives offered to individuals, households and firms encourage behaviour that does not spur socially optimum outcomes. When public goods, including public environmental goods and externalities, are present, incentive structures may lead to market failure. The market does not confront users with the full social costs of their actions. For example, markets that fail to reflect environmental values fully can lead to excessive environmental degradation. Some form of public or collective action involving regulation, market-based incentives or institutional measures is required if market failures are to be corrected.

Forests may be affected in several ways. For example, the market prices of widely traded timber products typically do not reflect the environmental costs of their production. Market prices fail to account for indirect use values (e.g. watershed protection or nutrient cycling) as well as future use and non-use values (e.g. option value or existence value), which may be lost or degraded by the production or consumption of forest products. Many environmental benefits are public goods and thus have no market price.

The agriculture sector too contributes less tangible, non-market-mediated services and benefits. The classic environmental externality is a good example. Most of us are familiar with the negative externalities associated with agriculture. A wide range of policy measures are used to internalize externalities, making the agricultural polluter pay for harming soil and water with pesticides and fertilizers, or even for causing bad odours or too much noise.

Agricultural activities also generate positive externalities. Important examples include watershed services, wildlife habitat, biodiversity benefits and carbon sequestration. The research project identified a range of these non-market-mediated agricultural benefits, from environmental services to food security and social viability benefits.

Take food security, for example. We know that when children are consistently well fed, they miss fewer days of school and they have much better learning outcomes. When their parents are well fed, they have more productive days at work, are healthier, stronger and can earn more income. Aside from the private, direct return to these households, increased earnings (and even more money to spend on health care, education, and productivity-enhancing investments), families with higher levels of food security provide indirect, public benefits. The health, well-being and income benefits that come with greater food security enable parents to be better mothers and fathers. Farm families are able to participate more in local organizations, contributing to a more stable and thriving rural community. These outcomes yield significant benefits to others, not the least of which is a better chance for prosperity, peace and stable institutions. These food security benefits contribute to society in ways that cannot be measured easily by economic growth.

Policy failures

Property rights also shape the system of incentives and disincentives for forest use. The structure of property rights defines the rules, rights and duties within which users of the forest operate. Economic policy-makers place great importance on property rights systems because they govern the efficiency of resource use throughout the economy as well as the distribution of income. Forest tenure systems range from exclusive rights to open access. Public forests in many developing countries are often open-access resources from which no one can be excluded. In countries with extensive tropical forests, the public sector's claims on tree-covered land far outstrip its ability to manage or control forest resources. In an open-access situation, one individual may want to conserve the forest or set it aside for future use, while another may decide to extract timber for personal gain. The risk and uncertainty associated with abstaining from current use creates an incentive to maximize short-term returns by harvesting forest products immediately. Thus, the opportunity costs of resource use (at long-term social prices) are not taken into account. This type of behaviour is not limited to open-access situations. If the owner of a forest parcel is so poor that the revenue from selling trees is needed immediately, the discount rate applied to tomorrow's benefit is infinite and the parcel would therefore be logged today.

Institutional and legal arrangements governing land tenure and transactions can also have significant effects on forest land use. In the past, property laws and land reform legislation in many countries required settlers to clear land in order to secure title to forested areas. Large areas that could have been exploited for commercial timber and non-wood products on a sustainable basis were lost in this way. With the exception of remote frontier areas of tropical forests, very little forested land exploited for timber is subject to pure open-access conditions. However, the failure to design appropriate concession arrangements for public forest lands and insecure tenure rights to plantations can create conditions similar to an open-access situation. Private individuals and businesses make harvesting decisions on the basis of short-term profit-maximizing objectives and have little regard for the potential for greater returns from timber stands in the future.

Public investment often has direct effects on forest-based activities, particularly where transport infrastructure and public services are extended to previously inaccessible forested areas. This type of investment may be an important subsidy for the logging and wood-processing industry, because it reduces the costs of gaining access to forest resources. Likewise, it represents a subsidy for consumers, as it brings forest products to market less expensively. Public investment in remote forested areas also acts as an impetus to human migration and agricultural expansion, which is the major cause of forest clearing in many countries. Examples of forest sector–specific policies that aim directly at forest management include tax credits or subsidies for forest conversion, afforestation and wood production. Forestry is also affected by policies that alter incentives and impede competition in downstream industries or related sectors, such as wood processing and construction.

Domestic market and policy failures also have a major influence on the conversion of forest land to agriculture and other uses. As this is the single largest cause of deforestation in the world, addressing only market and policy failures that directly affect the forestry sector will by no means be sufficient to halt deforestation and forest degradation in most countries. All these domestic market and policy failures have important implications for sustainable agriculture and rural development, not just forest management. If public policies are to be redirected to achieve efficient and sustainable resource management, changes are required. The research project argues that the economic valuation of current policies plays an important role in determining the appropriate policy responses. Often, however, insufficient economic data and information exist to allow a precise estimation of the economic costs arising from domestic market and policy failures.

Conclusions

Experience to date suggests that agriculture's non-commodity benefits are too easily overlooked. The reason is that producers and firms tend to make most decisions on the basis of private returns; hence, any external effects of their production activities, either positive or negative, lead to underprovision or overprovision of non-commodity outputs. Market prices fail to convey proper signals to secure an optimal provision of the multiple services and benefits that agriculturalists provide.

The various policy and market failures facing agriculture and forest landscapes are related in large part to the lack of information concerning these evolving market and non-market roles. Ignoring the whole range of environmental, social and economic contributions of agriculture and forests underestimates the returns to investment in these sectors and the trade-offs between them. Ignoring these contributions also underestimates their ability to shape the conditions in which the rural poor are able to enhance environmental outcomes, to raise their incomes and to live longer, healthier and more productive lives. Finally, ignoring the good and bad consequences of cross-sectoral policy impacts too often leads to undesirable outcomes.

Given the importance of agriculture's and forestry's non-commodity benefits in developing countries as well as the trade-offs between them, one of the most pressing policy questions is how best to target and allocate domestic and international investment for rural and agricultural development in ways that enhance and maintain the non-market roles. By calling attention to the externalities and public benefits generated by agriculture in the developing world, the research study raises the question of how governments can best support and facilitate an incentive framework that ensures a more optimal level of market and non-market agricultural goods and services.

The research project results present evidence of a diverse set of indirect social and economic contributions by agriculture. The evidence suggests that these indirect contributions are not well understood, seldom analysed in the context of cross-sectoral policy development, and rarely reflected in national and rural development policy formulation. The various policy and market failures facing agriculture are related in large part to the lack of information concerning the sector's evolving market and non-market roles.

Some policy signs and trends are encouraging. For instance, reversing past discrimination and policy bias against rural sectors in developing countries is evolving gradually. The country case studies do demonstrate a strong interdependence between the rural and urban sectors, and the many cross-sectoral links through which agricultural growth supports overall economic growth is more widely recognized. However, the less visible roles of agriculture in providing environmental services, and reducing poverty and hunger are increasingly recognized, yet still insufficiently addressed by the policy community.

The research presents a compelling case for exploring further the many additional economic and social benefits of agriculture. Agriculture does contribute to social and economic development in various ways, providing environmental, social and cultural benefits and services that are neither the object of market transactions nor reflected in agricultural prices or other forms of payments. As a result, these other roles are often ignored or overlooked in policies and development efforts. These contributions do not appear in national accounts. Their interactions and the interlinkages and cross-sectoral implications imply trade-offs that cannot be assessed properly because of the lack of data and analysis.

References

FAO (Food and Agriculture Organization of the United Nations) (1994) *The State of Food and Agriculture, Forest Development and Policy Dilemmas*. FAO, Rome.

FAO (2004) *FAO Research Programme Summary Report*. October. FAO, Rome. [This report and all the country module reports are available at: www.fao.org/es/ESA/Roa/pub_docs_en.asp]

Stringer, R. and Pingali, P. (2004) Agriculture's contributions to economic and social development. *Journal of Agricultural and Development Economics* 1(1), 1–5.

6 Policy Impact of the Kyoto Protocol on Sustainable Forest Management

Dieter Schoene

Introduction

Evidence of global warming is mounting, and public awareness of it is rapidly increasing. Forests respond with great sensitivity to even minute differences in temperature and moisture regimes. The rise in the global average surface temperature by 0.6°C since 1900 is already killing trees in boreal forests (FAO, 2003), and major shifts in the geographic distribution of forest vegetation and some dieback are expected in the future. In most instances, forest decline will not visibly be caused by climate change, but by climate-influenced stress effects such as fire, disease, deficiencies of nutrients or water, and insect pests. Recent harbingers of global warming – warmer winter temperatures and more arid Canadian summers – led to an outbreak of the mountain pine beetle, which killed forests in an area of almost 9 million ha, comparable to the entire forested area in Poland. On the other hand, some forests may actually benefit from longer growing seasons, warmer temperatures and enhanced growth.

Potential Roles of Forests in Climate Change

Planting new forests to absorb excess carbon dioxide (CO_2) in the atmosphere is the option that usually comes to mind first in connection with using forests to curb climate change. The idea of carbon-offset plantings, originally proposed by Dyson (1977), is now being implemented throughout the world under the Kyoto Protocol. However, beyond the technique of planting trees in productive or protective plantations, agroforests or urban forests, an entire palette of silvicultural and management options exists for enhancing carbon uptake and storage in forest ecosystems – such as restoring degraded forests, enrichment plantings, extending rotations in even-aged forests, thinning lightly, favouring species with high sequestration rates, reduced-impact logging, underplanting open forests, and fertilizing or irrigating deficient stands.

Outside forests, wood products can store carbon for decades and centuries; in addition, producing them requires less energy from fossil fuel than competing products made from steel or aluminium.[1] In industrial countries, the carbon pool in wood products amounts to 20–40 t of carbon per hectare of forest (Dewar, 1990; Schoene and Schulte, 1999). In some circumstances, managed forests and their

[1]On average, wood production and harvest cause emissions of approximately 3 kg of carbon per cubic metre (Wegener *et al.*, 1996), and every cubic metre of timber substituting for high-energy raw materials, such as aluminium and steel, avoids emissions of 0.3 t of carbon on average (Burschel *et al.*, 1993).

products may store more carbon than unmanaged natural forests (Dewar and Cannell, 1992).

Immature forests – widespread in Europe, North America and East Asia – act as carbon 'sinks' naturally and without deliberate human intervention when they accumulate biomass, half of which is carbon. For example, the 2005 Global Forest Resource Assessment (FAO, 2006) estimates an annual carbon sink of over 0.5 Gt carbon in countries in which the biomass carbon stocks increased from 1990 to 2005 as a result of afforestation or carbon accumulation in growing stock.

Unfortunately, this gain is currently overcompensated by an annual drain of approximately 1.6 Gt of carbon in countries reporting decreasing carbon stocks due to deforestation and degradation (FAO, 2006). Thus, if global warming is conceived allegorically as representing a fever of the planet, forests not only function as the potential remedy but they also contribute indirectly to the illness. Deforestation and forest degradation contribute 24% of all anthropogenic carbon emissions, and 18% of all greenhouse gas emissions combined (Watson *et al.*, 2000; Baumert *et al.*, 2005).

As many as 60 million indigenous forest dwellers depend fully on forests and their products; 1.2 billion people in developing countries obtain food from trees, and at least 70% depend on forests as their sole source of medicine (Bundesministerium für Wirtschaftliche Zusammenarbeit und Entwicklung, 2004). How forests fare during climate change will therefore strongly influence human well-being and will affect progress towards the United Nations' Millennium Development Goals.

Forest Policy Guidance from the Kyoto Protocol

Both the United Nations Framework Convention on Climate Change and the Kyoto Protocol list general obligations regarding forests that apply to all member countries. They are called upon to manage forests sustainably and to conserve and enhance them as sinks and reservoirs of carbon. They are obligated to promote afforestation and reforestation, as well as renewable energy. Finally, they are to include forests in national inventories of greenhouse gas emissions and removals, in technology transfer and in national programmes of adaptation to climate change. The Kyoto Protocol assigns different roles to forests in developed and developing countries.

Role of developed countries

The developed countries have to assess greenhouse gas emissions and carbon removals due to afforestation, reforestation and deforestation since 1990 that occur during the first commitment period 2008–2012, and incorporate them into their net greenhouse gas emissions. They have to decide by the end of 2006 whether they wish to select forest management as an optional domestic activity under the protocol and include greenhouse gas removals or emissions up to country-specific limits in their national accounts (FAO, 2003). Developed countries that are members may carry out forestry carbon offset projects jointly, involving afforestation, reforestation or forest management under the Protocol's instrument of joint implementation. Carbon offsets achieved in the host country's forests are subsequently transferred to the investor country as a whole or in part, on the basis of contractual arrangements between the partners; they reduce the host country's emission rights. This arrangement represents a zero-sum game for emission rights. Joint implementation is of primary interest to industrialized countries and countries with economies in transition.

Through the climate change focal area of the global environment facility trust fund and three new funds established under the convention and the Kyoto Protocol, industrialized countries are obliged to finance preparation for, and implementation of, adaptation measures in developing countries to varying degrees. For instance, they fully finance national action plans for adaptation that all of the least-developed countries prepare, listing their priorities for adaptation.

Role of developing countries

The United Nations Climate Change Convention and the Kyoto Protocol refer in several articles specifically to forests in developing countries. Article 4 (1e) of the Protocol includes a mandate to all members to protect and rehabilitate areas affected by drought and desertification, particularly in Africa. According to Article 4 (8), developed countries have to give full consideration to meeting the needs relating to climate change of developing countries with forested areas and areas liable to forest decay; forests may be included in vulnerability assessments and adaptive measures funded by the financial mechanism of the convention and the Kyoto Protocol (Verheyen, 2003; Robledo and Forner, 2005).

The Convention requires developing countries to submit periodic national inventories of greenhouse gas emissions by sources, and removals by sinks, as part of their national communications. Industrialized countries are obligated to finance the full costs of these communications. For many developing countries, particularly in Africa, forests constitute a major source of emissions. Yet many of these countries' national forest assessments are obsolete or of poor quality, or both (Saket, 2003), making their greenhouse gas inventories very unreliable. While forests have so far received only scarce attention in national communications from developing countries, there is every reason to take them into account, in view of their importance for food security and rural livelihood in some countries. The most prominent role of forests in developing countries occurs in the context of the Kyoto Protocol's clean development mechanism.

Clean Development Mechanism

The clean development mechanism (CDM) allows industrialized member countries to meet a part of their greenhouse gas reduction obligations through offset projects in developing countries. These projects also have to promote sustainable development through investment in host countries, as well as through the transfer of knowledge and technology. Unilateral projects in the host country and subsequent sale of credits are also feasible.

The CDM is a market-based mechanism, driven by demand for credits – certified emission reductions – from private or public entities in developed countries, and by supply from offset projects in developing countries. Projects that reduce emissions from sources can be carried out in many sectors, particularly energy, including wood energy. However, only afforestation and reforestation qualify as projects that remove carbon from the atmosphere. Carbon sequestration in agricultural crops and soils is not eligible under the CDM during the first commitment period (2008–2012).

To avert criticism of large-scale forest plantations and to help meet the goals of food security and rural development, the mechanism contains a small-scale category with simplified conditions and reduced costs. Projects cannot obtain credit for more than 8000 t of CO_2 or, equivalently, 2200 t of carbon sequestered annually on average and must be undertaken by low-income communities and individuals. Projects may include agroforests or urban forests and – depending on productivity and envisaged stocking levels – may encompass areas between 200 and 4000 ha.

By June 2006, over 800 CDM projects – expected to yield over 1.5 Gt of CO_2 reductions – were in the pipeline, of which 222 had achieved registration. Only about 20 were afforestation and reforestation projects, and of these all but two were rejected due to methodological flaws involving either the rules of the mechanism or forestry issues. The Food and Agriculture Organization of the United Nations (FAO) has analysed forestry projects submitted to the convention and compiled a list of the lessons learned (Kägi and Schoene, 2005). While recent top prices of over US$100/t of carbon apply in European Union emission trading, carbon credits from afforestation projects are currently discounted by buyers for risk to prices of US$10–15/t of carbon. Clearly, unless hurdles to widespread implementation and some misconceptions about carbon credits from forestry projects can be overcome and a market developed, a potentially huge stream of investments in the CDM could bypass forestry.

Future Negotiations

Negotiations for the second commitment period under the Kyoto Protocol started in 2005. Previous negotiations notably lacked forestry expertise in general, particularly from developing countries. Countries have to decide whether they wish to include other forestry activities under the flexible mechanisms of the Protocol. Reducing carbon emissions due to deforestation and forest degradation is a prime contender for inclusion; other feasible options are rehabilitation of degraded forests, reduced-impact logging, and reducing losses in converting timber to forest products, as well as enhancing wood energy use.

Being aware of demands from some developed countries that they should accept emission reduction commitments, many developing countries are taking a new look at their forests from the perspective of the Protocol. Developing countries currently cause approximately 60% of all anthropogenic greenhouse gas emissions, including those from land use change and forestry. One-third of greenhouse gas emissions from developing countries originate in land use change and forestry, primarily from deforestation; in the least-developed countries, this sector contributes 62% (Baumert *et al.*, 2005). Emissions due to deforestation are no less harmful than those due to fossil fuels; on the contrary, they not only affect the earth's atmosphere immediately after their release, but they also signify a reduced capacity to sequester excess carbon in the future. New proposals from developing countries, led by Papua New Guinea and Costa Rica, for industrialized countries to compensate for forest conservation in developing countries appear understandable from this perspective (Moutinho and Schwartzman, 2005). Opponents of this approach point to technical difficulties in assessing carbon savings and doubt whether monetary rewards for carbon conservation can succeed in markedly reducing deforestation.

Future Prospects

Forests and forestry are intricately intertwined with climate change. The United Nations Framework Convention on Climate Change and the Kyoto Protocol explicitly acknowledge this link. The international treaties seek to protect forests against the effects of global change and to harness their unique powers for mitigating it and safeguarding human societies. The CDM singles out afforestation and reforestation projects for carbon sequestration in developing countries. With this flexible instrument, the Kyoto Protocol has created a striking example of a way of achieving the United Nations' Millennium Development Goals through global partnerships for development.

In industrialized countries that opt to include forest management as an elective activity in accounting under the Protocol, growing stock values will increase on 1 January 2008 through the monetary equivalent of carbon stocks. After that date, converting forest to a highway, housing development or golf course will become more expensive, as the country will have to offset the lost carbon.

Remote sensing, carbon flux measurements and inverse atmospheric modelling are contributing to an improved understanding of the earth's carbon cycle and the role of forests in it. However, terrestrial forest inventories are indispensable in order to complement or substantiate estimates and models for quantifying the vast carbon stocks and flows involved in forest ecosystems. In view of the unsatisfactory state of national forest inventories in many countries (Saket, 2003), better and more frequent national forest assessments have become a more urgent matter, with the advent of obligatory reporting on carbon stock changes by countries (FAO, 2003).

Few forestry faculties appear to have integrated climate change into their programmes, although new professional opportunities are arising in the field of forestry and climate change (FAO, 2001–2005). Proposed afforestation and reforestation projects under the CDM are floundering at times because they are prepared with insufficient forestry expertise. With arrangements for implementing the Kyoto Protocol for the first commitment period and negotiations for the second commitment already underway, expertise in both forestry and climate change will be vital in order to define the future role of forests in it.

Assessing vulnerability and adapting human societies to climate change have become a distinct field with regard to negotiations, research, funding and implementation (Smith *et al.*, 2003). However, despite the fact that forests established today will grow for decades or centuries and will undoubtedly undergo climate change, the tasks of assessing forest vulnerability, adapting such tasks and developing new management methods have only recently received more attention (Spittlehouse and Stewart, 2003; Spittlehouse, 2005).

Forestry research may have to incorporate the novel aspect of climate change. Better and more specific data on carbon in wood products and about its fate in the life cycle of products may help countries to overcome any hesitancy regarding the inclusion of wood products as accountable carbon stores in future commitment periods. Is fuelwood for bioenergy use optimally produced by the ancient practice of coppicing? Could coppice forests with standards fulfil the goals of bioenergy production and carbon storage simultaneously? Can the fledgling art of carbon inventories and assessing sequestration rates in forests be developed in such a way as to make it as simple as using a yield table today?

Climate change and the international treaties dealing with it have created a plethora of new challenges, opportunities and tasks for societies in order to achieve sustainable forest management and the millennium development goals. Meeting these goals successfully will require a significant paradigm shift; fresh perspectives, modified priorities, new information, skills and creativity are needed.

References

Baumert, K.A., Herzog, T. and Pershing, J. (2005) *Navigating the Numbers: Greenhouse Gas Data and International Climate Change Policy*. World Resources Institute, Washington, DC.

Bundesministerium für Wirtschaftliche Zusammenarbeit und Entwicklung (BMZ)/Referat für Entwicklungspolitische Informations- und Bildungsarbeit (2004) Fortschrittsbericht zur deutschen bilateralen Entwicklungszusammenarbeit im Waldsektor. Bundesministerium für wirtschaftliche Zusammenarbeit und Entwicklung, Bonn. Available at: http://www.bmz.de/de/service/infothek/fach/materialien/materialie142.pdf

Burschel, P., Kürsten, E. and Larson, B. (1993) Die Rolle von Wald und Forstwirtschaft im Kohlenstoffhaushalt. Eine Betrachtung für die BRD. University of Munich, Munich. (Schriftenreihe der forstwissenschaftlichen Fakultät der Universität München und der forstlichen Versuchs- und Forschungsanstalt.)

Dewar, R.C. (1990) A model for carbon storage in forests and forest products. *Tree Physiology* 6, 417–428.

Dewar, R.C. and Cannell, M.G.R. (1992) Carbon sequestration in the trees, products and soils of forest plantations: an analysis using U.K. examples. *Tree Physiology* 11, 49–71.

Dyson, F.J. (1977) Can we control the carbon dioxide in the air? *Energy* 2, 287–291.

FAO (Food and Agriculture Organization of the United Nations) (2001–2005) Clim-FO-L electronic newsletter. FAO, Rome. Available at: http://www.fao.org/forestry/site/17827/en

FAO (2003) *State of the World's Forests, 2003*. FAO, Rome. Available at: http://www.fao.org/docrep/005/y7581e/y7581e00.htm

FAO (2006) *Global Forest Resources Assessment 2005: Progress towards Sustainable Forest Management*. FAO, Rome.

Kägi, W. and Schoene, D. (2005) *Forestry Projects under the CDM: Procedures, Experiences and Lessons Learned*. Forests and Climate Change Working Paper 3. FAO, Rome. Available at: ftp://ftp.fao.org/docrep/fao/008/j7017e/j7017e00.pdf

Moutinho, P. and Schwartzman, S. (eds) (2005) *Tropical Deforestation and Climate Change*. IPAM (Instituto de Pesquisa Ambiental da Amazônia), Belém, Brazil/Environmental Defense, Washington, DC. Available at: http://www.environmentaldefense.org/documents/4930_TropicalDeforestation_and_ClimateChange.pdf#search='Tropical%20Deforestation%20and%20Climate%20Change'

Robledo, C. and Forner, C. (2005) *Adaptation of Forest Ecosystems and the Forest Sector to Climate Change*. Forests and Climate Change Working Paper 2. FAO, Rome. Available at: http://www.fao.org/docrep/008/j6525e/j6525e00.htm

Saket, M. (2003) Gaps in national-level information on forest and trees in developing countries. *Unasylva* 53, 24–27.

Schoene, D. and Schulte, A. (1999) Forstwirtschaft nach Kyoto: Ansätze zur Quantifizierung und betrieblichen Nutzung von Kohlenstoffsenken. *Forstarchiv* 70, 167–176.

Smith, J.B., Klein, R.J.T. and Huq, S. (eds) (2003) *Climate Change, Adaptive Capacity and Development*. Imperial College Press, London.

Spittlehouse, D.L. (2005) Integrating climate change adaptation into forest management. *Forestry Chronicle* 81, 691–695.

Spittlehouse, D.L. and Stewart, R.B. (2003) Adaptation to climate change in forest management. *BC Journal of Ecosystems and Management* 4, 1–11.

Verheyen, R. (2003) Adaptation funding: legal and institutional issues. In: Smith, J.B., Klein, R.J.T. and Huq, S. (eds) *Climate Change, Adaptive Capacity and Development*. Imperial College Press, London, pp. 191–216.

Watson, R.T., Noble, I.R., Bolin, B., Ravindranath, N.H., Verardo, D.J. and Dokken, D.J. (eds) (2000) *Land Use, Land-Use Change, and Forestry: 2000 Special Report of the Intergovernmental Panel on Climate Change*. Cambridge University Press, Cambridge.

Wegener, G., Beudsert, M., Fruehwald, A., Dreiner, K. and Scharai-Rad, M. (1996) *Bewertung von Holz im Vergleich mit anderen Werkstoffen unter dem Aspekt der CO_2-Bilanz*. Fraunhofer-Informationszentrum Raum und Bau/IRB-Verlag, Stuttgart, Germany.

7 The United Nations Convention to Combat Desertification: a Global Framework for Cross-sectoral Policy Coordination Addressing Sustainable Forest Management

Camilla Nordheim-Larsen, Eleonora Canigiani and Paule Hérodote

Introduction

Land degradation, which includes both desertification and deforestation, is a worldwide phenomenon that severely affects the poorest rural communities. Estimates show that drought and desertification threaten the livelihood of over 1 billion people in more than 110 countries around the world.[1] Land degradation is mainly the result of climate variability and unsustainable human activity and is both a cause and effect of poverty. The World Summit on Sustainable Development held in September 2002 reaffirmed land degradation as one of the major global environment and sustainable development challenges of the 21st century. It called for action to 'address causes of desertification and land degradation in order to restore land, and to address poverty resulting from land degradation' (UN, 2002). Thus, combating desertification is essential for achieving a much broader objective: the sustainable development of countries affected by drought and desertification.

Recognizing the linkages between poverty and environmental degradation, the United Nations Convention to Combat Desertification (UNCCD) was established following the 1992 Earth Summit in Rio de Janeiro. To date, 191 countries have ratified the convention (UNCCD, 2004a) as a legally binding framework that contributes to providing a comprehensive answer to problems related to the environment and sustainable livelihood. This Convention is referred to as one of the Multilateral Environmental Agreements; however, it is also recognized as a global framework for sustainable development. Designed as an overarching and cross-sectoral legal instrument, the convention is therefore well positioned to support countries achieving their national development objectives and to contribute significantly to the millennium development goals of halving the number of people living in poverty by 2015 and ensuring environmental sustainability.

The Convention defines land degradation as reduction or loss, in arid, semiarid and dry subhumid areas, of the biological or economic productivity and complexity of rain-fed cropland, irrigated cropland, or range, pasture, forest and woodlands resulting from land uses or from a process or combination of processes, including those arising from human activities and habitation patterns. In terms of obligations under the Convention, country parties identified as developed pledge to implement relevant long-term, integrated strategies in their territory, as well as to support relevant actions in developing countries, provide funding, and facilitate access to technology and knowledge. In turn, affected developing countries commit to elaborating national action programmes that identify the factors contributing to desertification and establish

[1] Message of Kofi A. Annan, the then United Nations Secretary General, on World Day to Combat Desertification, 7 June 2001.

practical measures for implementing the convention (UNCCD, 1996a).

This chapter affirms that there is a close correlation between the challenge of promoting sustainable forest management and that of achieving sustainable land management. It advocates enhanced collaboration between the forestry sector and the Convention in achieving the Millennium Development Goals and reducing poverty. It emphasizes the importance of multisectoral, integrated approaches within a landscape perspective in creating an enabling environment for participatory rehabilitation practices. The strategy of the Global Mechanism of the convention is presented as an example of a cross-sectoral and comprehensive approach to resource mobilization. The chapter also draws on the forest landscape restoration approach to illustrate its potential in optimizing synergies between the forestry and soil conservation processes at national and local levels.

The UNCCD Framework and Forests

An analysis of the causes of deforestation and land degradation reveals a considerable convergence of solutions aimed at slowing deforestation and rehabilitating degraded land. Forests and forest ecosystems have been recognized by the signatories of the UNCCD, the convention on biological diversity, the United Nations Framework Convention on Climate Change and the United Nations Forum on Forests as areas in which synergetic action can be particularly effective. Moreover, the parties have stressed the need to focus on a broader development framework that includes issues such as desertification and land management, biological diversity, climate change and socio-economic development. In particular, they recognize the need to join efforts to address the interlinked issues of poverty eradication, sustainable development and environmental security (UNCCD, 2004b).

The Convention pays particular attention to countries with low forest cover and advocates strengthening the capacity of these countries to combat desertification, land degradation and deforestation. Forests in such countries are, by definition, a scarce resource and therefore likely to be under greater pressure than those elsewhere. In addition, they are little known and generally unprotected, and the forest conservation and protected area management strategies required may therefore be different from those of countries with large forest estate (UNCCD, 2005). Given that aridity and drought are among the main causes of the reduction in forest cover in these countries, common goals and objectives can be identified for the UNCCD and the countries with low forest cover. These include integrated land use planning, controlled resource use in forests, and afforestation or re-afforestation to alleviate the pressure on natural forests. Accordingly, in these countries, holistic and cross-sectoral participatory approaches to the development of national forest programmes and policies are highly relevant.

Promoting a landscape perspective

Forest landscape restoration, as it is being developed within the forestry sector, is considered an effective approach for sustainable land management and an operational framework for the Global Mechanism's cross-sectoral and integrated approach. The focus is on the landscape as a whole and its multiple functions for delivering ecosystem services, rather than on site-based interventions. A landscape approach allows for site-level decisions to contribute to an integrated restoration strategy and a better understanding of the factors that determine whether different land uses and land use policies are mutually reinforcing or in conflict with each other (ITTO/IUCN, 2005). Forest landscape restoration seeks to balance biodiversity conservation with poverty alleviation and socio-economic needs in order to achieve sustainable development, which inevitably entails negotiations and trade-offs between stakeholders at the landscape level (Aldrich *et al.*, 2004).

Such an approach allows the connectivity between systems on different scales to be addressed, including the link between local and global environmental benefits (GEF, 2005a). Accordingly, the global environmental facility focal area on land degradation has also adopted the landscape approach (GEF, 2005b). Its operational programmes for sustainable land management and integrated ecosystem management, as well as operational programmes of the biodiversity and climate change focal areas, can be useful tools for optimizing synergies between different sectors and mobilizing resources in support of the UNCCD and other relevant international agreements.

Global Mechanism cross-sectoral approach

As part of the solution to the problem of desertification, the Global Mechanism was established in 1998 under the authority of the conference of the parties of the UNCCD. It is an instrument to facilitate the rationalization of resource allocation and the mobilization of additional resources to combat land degradation and poverty. During negotiations, it emerged as a consensus option to optimize resource flows to combat desertification as a cross-sectoral concern. The International Fund for Agricultural Development, as the global agency at the forefront of tackling land degradation and reducing rural poverty in countries most affected by desertification, was selected in 1997 to house the Global Mechanism. It is guided by decisions and recommendations of the parties to the UNCCD and receives support from a facilitation committee.[2]

The Global Mechanism's support for affected developing countries is guided by its consolidated strategy and enhanced approach to resource mobilization. This was prepared in response to the changing resource allocation environment, particularly at the national level. As donors to an increasing extent align their priorities with those of recipient countries, articulating land degradation as a development priority becomes more important. The significance of domestic public budget allocations is increasing considerably through new approaches such as basket funding and direct public budget support. As a consequence, the level of finance for implementation depends on the political will of governments to identify land rehabilitation and sustainable land management as national priorities. Given that national development frameworks such as poverty reduction strategies often place strong emphasis on sectors such as education, public health and infrastructure, the objective of combating desertification faces strong competition with other development priorities.

The cross-cutting nature of the UNCCD mandates the Global Mechanism to mobilize resources in a wide range of cross-sectoral areas. It is therefore currently exploring a number of strategic areas as a means to increase the resources available for the implementation of the convention. These areas include: economics and financing instruments, private sector investments, effective civil society participation, improved market access and trade, forestry in low forest cover countries and education. The Global Mechanism does not intend to build in-house expertise on these issues but to focus on their financial aspects.

Optimizing synergies at national and local levels

Moving from policy dialogue to effective action through the implementation of forest landscape restoration at national and local levels and across sectors could contribute to, and enrich, multilateral environmental agreements and policy processes (Saint-Laurent, 2005). Forest landscape restoration and sustainable land management require supportive national policy frameworks that provide incentives for long-term investment and acceptable returns. The challenges and constraints of implementing sustainable forest management at national level coincide to a great extent with those posed by UNCCD implementation, such as integrating programmes into national development frameworks, scarcity of financing, institutional weaknesses and insufficient interaction with other development sectors. Whilst the national action programmes of the Convention have been able to broadly capture the technical aspects of desertification and certain strategic elements, many of them have been unable to effectively translate the principles into strategic and fundable work programmes aimed at mitigating the root causes of land degradation. The integration of sustainable land management practices into national development frameworks, such as poverty reduction strategies, can facilitate the coordinated mobilization of funding for

[2]Members of the facilitation committee comprise IFAD, UNDP, World Bank, FAO, UNEP, the African, Asian and Inter-American Development Banks, the Consultative Group on International Agricultural Research, and the UNCCD and the Global Environment Facility Secretariats.

a successful implementation of cost-effective and sustainable programmes (GEF, 2003).

To increase the resources allocated to sustainable land management, the Global Mechanism contributes to national policy processes, working with governments to mainstream sustainable land management issues. Mainstreaming implies changes in the way of doing business, for instance, through policy reform, institutional change, enhanced coordination arrangements, and planning/budgeting/resource allocation modalities. At the core of the resource mobilization strategy is the promotion of financing strategies that have proven successful in other sectors. Such financing strategies will provide country partners with tools to align UNCCD priorities with those of other sectors and to compete for the allocation of resources in order to mobilize financial resources in a systematic, coherent and predictable manner. A national financing strategy, with a clear objective of mobilizing resources, will be an integral part of national action programmes and should lead to country financing partnerships. The strategies to be implemented under country leadership must be based on an analysis of the investment climate, identify financing instruments and sources of finance, and contribute to increased financing from complementary sources, including domestic budgets. Increasing financial resources is, however, dependent on the mobilization of a range of resources. The Global Mechanism's definition of 'resources' includes not only funds but also:

- Instrumental resources: strategic frameworks, policy instruments, legislation, programmes and plans;
- Human resources: stakeholders, organizations and institutions;
- Knowledge and information resources;
- Financial resources.

The UNCCD puts particular emphasis on ensuring 'the full participation at all levels of local people' in decision-making processes and assumes that 'decisions on the design and implementation of programmes ... are taken with the participation of populations and local communities and that an enabling environment is created at higher levels to facilitate actions at national and local levels' (UNCCD, 1996b). Similarly, forest landscape restoration seeks to take into account the needs and priorities of the numerous stakeholder groups and to involve local people in decision-making processes. In an attempt to increase the involvement of local communities, the Global Mechanism and forest landscape restoration promote local actions across sectors that could contribute to the creation of incentive mechanisms for investment in the rehabilitation of degraded landscapes.

Recent studies indicate that environmental and social forces may not be sufficient to provide these incentives and that market and economic mechanisms should be a driving force for change (Hazell, 2001; GM, 2004). In this context, the development of community-based trade is viewed as a potential means of increasing local participation in rehabilitation activities. This draws on the assumption that local communities and households are those ultimately responsible for the sustainable use and management of natural resources, and that their involvement in profit-making activities through trade and business development would motivate them to increase investments in preserving these resources and restoring degraded ones. The promotion of community-based trade could also serve to mobilize additional resources and financial flows by involving stakeholders so far excluded, or only marginally involved, in restoration activities. For example, evidence of potential economic and financial benefits may attract increased investments from the private sector and contribute to the sustainable development of degraded landscapes by triggering business development, increasing market opportunities and generating values for local products and services.

Under current agricultural pricing schemes it is difficult for rural producers to cover both the total costs of production and those of replenishing the natural capital. This in turn creates pressures on land use leading to the expansion of the agricultural border and thus generating land degradation and desertification problems. If local producers were to be compensated for part of the costs of replenishing their natural capital, the degradation processes could be reversed. Benefit-sharing could be enhanced through compensation for ecosystem services. These are innovative mechanisms aimed at mobilizing fresh resources so as to allow more efficient use of rural energy, the improvement of agricultural and forestry practices, as well as the conservation and expansion of forested areas through contractual arrangements between local producers and the parties benefiting from these actions. Forest ecosystems provide, in fact, a number of ecological services that can serve as a potential platform for synergy: soil stabilization, biological diversity, carbon sequestration, water resources management, etc. While the value of the services provided by nature is being recognized, it is still a challenge to bring them into the mainstream of existing markets. A key role in enhancing benefit-sharing is to be played by civil society and non-governmental organizations (NGOs). By reaching out to the most disadvantaged and isolated groups and engaging in advocacy- and capacity-building activities, they can promote effective participation in decision-making processes. This will contribute to provide them with knowledge and practical tools for taking advantage of economic opportunities at the landscape level.

All local actions are ultimately aimed at building synergies between local, national and international levels, and fostering political commitment by informing policy-making processes and related frameworks at national and international level. In partnership with the Small Grants Programme of the Global Environment Facility of the United Nations Development Programme, the Global Mechanism seeks to design interventions within the forest landscape restoration framework for engaging local households and communities in land rehabilitation activities in order to inform national and international policy processes. Local actions are aimed at triggering a process leading to the mobilization of all types of resources – human, knowledge and information, instrumental and financial – at the different levels.

Conclusions

The close relationship between land degradation and forestry demonstrates the potential for optimizing synergies between the soil conservation and forest management processes by adopting a holistic approach towards sustainable development and poverty alleviation. Until these issues are addressed in a more integrated manner, it will be difficult to formulate and implement an optimal set of policies, practices and technologies, and to develop effective financing mechanisms. In line with the cross-cutting nature of the UNCCD, the Global Mechanism advocates a multifaceted and cross-sectoral approach to resource mobilization that broadens the scope of sustainable land management. This approach allows greater flexibility in mobilizing additional resources for the implementation of the convention and provides increased opportunities for synergetic actions with a broad range of stakeholders.

Creating linkages between various sectors is crucial for tapping potential flows of additional financial resources and for facilitating the involvement of multiple stakeholders in the implementation of the Convention. In order to effectively mobilize resources, the Global Mechanism considers a landscape approach for its interventions and draws on various international, national and local processes for sustainable land management. Forest landscape restoration can thus be an effective approach for optimizing synergies between relevant processes at all levels and for facilitating multiple-stakeholder involvement in land rehabilitation and restoration activities. With a view to contributing to the achievement of the Millennium Development Goals and reducing poverty, the UNCCD and the forestry sector will mutually benefit from promoting joint development objectives that offer enabling conditions for household and community investment in land rehabilitation and in the sustainable management of natural resources.

References

Aldrich, M. *et al.* (2004) *Integrating Forest Protection, Management and Restoration at a Landscape Scale.* WWF, Gland, Switzerland.

GEF (Global Environment Facility) (2003) GEF Operational Programme on Sustainable Land Management (OP, no. 15). Available at: http://www.gefweb.org

GEF (Global Environment Facility) (2005a) Working Draft: GEF Land Degradation Strategy. Available at: http://www.gefweb.org

GEF (Global Environment Facility) (2005b) Working Draft: GEF Land Degradation Strategy. Annex 1. Available at: http://www.gefweb.org

GM (Global Mechanism) (2004) Why Invest in Drylands? Draft working paper. Rome, Italy.

Hazell, P. (2001) *Strategies for the Sustainable Development of Dryland Areas*. International Food Policy research Institute (IFPRI), Washington, DC.

ITTO/IUCN (2005) *Restoring Forest Landscapes: An Introduction to the Art and Science of Forest Landscape Restoration*. ITTO Technical Series, no. 23. International Tropical Timber Organization, Yokohama, Japan.

Saint-Laurent, C. (2005) Optimising synergies on forest landscape restoration between the Rio Conventions and the UN Forum on Forests to deliver good value for implementers. *Review of European Community and International Environmental Law (RECIEL)* 14(1), 39–49.

UN (2002) Report on the World Summit of Sustainable Development, 26 August–4 September 2002, Johannesburg, South Africa. United Nations, New York. Available at: http://www.adb.org/Documents/Conference/WSSD/WSSD.pdf

UNCCD (1996a) Text of the Convention. Available at: http://www.unccd.int, Art. 1(f).

UNCCD (1996b) Text of the Convention. Available at: http://www.unccd.int, Par 1(a) and 3(a).

UNCCD Secretariat (2004a) Update on Ratification of the UNCCD. Available at: http://www.unccd.int

UNCCD Secretariat (2004b) Workshop on Forests and Forest Ecosystems: Promoting Synergy in the Implementation of the Three Rio Conventions. Concept Note. Available at: http://www.unccd.int

UNCCD Secretariat (2005) Review of the Activities for the Promotion and Strengthening of Relationships with Other Relevant Conventions and Relevant International Organizations, Institutions and Agencies. Note by the Secretariat. ICCD/COP/(7)/5. Available at: http://www.unccd.int

8 Bringing Forest Ecosystems into National Income Accounts: an Environmental Accounting Approach

Glenn-Marie Lange

Introduction

Deforestation is a serious problem that has local, national and global consequences. Yet efforts to address this problem through sustainable forestry have often failed – in part because of a lack of information about the multiple economic contributions of forests: how forest goods and services are linked to the rest of the economy, and who benefits from them. While there is information about the economic value of commercial timber, many other contributions made by forests – such as tourism, biodiversity and ecosystem protection services – are usually missing from the national accounts, our primary source of information about the economy. This lack of information distorts policy analysis, but perhaps more importantly, it hampers the development of cross-sectoral institutional alliances to promote sustainable forestry, because stakeholders outside the forestry sector are not fully aware of how much they depend on forest ecosystem services.

Sustainable forestry requires addressing forest policy in a cross-sectoral context, which can be achieved only when there is a clear incentive to build an alliance among stakeholders in different sectors benefiting directly or indirectly from forests. A powerful incentive for building such alliances is information about the economic benefits of forest ecosystems to non-forestry sectors, including agriculture, fisheries, tourism, municipal water supplies, livelihood of rural communities, and other fields. Potential stakeholders include government, communities, business and civil organizations, as well as private citizens.

This chapter describes the System of Environmental and Economic Accounting (SEEA) forestry accounts and reports on the status of current efforts to develop forestry accounts at the global and national levels. It concludes that the SEEA can be used to integrate non-market forest goods and services with conventional economic information, and also how this information can contribute to better sectoral and macroeconomic policy-making by governments, more effective cross-sectoral alliances and, ultimately, to more effective forest conservation (Dubé and Schmithüsen, 2003; Lange, 2003; Dubé *et al.*, 2006).

Overview of Environmental Accounts

A great deal of work has been done by forest economists and others to measure the multiple benefits from forests. New initiatives such as 'payment for ecosystem services' have emerged to promote recognition of these services by beneficiaries and their internalization into costs (Campbell and Luckert, 2002; Pagiola *et al.*, 2002; Gutman, 2003). However, this information has not found its way into the System of National Accounts, which provides the most widely used measures of economic performance, such as gross domestic product (GDP).

Environmental accounts, formalized in the SEEA (United Nations *et al.*, 2003), are an extension of the System of National Accounts. The SEEA puts the total economic value of forest resources within the larger context of accounts for the national (or regional) economy. Thus, the SEEA forestry accounts provide a tool for quantitative analysis of the total economic value of forests, as well as a framework for linking information about forestry to the status of other resources (water, land and soil) and to the broader economy, integrating forestry policy with national development and macroeconomic and sectoral policies.

Over the past few decades, most countries have come to embrace the notion of sustainable development. The search for ways to operationalize this notion has focused, in part, on national economic accounts: incorporating the role of the environment and natural capital more fully into the conventional System of National Accounts through a system of satellite accounts for the environment. This is particularly important, as it constitutes the primary source of information about the economy and is widely used for analysis and decision-making in all countries (United Nations *et al.*, 1993). However, the System of National Accounts has had a number of well-known shortcomings with regard to the treatment of the environment, including forests.

First, it has treated cultivated forests and natural forests quite differently. For cultivated forests, the System of National Accounts records both production and changes in the forest stock, so that the consequences of depletion or afforestation are accounted for. For natural forests, however, it records only the income from logging, but not changes in natural forest stocks. This can result in quite misleading economic signals about changes in a natural forest: income from overexploitation would be recorded as part of GDP, but the corresponding depletion of the forest stocks would not be recorded. Similarly, the benefits of afforestation would not be recorded.

The 1993 revision of the System of National Accounts addresses some of these problems, notably by expanding the asset boundary to include a broader range of assets such as natural forests. Even with this expanded coverage, significant gaps remain:

- Non-marketed forest products are critical to rural livelihood in developing countries, yet they are often not included in the national accounts. In principle, the System of National Accounts includes such products, but measurement difficulties have limited implementation in many countries.
- Many of the non-market services from forests are wrongly attributed to other sectors of the economy. The value of forest services provided as intermediate inputs to other sectors, such as livestock grazing or tourism, is attributed to the using sector, not to forestry, thereby underestimating the economic value of forests.
- Ecosystem services such as watershed protection and carbon storage may not be represented at all.

The SEEA was developed as a set of satellite accounts to the System of National Accounts to address these gaps. Environmental and natural resource accounts have evolved since the 1970s through the efforts of individual countries and practitioners, each developing their own frameworks and methodologies to represent their environmental priorities. Since the late 1980s, a concerted effort has been underway through the United Nations Statistics Division (Eurostat), the Organization for Economic Cooperation and Development (OECD), the World Bank, national statistical offices, and other organizations to standardize the framework and methodologies used. The United Nations published an interim handbook on environmental accounting in 1993 and a revised, expanded version in 2003 (United Nations *et al.*, 2003).

As a system of satellite accounts, the SEEA has a similar structure to that of the System of National Accounts. The SEEA consists of stocks and flows of environmental goods and services. It provides a set of aggregate indicators for monitoring environmental-economic performance at the sectoral and macroeconomic level, as well as a detailed set of statistics to guide resource managers towards policy decisions that will improve environmental-economic performance in the future. The definition of environmental goods and services in the SEEA is much broader than in the System of National Accounts, in principle attempting to measure total economic value, not just market values.

Significantly, the SEEA includes two parallel sets of accounts, one in physical units and the other in monetary values. The SEEA forest accounts have four major components:

- *Asset accounts* record stocks and changes in stocks of forest resources (land cover, timber and carbon).
- *Flow or production accounts* for forest goods and services include:
  - Market and non-market goods (timber and non-timber forest products);
  - Forest ecosystem services such as tourism and biodiversity protection, services to agriculture (pollination by wild bees, livestock grazing, etc.), and soil and water protection services to a wide range of industries (agriculture, hydroelectric power, industry and municipal water supply).
  - Environmental degradation and externalities, such as soil erosion from logging.
- *Environmental protection and resource management expenditure accounts* include forest management expenditures by government, environmental protection expenditures by public and private sectors, and user fees and taxes paid by forest users to the government.
- *Environmentally adjusted macroeconomic indicators* include indicators of sustainability such as adjusted net savings, environmental income, and total wealth including natural capital. Forestry accounts show the addition to GDP of unvalued forest goods and services, the costs of deforestation or loss of forest services, and the contribution of forest assets to national wealth.

The Earth Institute, headed by the well-known economist Jeffrey Sachs, is at the forefront of environmental accounting and, together with the World Bank, is launching a Global Initiative for Environmental Accounting. A priority of the global initiative is to develop a framework for ecosystem accounting and test it in several countries, both developing and developed. Forest ecosystem accounts will be one of the pilot studies for this work.

The European Environment Agency is also pioneering work on land and ecosystem accounts, of which forest accounts are an important component (Weber, 2007). The purpose is to provide an information system to support ecosystem assessment, especially their interaction with economic and social developments. It is leading a subgroup on ecosystem accounting for the United Nations Committee of Experts for Environmental-Economic Accounting[1]. Ecosystem accounting is considered one of the experimental components of the SEEA, but interest in the ecosystem approach and demands for an economic perspective on ecosystems are growing – for example, the Millennium Ecosystem Assessment Report (Millennium Ecosystem Assessment, 2005). Forest ecosystem accounting is one of the most promising areas for progress.

Status of Current National Efforts

Environmental accounts have been constructed for forest resources more often than for most other resources. The earliest set of forest accounts was constructed by Norway in the late 1970s. Since that time, many other countries have constructed forest accounts, and the accounts have expanded to include monetary asset accounts for standing timber as well as non-timber goods and services.

Table 8.1 shows countries that have formal forest accounting programmes sponsored by government agencies and the type of information included in their accounts. New efforts are currently underway in Tanzania, Uganda and Guatemala. The Food and Agriculture Organization of the United Nations (FAO) manual for forest accounting was recently tested in Jilin province of China (Guangcui and Huang, 2006). Forestry accounts for all countries include timber asset accounts in physical and monetary terms. Forestry accounts are more common in developed countries than in developing countries. Eurostat has had an ongoing programme to develop forest resource accounts since 1995, and many of the participating countries have developed extensive accounts. There are also many additional academic studies and one-off studies by governments and non-governmental organizations (NGOs) that are not shown here.

[1]This is a group established in 2005 by the United Nations Statistical Commission to elevate the SEEA from the level of recommendations to statistical standards, and to promote the implementation of the SEEA worldwide.

Table 8.1. Forest accounts constructed by selected countries. (Based on FAO, 2004.)

	Forest accounts							
	Timber		Non-timber goods and services		Forest-related accounts			
	Asset accounts, physical and monetary	Supply and use table for wood products	Carbon storage	Other goods and services	Land	Energy	Water	Pollution and environmental degradation
Developing countries								
Brazil	X							
Chile	X							
China	X		X	X				
Costa Rica	X							
Indonesia	X		X					
Mexico	X		X		X	X		X
Philippines	X		X	X	X	X		X
Thailand	X							
South Africa	X		X	X			X	
Swaziland	X		X	X				
Developed countries								
Austria	X	X	X	X	X	X		X
Finland	X	X	X	X	X	X		X
Denmark	X	X	X	X	X	X	X	X
France	X	X	X	X	X	X	X	X
Norway	X	X	X	X	X	X		X
Sweden	X	X	X	X	X	X	X	X
Spain	X	X	X	X	X	X		X
Germany	X	X	X	X	X	X	X	X
Italy	X	X	X	X	X	X		X
Canada	X	X	X	X	X	X	X	X
Australia	X	X	X	X	X	X	X	X
New Zealand	X	X	X	X	X	X	X	X

Note: Countries included here have ongoing accounting programmes run by government agencies, or by non-governmental agencies in cooperation with governments. There have been many additional academic studies and individual studies by governments or international agencies.

Forests are disaggregated in different ways, depending on the policy issues and characteristics of forests in each country. Virtually all forestry accounts distinguish between cultivated and natural forests, and disaggregate forests by major tree species. Many developing countries limit timber accounts to commercial timber production, but are beginning to add non-commercial timber production and the use of non-wood products as data become available. In all countries, there is an increasing number of spatially disaggregated forest accounts, compiled at the regional level, which are often better suited to forest management than purely national forest accounts. Carbon accounts are also compiled by most countries. This practice has been less widespread in developing countries than in developed countries, but is likely to increase with the growing potential for markets in forest carbon mitigation.

Countries have used forest accounts in many ways. A comprehensive survey of policy uses of forest accounting is provided by the manual for environmental and economic accounts for forestry (FAO, 2004), which places special emphasis on the contribution of forest accounting to cross-sectoral policy analysis. The accounts improve policy by providing more accurate and comprehensive information about the total economic value of forest goods and services, both market and non-market. The emphasis on systematic measurement of non-market forest goods and services is a critical advance over current national income accounts, which often include only timber values. The Millennium Ecosystem Assessment reports that in a number of countries, the timber value of forests accounted for less than one-third of the total economic value of the forest ecosystem; the remaining two-thirds of forest value was provided by non-market goods and services, particularly environmental services provided to other sectors such as agriculture and tourism.

More comprehensive and accurate measures of all forest values improve monitoring and the coordination of decision-making between the forest and other sectors, among government ministries and public administrations, and in the private sector. Within the forestry sector itself, improved measures for forest values reduce the incentives for deforestation and provide more accurate economic assessments of the trade-offs among competing users of forest land and forest resources. For example, a forest ministry contemplating the award of logging concessions will know how the volume and method of logging will affect water supplies to downstream cities, production of non-timber forest products that are critical to the livelihood of poor households, and opportunities for agriculture and tourism.

At the macroeconomic level, forest accounts contribute in the following areas:

- Forest accounts measure the costs of deforestation and indicate at the macroeconomic level whether economic growth is sustainable or a country is 'living off its natural capital'. Forest accounts contribute to environmentally adjusted macroeconomic indicators of sustainability, such as adjusted net saving or changes in total wealth (including natural capital). Such indicators of (weak) sustainability monitor whether loss of forests is compensated for by investment in other forms of wealth. In the World Bank estimates of adjusted net savings (World Bank, 2005), forest depletion reduced net domestic savings by 20% in low-income countries, mostly in Asia.
- Budgetary support for sustainable forest management. Ministries of finance often make sectoral budgetary allocations based on information from national income accounts. But national income accounts underestimate the true contribution from forests to the national economy and poverty reduction, resulting in misguided government policies and poor investment decisions. More accurate information about the value of forests can influence budgetary decisions.
- The accounts also provide information about the distribution of forest benefits among different groups in society (e.g. commercial operators versus poor rural households; local versus regional/global), an important element in decision-making about forest management. Forest accounts are critical in achieving poverty reduction under the millennium development goals (MDGs) and the poverty reduction strategy programmes (PRSPs) that many countries have adopted to guide their development. Furthermore, understanding the distribution of benefits is essential for generating revenue for sustainable forest management. It is often only the local beneficiaries who are recognized, but assessment of forest services, such as recreation and carbon storage, reveals non-local beneficiaries who may be requested to pay for the benefits they receive.

For the private sector, more accurate information about forest values can improve private-sector decisions. Payments for environmental services, mentioned earlier, are an example of private-sector solutions to maintaining forests through market mechanisms, based on recognition of the value of forest services to downstream beneficiaries such as hydroelectric power, fisheries and irrigated agriculture, tourism and municipal water supplies. The result is an improvement in economic efficiency for all parties.

Finally, forest accounts can help build more effective cross-ministerial/multiple-stakeholder alliances by demonstrating the economic contribution of forests to other sectors such as agriculture, tourism and recreation, and hydroelectric power.[2] When stakeholders understand

[2]See Schmithüsen (2003) for a more detailed discussion of the policy aspects.

the economic benefits they receive from forests, they are more likely to support sustainable forestry initiatives.

Conclusions

Eventually, forest accounting and environmental accounting more generally should be viewed as simply a more comprehensive way of compiling national accounts that will provide policy-makers with a more accurate set of tools to manage both the economy and the natural capital on which the economy depends. To promote implementation, the Global Initiative of the Earth Institute and the United Nations Committee of Experts for Environmental-Economic Accounting have identified several important issues that need to be addressed (Lange and Hamilton, 2005):

- Implementation in developing countries would benefit greatly from strong endorsement by, and mainstreaming in the programmes of, international agencies. Forest accounting should be recognized as a tool for achieving goals of poverty reduction strategy papers and the MDGs targets. A collection of case studies and a detailed manual (more detailed than FAO, 2004) would help build support for forest accounting.
- Data collection is resource-intensive, and countries should use existing data wherever possible, especially combining efforts with other programmes such as the National Forest Assessments. As part of its research agenda, the United Nations Committee of Experts for Environmental-Economic Accounting has adopted the development of an 'umbrella framework' that combines environmental accounts and environmental statistics in order to streamline data requests.
- In many developing countries, a major obstacle to implementing forest accounting is a lack of support and technical cooperation. The expertise required is scarce in many developing countries; technical cooperation and extensive training is necessary, especially in the policy applications of forest accounts. An extensive programme of technical support is needed that includes three components: compilation of the accounts, using the accounts for policy analysis and capacity building. The Eurostat has taken a step towards providing such support by establishing an easily accessible web site, with national case studies that cover not only the compilation of accounts, but also the policy applications.

References

Campbell, B. and Luckert, M. (2002) *Uncovering the Hidden Harvest*. Earthscan, London.

Dubé, Y.C. and Schmithüsen, F. (eds) (2003) *Cross-sectoral Policy Impacts between Forestry and Other Sectors*. FAO Forestry Paper, 142. Food and Agriculture Organization of the United Nations, Rome. Available at: http://www.fao.org/docrep/006/y4653e/y4653e00.htm

Dubé, Y., Lange, G. and Schmithüsen, F. (2006) Cross-sectoral policy linkages and environmental accounting in forestry. *Journal of Sustainable Forestry* 23(3), 47–66.

Guangcui, D. and Huang, D. (2006) *China Environmental and Economic Accounts for Forestry: Jilin Province, Jiaohe County, and Fengman District Case Studies*. Report to the Forestry Department, FAO, Rome.

Gutman, P. (ed.) (2003) *From Goodwill to Payments for Environmental Services: A Survey of Financing Options for Sustainable Natural Resource Management in Developing Countries*. World Wildlife Fund Macroeconomics Program Office, Washington, DC. Available at: http://assets.panda.org/downloads/fin_alt.pdf#search='goodwill%20to%20payments%20WWF'

Lange, G. (2003) Bringing forest ecosystem services into the national economic accounts: an environmental accounting approach. Paper presented at the Conference on Ecosystem Services in the Tropics, International Society of Tropical Foresters, Yale University, 4–5 April 2003.

Food and Agriculture Organization of the United Nations (FAO) (2004) *Manual for Environmental and Economic Accounts for Forestry: A Tool for Cross-sectoral Policy Analysis*. By G.M. Lange, FAO Forestry Department, Policy and Institutions Service, Working Paper. FAO, Rome. Available at: ftp://ftp.fao.org/docrep/fao/007/j1972e/j1972e00.pdf

Lange, G. and Hamilton, K. (2005) Implementation of environmental accounts. Paper presented at the Preliminary Meeting of the UN Committee of Experts on Environmental-Economic Accounting, August 2005, New York.

Millennium Ecosystem Assessment (2005) *Ecosystems and Human Well-being*. Island Press, Washington, DC.

Pagiola, S., Bishop, J. and Landell-Mills, N. (eds) (2002) *Selling Forest Environmental Services: Market-based Mechanisms for Conservation and Development*. Earthscan, London.

Schmithüsen, F. (2003) Understanding cross-sectoral policy impacts – policy and legal aspects. In: Dubé, Y.C. and Schmithüsen, F. (eds) *Cross-sectoral Policy Impacts between Forestry and Other Sectors*. FAO Forestry Paper, 142. Food and Agriculture Organization of the United Nations, Rome, pp. 5–44.

United Nations, European Commission, International Monetary Fund, Organization for Economic Cooperation and Development, World Bank (1993) *System of National Accounts 1993, Prepared under the Auspices of the Inter-Secretariat Working Group on National Accounts*. Commission of the European Communities, Brussels/Luxembourg; International Monetary Fund, Washington, DC; Organization for Economic Cooperation and Development, Paris; United Nations, New York; World Bank, Washington, DC. Available at: http://unstats.un.org/unsd/sna1993/toctop.asp

United Nations, European Commission, International Monetary Fund, Organization for Economic Cooperation and Development, World Bank (2003) *Handbook of National Accounting: Integrated Environmental and Economic Accounting 2003*. United Nations, New York. Avaialble at: http://unstats.un.org/unsd/envAccounting/seea2003.pdf

Weber, J.L. (2007) Implementation of land and ecosystem accounts at the European Environment Agency. *Ecological Economics*.

World Bank (2005) *Where is the Wealth of Nations? Measuring Capital for the 21st Century*. World Bank, Washington, DC.

9 The Model Forest Experience in Using Broad-based Partnerships for Sustainable Forest Management

Peter Besseau, Washington Alvarado, Ramon Calumbay, Purificación Daloos, Rosalie Imperial, Lisa Jones, Cyprain Jum, Christa Mooney, Johan Svensson, Genrikh P. Telitsyn and Lourdes Wagan

Introduction

The question of how to involve different interest groups as active partners in sustainable forest management inspired the creation of Canada's Model Forest Program in the early 1990s. At the time, major conflicts within the Canadian forest sector were impacting communities, industry, government, indigenous groups and international trade in Canadian forest products. At the heart of the conflict were competing perceptions about forest management practices, about decision-making processes and about the increasing importance being placed on the value of non-traditional forest-based goods and services. In short, it was a period when the paradigms that had long grounded and directed forest planning and management were being forcefully challenged and new values were vying for voice and accommodation. This transition continues in different degrees around the world today.

The model forest concept is based on the need to create a forum in which all views can be considered within forest resource planning and management processes. At the time of its inception, the model forest idea was novel: it proposed a functional way to fully and constructively engage civil society, together with government, industry, indigenous groups, research organizations and non-governmental organizations (NGOs) on a large landscape scale. In so doing, the initiative sought to create partnerships that were as complex, interdependent and multifunctional as the landscapes they inhabited. It would be the task of these partnerships to work together to understand what sustainability means and how it could be applied in their landscape and then work collaboratively towards it (Besseau *et al.*, 2002). At the United Nations Conference on Environment and Development Summit in 1992, and on the basis of initial success with the approach in Canada, the Canadian government announced an international model forest network initiative and invited other countries to join. Initially, Russia and Mexico worked with Canada to create their own model forests. The network today includes nearly 20 participating countries and close to 40 model forest sites.

Core Attributes of Model Forests

Model forests are located in highly diverse landscapes – from the boreal forests of Sweden, to the temperate forests of Patagonia, to the tropics of Indonesia. Each of these landscapes is unique to its setting in terms of history of use, population pressures and impacts, culture and values, and legislation. Despite these significant differences between sites and regions, the model forest idea has shown itself to be flexible and adaptive to its setting without compromising the core attributes shared by all model forests:

- Broad-based, inclusive, voluntary partnerships;
- A commitment by all partners to work collaboratively in support of sustainable forest management;
- A land base large enough to incorporate the main land uses and values;
- A governance structure that is representative, transparent and accountable to its members;
- A programme of work reflective of its partners' needs and values;
- A commitment to networking: sharing their know-how and expertise with each other and the broader international network.

The sixth attribute underpins the justification for a network of sites. Network affiliation is designed to allow for the efficient movement of knowledge and know-how to accelerate innovation between model forests, and otherwise create opportunities for learning. Overseeing development of the network is the role of the International Model Forest Network Secretariat, based in Ottawa, Canada. One of the secretariat's key objectives is to support development of an international network that, in effect, constitutes an integrated global community of leading-edge practice for sustainable forest management (Besseau *et al.*, 2002).

This chapter illustrates the work of a number of model forest partnerships. As Table 9.1 shows, these sites are highly varied geographically, programmatically and in other significant ways. But what each of their experiences illustrates is that broad-based functional partnerships have indeed been created. The partnerships represented in the examples below have leveraged and used substantial resources to address priorities that are important to stakeholders in those landscapes: from watershed management to biodiversity conservation, and from governance and conflict resolution issues to inter-regional collaboration for more resilient resource-based economies.

Such partnerships are not easily or quickly established, but to the extent that ecosystems continue to be compromised by pressure for more intensive use, and to the extent that non-traditional players have concrete skills and resources to contribute, they must be seen as indispensable to making tangible and durable progress towards sustainable forest management. The International Model Forest Network experience backs this up. For policy-makers, particularly with limited resources, these partnerships are an efficient way to test, adapt, implement and inform the policy-making process.

In considering the six country case studies dealing with different management aspects that are presented below, it is important to note that model forest partnerships do not exercise control over the land bases in question. Rather, they typically represent those with legal title or authority over the land in addition to research, academic, community, indigenous and other stakeholders. The objective of the partnership is to explore innovative management options that tenure holders may, at their discretion, choose to adopt, generally with the assistance of the model forest organization through its experts and programme budget. Participation in a model forest partnership is generally an acknowledgement that there are shared problems that require shared solutions, as well as a willingness on the part of tenure holders to try them.

Interestingly, this lack of executive authority over the land base, which might suggest that the model forest is impotent, is in fact one of the features that gives it strength: it provides a genuinely risk-free and neutral forum in which tenure holders and non-tenure holders alike can sit at the same table and openly ask the question 'what if?' – and then, in getting an answer, do something about it.

Canada: Grizzly Bear Research Project

The 2.75 million-ha Foothills Model Forest in western Canada includes Jasper National Park, oil, timber and coal industries, recreation outlets and more. In 1999, the Foothills Model Forest and its partners initiated the Grizzly Bear Research Program to provide resource managers with the necessary knowledge and planning tools to ensure the long-term conservation of grizzly bears in west-central Alberta. Eventually, the programme expanded its study to include all grizzly bear habitat in Alberta (100,000 km^2). Research of this nature, and developing management tools of this complexity over a large scale, is unprecedented in wildlife research (Foothills Model Forest, 2004).

Table 9.1. Comparison of model forests.

Model forest	Araucarias del Alto Malleco (Chile)	Ulot Watershed (Philippines)	Foothills Model (Canada)	Gassinski (Russia)	Vilhelmina (Sweden)	Dja et Mpomo (Cameroon)
Area	• 400,000 ha	• 87,536 ha	• 2,756,692 ha	• 400,000 ha	• 850,000 ha	• 700,000 ha
Joined IMFN	• 2002	• 2000	• 1993	• 1994	• 2004	• 2005
Land ownership	• Park and national reservation 17.2% • Private lands, indigenous 29.0% • Provincial areas • Private lands, not 7.0% indigenous 46.8%	• Private lands 22.0% • Community-based forest management agreements 6.0% • MPSA 8.0% • SINP 64.0%	• Forest management agreements 36.0% • Provincial management units 4.0% • National Park • Provincial 40.7% protected areas 19.3%	• Forest of Federal Government 56.0% • National Park 44.0%	• Private owners (family) 36.0% • Private forest companies 22.0% • State forest companies 29.0% • Vilhelmina common 10.0% • Church, Vilhelmina Municipality (for forested area only, 530,000 ha) 3.0%	• Government-owned forests 75.0% • Community forests 25.0%
Population	• 26,899 (2000) • Urban 58.6% • Rural 41.4% • Indigenous 21.8% • Non-indigenous 78.2%	• 12,632 (2000)	• 23,466 (2001)	• 12,180 (1995) • Urban • Rural 0.1% • Indigenous 99.9% • Non- 21.2% indigenous 78.8%	• 8000	• 25,000 • Urban 3.0% • Rural 97.0% • Indigenous 95.0% • Non-indigenous 5.0%
Land cover and use	• Agricultural land 6.6% • Grasslands and shrubs 29.2% • Native forests 48.5% • Plantations 0.9% • Mixed forests 0.1% • Wetlands 0.7% • Areas devoid of vegetation 4.9% • Snow and glaciers 8.5% • Water bodies 0.6%	• Bare/crop/grassland 6.0% • Brushland/coconuts 48.0% • Built-up areas/roads 0.5% • Water bodies 1.0% • Fishpond 0.4% • Forest/mangroves 44.0% • Swamp/wetlands 0.1%	• Alpine natural subregion 25.1% • Subalpine natural subregion 36.9% • Montane natural subregion 4.8% • Upper foothills natural subregion 21.7% • Lower foothills natural subregion 11.5%	• Agricultural land 0.1% • Grasslands and shrubs 6.2% • Native forests 6.1% • Plantations 0.4% • Mixed forests 64.7% • Wetlands 22.5% • Areas devoid of vegetation 0.01% • Water bodies 2.0%	• Forest management (350,000 ha) 41.0% • Protected forest area (180,000 ha) 21.0% • Non-forested mountains, water, etc. (320,000 ha) 38.0%	• Agricultural lands 10.0% • Forest management units 40.0% • Community forests 15.0% • Protected areas 23.0% • Mining 10.0% • Municipal forests 2.0%

Continued

Table 9.1. *Continued*

Model forest	Araucarias del Alto Malleco (Chile)	Ulot Watershed (Philippines)	Foothills Model (Canada)	Gassinski (Russia)	Vilhelmina (Sweden)	Dja et Mpomo (Cameroon)
Key partners	• 22 stakeholder groups • National government • Communities and municipalities • Indigenous peoples • NGOs • Private sector (farming) • Private sector (forestry) • Conservationist • Religious groups	• 33 stakeholder groups • National government (multiple agencies) • Local government units • NGOs • People's organizations and cooperatives • Municipalities and communities • Philippine Army • Academia	• National and provincial government agencies • Private sector (forestry) • Private sector (mining/oil and gas) • Indigenous peoples • Municipalities and communities • NGOs • Academia and research institutions	• 28 stakeholder groups • State government • Indigenous peoples • Academia and research institutions • Private sector (forestry) • NGOs • Schools • Conservationists • Consumer groups	• National government (State Forest Agency) • Indigenous peoples • Municipalities and communities • Academia and research institutions • Private forest owners and associations • Private and state forest companies • NGOs, schools, wildlife tourism, etc. • Local entrepreneurs	• National government • Communities and municipalities • Indigenous peoples • NGOs • Private sector (mining) • Private sector (forestry) • Conservationists • Religious groups
Governance	• Board of Directors (25 persons) • Various committees / working groups • Dedicated staff • Financial administration through national foundation	• Board of Directors (15 persons) • Various committees / working groups • Legally registered as a Federation • Staffing provided by government	• Board of Directors (19 persons) • Various teams and committees • Legally registered not-for-profit organization • Dedicated staff	• Council of Partners • Partners' meeting • Technical council • Coordination group	• Steering committee • Various teams and networks • Integrated with local Municipality Board • Integrated with State Forest Agency • Dedicated staff	• General Assembly (all actor-groups) • Board of Trustees (15 members) • Executive Board • Stakeholders' platforms (simple and complex) • Legally registered as an association
Goals	• Strengthen democracy and community participation in the sustainable management of natural resources • Contribute to the development of a diverse, sustainable and innovative economy which improves the standard of living for local residents	• To develop and enhance lasting partnerships among stakeholders in Ulot Watershed Model Forest • To promote and develop conservation-compatible sustainable livelihood and enterprises	• Demonstrate sustainable forest management • Develop and implement mechanisms that result in wider understanding and application of accrued knowledge and technology for sustainable forest management	• To create an association of partners with the ability and interest to pursue the goal of sustainable forest management in a forest area (populated fully or partly by indigenous peoples) large enough to increase living standards by using	• To promote strategic approaches for realization of SFM principles in practical forestry and land use, by using the MF as the setting and the interface between the research community, the authorities on the national, regional and local level, and the forest practitioners	• To build capacity of local communities and site actors to engage in negotiation processes, project identification and implementation • To strengthen local capacity to engage in economic activities and in sustainable management of projects

	• Contribute to sustainable natural resource management, increasing the social, environmental and economic value of local natural resources • Improve coordination between the various agencies operating in the model forest area, in order to increase collaboration and the effectiveness of local development initiatives • Improve the overall quality and effectiveness of model forest activities	• To harmonize the production and protection functions of the forest ecosystem through sustainable forestry practices while providing benefits to the communities/stakeholders • To contribute to improving policies geared towards SFM • To strengthen the capabilities of stakeholders in developing and managing the model forest • To actively share lessons, best practices and experiences in promoting the model forest concept at the national, regional and international level	• Deliver communications and outreach programmes that improve understanding of, and support for, sustainable forest management • Support and influence policy that improves the practice of sustainable forest management	forest resources on the basis of scientifically justified quotas, value-added processing and better technologies; and through the development of an appropriate social and industrial infrastructure • To promote the economic and social development of indigenous communities • To promote the conservation of biodiversity and the protection of rare and endangered species • To achieve and support sustainable forest management through decision-making processes that take into account the interests of the local people living in and around the Gassinski Model Forest, and which are based on monitoring of the condition of forest and water ecosystems	• To explore best management, land use and conservation principles in natural coniferous landscapes by creating thematic landscapes using modern techniques and up-to-date knowledge • To elaborate principles for integrated and balanced management objectives (production, ecology, sociocultural) to make forest owners feel secure in their right to use and manage their forests, and to aid the national, regional and local authority decision-making procedures • To develop and demonstrate forestry methods that are functional, with emphasis on giving consideration to land use by indigenous Sami people and to hydrological integrity in riparian ecosystems	• To develop participatory approaches and support for policy dialogues and capacity-building for multiple-stakeholder processes • To enhance the capacity of site actors and local communities to work constructively to implement key SFM policies and objectives • To promote a governance structure that is transparent and consensus-based
2006 activities	• Implementing sustainable cattle proposal for small owners • Coordinating a leadership training program for local community • Carrying out a needs assessment of local community organizations in the region • Supporting community-driven projects through a small grants programme	• Development of a watershed management plan for the Ulot Watershed • Ecotourism development in selected areas of the model forest • Skills training in coco coir and coco peat processing • Market analysis and development • Strengthening of information and education campaign	• Fish and watershed research Foothills stream crossing program • Hardisty Creek restoration project • Natural disturbance programme • Grizzly bear research programme • Aboriginal involvement • Caribou landscape management association • Foothills growth and yield association		• Integrated and balanced management objectives • Reindeer husbandry and forest management • Forest, forestry and water	

Continued

Table 9.1. *Continued*

Model forest	Araucarias del Alto Malleco (Chile)	Ulot Watershed (Philippines)	Foothills Model (Canada)	Gassinski (Russia)	Vilhelmina (Sweden)	Dja et Mpomo (Cameroon)
	• Creating micro-credit and credit programmes that support local entrepreneurship • Organizing regional festivals that celebrate and increase awareness of local biodiversity • Piñon Project (seed of *Araucaria araucana*) technology meeting • Increasing strategic partnerships, resources and organizational capacity • Implementing Global Environment Fund proposal. Focal area: soil degradation	• Partnership development and maintenance • Training in financial management and resource mobilization • Training in project proposal preparation and negotiation skills	• Adaptive forest management/history • Chisholm, Dogrib and Lost Creek post-fire research • Communications and extension • Geographic information systems • Local level indicators for sustainable forest • Social sciences: economic and social values of the forest			
Web site	www.imfn.net	www.imfn.net	www.fmf.ca www.modelforest.ca	www.gassi.khv.ru www.imfn.net	www.balticforest.net www.imfn.net	

IMFN = International Model Forest Network; MF = model forest; MPSA = mineral production sharing agreements; NGO = non-governmental organization; SFM = sustainable forest management; SINP = Samar Island Natural Park.

The study involves capturing grizzly bears and equipping them with Global Positioning System collars. These collars collect several locations over a 24-h period and are the basis for the innovative management tools developed through the Grizzly Bear Research Program and its university partners. Essential to the study are the habitat cover maps created from satellite imagery for Alberta's grizzly bear range that illustrate the likelihood of bears using different areas of the landscape at different times of the year. Industry and government can use these maps in their planning to minimize disturbance to important grizzly bear habitats (Foothills Model Forest, 2004).

The model forest partnership[1] provided base funding to get the research project to the point at which it could leverage funds from a much larger group of stakeholders. Partners also provided credibility for the programme, which made additional fund-raising possible (Foothills Model Forest, 2004). University and academic partners allowed innovative and new techniques to be tested and developed. These, in turn, can be used by other research programmes, thus advancing the discipline of wildlife conservation. The Foothills Model Forest Grizzly Bear Research Program concluded that grizzly bears, mining and forestry activities have coexisted, and continue to coexist, in the study area. This finding is somewhat contrary to popular belief. It suggests that landscape change that has occurred because of these activities has not resulted in habitat loss detrimental to grizzly bear populations. In fact, human-caused mortality is the biggest threat to grizzly bear conservation.

Sound science is important, but the application of science is what truly makes a difference in the long run. Many of the model forest's partners are now applying the knowledge and tools generated from the research programme. For example, Weldwood of Canada Ltd (a forest industry company) developed an access plan for the Athabasca West area that includes a reduction in the permanent road footprint by about 30% over traditional road planning and building (Foothills Model Forest, 2004). Permanent roads will avoid high-quality grizzly habitat and will not run parallel to creeks. These changes mark a transformation in forest industry practice, and others are making similar plans. The new procedures and techniques for the capture and handling of grizzly bears for research and management purposes are now being adopted as leading-edge standards for grizzly bear handling in Alberta and other jurisdictions in North America.

The diversity of partnerships in the Foothills Model Forest was an important factor in the success of the project. Stakeholders recognized the need for grizzly bear management to occur across jurisdictions and on a regional basis, three of the four sponsoring partners led the initiative, and decision-makers were able to apply the research.

Cameroon: Participatory Governance and Poverty Alleviation

The Congo Basin is home to the second-largest area of humid tropical forest on the planet. Deforestation rates in Africa, which lost 5.3 million ha of forest per year in the 1990s, are also among the highest in the world (World Summit on Sustainable Development, 2002). Population growth, poverty and the lack of an integrated management framework are three of the most pervasive underlying causes of forest degradation in the region.

There is a recent history of networking in Cameroon that promotes the sharing of experiences and allows members to work cooperatively, reconcile sustainable use and forest conservation with the development needs of local populations, and build institutional and organizational capacities. However, efforts have been disjointed and results poor (Pa'ah and Ondo, 2005). The contextual flexibility of the networking concept, as put forth by the International Model Forest Network, recently motivated the government of Cameroon to support the creation of two model forests – Dja et Mpomo and Campo Ma'an – and to join the international network.

By encouraging a transparent and participatory governance structure, the partnership approach provides a framework on which to build

[1]There were close to 80 contributors to the grizzly bear project. They included Weldwood of Canada Ltd, Jasper National Park, Alberta Sustainable Resource Development, Canadian federal government departments, Petro Canada, Center for Wildlife Conservation, University of Alberta, University of Washington and Mountain Equipment Co-op, among others.

a foundation of cooperation, understanding and respect between stakeholders representing a wide range of values. With access to decision-makers, stakeholders hope to work towards a common vision of sustainable forest management while addressing long-standing poverty issues faced by many forest-dependent groups, particularly indigenous peoples. It is also hoped to expand the Dja et Mpomo experience into a Congo Basin model forest network, representing the first trans-boundary, trans-sectoral sustainable forest management initiative in Africa (Pa'ah and Ondo, 2005).

Chile: Conflict Resolution

The situation of the Mapuche-Pahuenche indigenous peoples in the Araucarias del Alto Malleco Model Forest area has been historically marked by inequality, discrimination and limited participation in the decisions that affect them. Conflict over long-standing ancestral land claims has been exacerbated by agricultural settlers pushing the Mapuche off their traditional territory on to increasingly marginal land (International Model Forest Network Secretariat, 2005). The Mapuche have tended to rely on various provincial projects or NGOs to intervene on their behalf, but these groups tend to operate in isolation from one another, and in a top-down manner. In addition, a culture of work centred on individual gain translates into a resistance to shared effort.

Established in 1999, this model forest has become a negotiating basis that acts as a forum for working towards a future of consensus and away from a history of conflict. This forum has been legitimized through a transparent governance structure that is respectful of diversity. The model forest's 22-member board of directors engages 10 rural inhabitants, 7 of whom are indigenous, in a permanent dialogue and decision-making process representing a form of direct democracy they have never before experienced (International Model Forest Network Secretariat, 2005).

In concrete terms, the model forest has undertaken capacity-building for local NGOs, rural inhabitants and indigenous peoples, developed a strategic plan, and coordinated meetings and discussions on themes that are of interest to all and in which stakeholders want to participate. The sharing of resources has led to greater efficiencies. However, with each success comes increased attention. With more stakeholders taking an interest in the model forest, and earning a place on the board, reaching consensus is becoming increasingly difficult. Differing levels of education and different cultural backgrounds and interests complicate the principle of equal participation. The process of balancing short-, medium- and long-term goals is complicated by a lack of adequate support.

Despite these difficulties, individual interests motivating stakeholder involvement have gradually come to be seen as less important than the feeling of collective success on a given issue. Furthermore, the board of directors is now recognized nationally and internationally as a model for inclusiveness and indigenous participation. To achieve these and other successes, it was necessary to have people and organizations dedicated to network-building. The Araucarias del Alto Malleco Model Forest provided the forum for this fundamental role.

Philippines: Watershed Management Planning

The Ulot Watershed Model Forest in the Philippines was established in 2000 as one of the initial sites in the Regional Model Forest Network for Asia. The 87,536 ha site, as its name implies, includes the Ulot river watershed, Samar island's third largest. As the model forest lies within the boundaries of Samar Island Natural Park (SINP), any management strategy must also reflect the larger protected-area management plan. Portions of the model forest area have been deforested. Small-scale timber poaching, rampant slash-and-burn agriculture, and unregulated collection of wildlife species have been identified as some of the causes of forest degradation.

The diversity of stakeholders (including the state forestry department, the army, NGOs, academics and local communities, among others) and diverging issues and concerns in the model forest partnership group have tended to result in a fragmented approach to the implementation of activities and projects. Recognizing a need to address this issue, stakeholders prepared an integrated watershed plan, through participatory processes and in collaboration with the Global Environment Facility's Samar Island Biodiversity Project. Stakeholder representatives developed a planning framework, based on the watershed ecosystem management approach being adopted by the Philippine Department of Environment and Natural Resources, to guide the process. These activities included consultation with stakeholders to secure their support and commitment, watershed profiling, identification and analysis of issues, a public hearing and finalization of the plan (P.S. Daloos, Samar Island, 2006, personal communication).

Bringing together the various stakeholders in the planning process made it possible for them to examine and address the sustainability of watershed resources and values actively. It enhanced their understanding and widened their perspectives on the landscape and their roles in its overall management. However, while participatory management planning elicits a holistic and practicable result, it is also a time-consuming process. Concerns and issues had to be discussed thoroughly by the planning team representing various stakeholder groups till a decision was reached, and that decision then had to be supported by all of the other model forest partners and stakeholders. Reaching a consensus was difficult. Motivation was also an issue, considering that there was no immediate tangible benefit for stakeholders. Recognizing future impacts on increasing incomes and improving farming systems and the resource base motivated the stakeholders to participate. Despite difficulties, because model forests strive to strike a balance between social, economic and ecological considerations, the partnership approach seems to be the best option for achieving sustainable forest management in the long run (P.S. Daloos, Samar Island, 2006, personal communication).

Important to note here is that development of the watershed plan has helped create a meaningful partnership that sees value in the sharing of resources, expertise, information and participatory decision-making. It will likewise increase the participation of each stakeholder in implementing the activities in an integrated and cost-effective manner, while at the same time benefiting the larger watershed for posterity.

Russia: Creation of the First National Park

With a well-developed road system, the Gassinski Model Forest area in Russia's Far East was easily accessible to local residents, who have a history of exploiting their forests for fuelwood and other natural resources. This intensive use of forest stands diminished the availability of the mature trees that provided a home for local wildlife.

Despite the fact that Russia had never before established a national park in the Far East, it was the model forest partnership's goal to do just that in order to mitigate the anthropogenic stress on local plants and animals, and to balance future economic development with conservation values. In the mid-1990s, an initiative to establish the Aniuiski National Park was launched by the model forest association and the chief of the Khabarovsk state forest service. The proposal was also supported by international partners such as the International Model Forest Network Secretariat and McGregor Model Forest, both in Canada, and the World Wide Fund for Nature (Russian Federal Forest Agency, 1996).

During a stakeholders' meeting in the early 2000s, several model forest partners came out strongly against the creation of the park. The forest industry, consumer groups, indigenous peoples and fuelwood collectors opposed the project out of a fear of lost income or access to resources. The proponents persuaded reluctant stakeholders of the benefits the park would bring – conservation of the forest ecology, improved water resources and spawning sites, and free gathering of medicinal plants and other non-wood forest products for indigenous people only. By 2002, the creation of the 427,405 ha Aniuiski park was a reality. The governor of Khabarovsk state announced by resolution that the national park area would be a zone for 'resilient' forests, protected from private use, wildfires, insects and disease in particular. Also, the resolution required improvement in the management and use of forests, wildlife and natural resources through scientific research, as well as technical and financial assistance (Resolution of the Governor of Khabarovsk krai, 2001).

In addition, the resolution appointed the Nanaiski Leskhoz (a federal forestry unit and Gassinski model forest partner) to organize and implement additional measures for wildfire control, forest health surveys and monitoring.

Consequently, wildfires in the park have been reduced by more than 95% in comparison with the 1998–2001 period (Russian Federal Forest Agency, 2001). Logging activities in the area were subsequently restricted and tree planting was doubled in comparison with figures dating from 1994 to 2001. Wildlife conservation focusing on the endangered Siberian tiger was successful, with tiger populations not just stable, but growing. These measures have been valued at three times better than those in the remaining area of the Nanaiski Leskhoz, which is not under the national park system (Russian Federal Forest Agency, 2001).

Since then, the Gassinski Model Forest has continued to play a moderating role between park-dependent groups and conservationists, principally through stakeholders' meetings. Other successful dispute mitigation strategies have included information and education campaigns, clear designation of sites in which gathering of non-wood forest products is permitted, monitoring social opinion on sustainable forest management efforts and increased participation by indigenous peoples in the model forest board of directors.

Sweden: Trans-national Networking for Sustainable Forest Management

The forest sector in Sweden is of critical importance to the national economy, particularly for the local and regional well-being of rural landscapes and communities. Because the proportion of privately owned forest is high (about 50%), private landowners and their associations are key actors in the newly established Vilhelmina Model Forest partnership (International Model Forest Network, 2006). At the same time, traditional land use is fundamental to the economy and culture of the indigenous Saami (Lapp) people. A history of reindeer husbandry that follows migration patterns irrespective of landownership necessitates involvement of the Saami in the model forest partnership. Other key partners include the research community, the state, forest companies, NGOs and forest-based entrepreneurs.

Conflict over access to the forest resources claimed by these actors, each with different spheres of influence and interest, has kept them isolated from each other. However, the successful application of sustainable forest management principles demands open channels of communication, as well as ongoing knowledge and capacity development for applied research. An incentive behind the model forest programme in Sweden, then, is the need to provide a common forum in which these divergent groups can voice their concerns about national and local forest management practices, agree on a common approach to tackling them and then collectively work towards that goal. International networking through the International Model Forest Network will help the model forest gain access to up-to-date approaches and solutions to conflict management in a sustainable forest management context. Indeed, it was access to international expertise and networking that attracted Swedish stakeholders to the international network.

In addition, in recognition of the need to promote holistic solutions beyond local boundaries alone, the European Union has recently launched the Interregional Tacis and Baltic 21 project, which includes Sweden and seven other countries surrounding the Baltic Sea – the region's first cross-sectoral and trans-boundary forest-based cooperation initiative. The Baltic forest participants hope to secure forest sector sustainability and enhance input to regional development and spatial planning by exploring ways in which the model forest concept can be used to meet Baltic Sea region goals (International Model Forest Network, 2006).

Conclusions

Whether driven by an interest in conflict reduction, participatory governance, conservation or networking, it is clear that model forest partnerships have been highly effective over the last decade. Less clear is why – if such broad partnerships are such a good thing – they are not already the standard for addressing sustainable forest management challenges. As several of the case studies suggest, the time and effort required for partnership building is significant. Donors, governments and other key supporters tend to operate within a fixed amount of time and seek material results – economic or otherwise – in the short- to mid-term period.

Because model forests first tackle the social aspects of sustainability, they have to be seen as processes, not projects, and tend to operate for longer periods. It also appears that, while much success has been achieved in the economic and biophysical sciences related to resource management, the social science of sustainable management – the way in which we manage ourselves and the demands we place on the environment – remains the weakest variable in the equation. Model forests devote considerable attention to this issue.

Finally, whether involving local, regional or trans-boundary sustainable forest management planning, model forests have demonstrated that both local stakeholders and decision-makers alike must be part of the process and take an active part in delivering solutions on the ground. Broadly speaking, the array of questions that model forests address are not issues for developed or developing countries alone; they are familiar in all our landscapes. However, both the range of issues considered and the options for addressing them are substantially enriched through broad-based partnerships such as those found in model forests.

References

Besseau, P., Dansou, K. and Johnson, F. (2002) The International Model Forest Network (IMFN): Elements of Success. *Forestry Chronicle* 78, 648–654.

Foothills Model Forest (2004) The Foothills Model Forest Grizzly Bear Research Program. In: *Footnotes: The Newsletter of the Foothills Model Forest*. Available at: http://www.fmf.ca/CE/FN_F04/Fn_F04_tc.html

International Model Forest Network (2006) New Baltic forest initiative to boost sustainable regional development through trans-national forest sector cooperation. Available at: http://www.imfn.net/en/ev-96240-201-1-DO_TOPIC.html

International Model Forest Network Secretariat (2005) Governance in a model forest context: giving people a voice in their future. In: International Model Forest Network Secretariat (ed.) *Partnerships to Success in Sustainable Forest Management*. International Model Forest Network Secretariat, Ottawa, pp. 10–13. Available at: http://www.imfn.net/uploads/user-S/11301799781Partnerships_to_Success.pdf

Pa'ah, P.A. and Ondo, B. (2005) Approaches to networking: experiences and opportunities across the IMFN (Africa). *Connections: International Model Forest Network News* January 2006, 4. Available at: http://www.imfn.net/uploads/user-S/11395815911Newsletter_Eng_final.pdf

Resolution of the Governor of Khabarovsk krai (2001) Concerning the establishment of the National Park 'Aniuiski' in the Nanai District.

Russian Federal Forest Agency (1996) Report: Decision of the Second Forum of Forest Rangers of Russia.

Russian Federal Forest Agency (2001) Report: Instruction of the RF Government on Approval of the Proposal on Establishment of National Forest Reserves and Parks in the RF during 2001–2010.

World Summit on Sustainable Development (2002) *Facts on Forests*. Available at: www.johannesburgsummit.org/html/media_info/pressreleases_factsheets/wssd10_africa.pdf

10 The World Agroforestry Centre's Experience in Cross-sectoral Policy Planning in Africa

Jan G. Laarman

Introduction

This chapter uses a case study approach to present the role and experience of the World Agroforestry Centre (officially known as the International Centre for Research in Agroforestry, ICRAF) in cross-sectoral policy. Rather than laying out a theoretical framework, the working model used by our institution is discussed in relation to our partners and clients, in order to offer a number of contextual observations. The scope is mainly confined to Africa, which is where about two-thirds of ICRAF's project portfolio is located.

ICRAF is explicitly defined by a cross-sectoral professional discipline, one that fuses agriculture with trees. None of the other 14 international agricultural research centres in the Consultative Group for International Agricultural Research has an equivalent cross-sectoral mandate, or certainly not one as clearly specified as that of ICRAF. In the early 1990s, the Consultative Group's decision-makers considered incorporating ICRAF with the Centre for International Forestry Research to form a single entity. However, they determined that agroforestry was important in its own right, and that research and development agendas would be blurred if forestry and agroforestry were to be merged under the same roof. Although the question of integration still comes up from time to time, ICRAF has to date forged its own identity as a centre wholly devoted to agroforestry.

A testament to ICRAF's cross-sectoral approach is captured in its slogan, 'Transforming Lives and Landscapes'. Our perspective is to take in the entire landscape as the conceptual framework for defining problems and solutions. Whether deliberately or sometimes implicitly, our view embraces all the stakeholders and clients who in one way or another influence what happens to the management of the entire landscape and its resources. This is our definition of a holistic view, and certainly also a cross-sectoral one.

The Policy Planning Framework

ICRAF believes that worldwide interest in agroforestry as a cross-sectoral organizing framework is growing and will continue to grow. At the same time, agroforestry practices are not well-known in ministries of finance, local government, health or other official bodies (Bakengesa *et al.*, 2004). How does an organization establish influence when its partners are not certain what it represents? Our name and cross-sectoral identity, therefore, do not necessarily open policy doors. An important first-order consideration is to explain and keep explaining aspects of the agroforestry mission, motivation and linkages to poverty. The policy planning framework begins with that humble position – the weak identity of agroforestry in policy contexts. Agroforestry can be embraced by nearly everyone in principle, but in reality there are comparatively few organizations and institutions that champion it.

That is why the essence of planning to achieve greater influence defines education and outreach as a priority, and cross-sectoral partnerships as another priority. Outreach involves disseminating key messages widely even if somewhat superficially, while partnerships are a means of stimulating cross-sectoral learning experiences at a deeper level. A few examples of the key planning elements are given below.

Planning element no. 1: external messages

ICRAF is like all other institutions in wanting to achieve a highly visible presence in the external world. The ideal is for our messages to be known and understood by a wide range of audiences, including policy-oriented individuals across sectors. On the one hand, this involves advocacy for agroforestry, but it is also our desire to establish a science-based framework for policy decisions. In a world crowded with competing types and sources of information regarding policy and development, what should be our targeting strategy? The main planning challenges include: (i) careful writing of science messages for non-scientific audiences, to make the content understandable but not oversimplified; (ii) encouraging regional field teams to develop their own messages and distribution lists for materials targeted locally, including materials in non-English languages; (iii) finding alternatives to Internet-based information platforms for outreach in Africa, where telecommunications continue to be poor and expensive; and (iv) inviting journalists and diplomats (e.g. ambassadors and consulate officers) to visit us, both in the office and in the field, as a way to extend our messages. In each of these areas, the intent is to deepen our knowledge of the impacts that we may be having. However, that requires deliberate planning in order to correctly design and implement an approach for monitoring and evaluation. To date, ICRAF continues to operate mainly on the basis of informal feedback through its web site, media days, and the choice and quality of news stories prepared for external audiences. However, our scientists and managers would like to be able to direct our strategy more formally towards delivering externally directed messages.

Planning element no. 2: personnel location

Institutions such as ICRAF take pride in their field projects. These are our laboratory for testing, demonstrating and evaluating agroforestry technologies. However, we are increasingly accepting the fact that field projects that are distant with ordinary transportation will not be good choices for policy influence in the usual set of circumstances. Either projects have to be taken to the capital cities by means of slide shows and other media approaches, or some of our professional staff need to be located in the capitals, where they can more effectively interact with the main policy actors. There is an important aspect of planning the balance between being well represented in the field versus being positioned close to government decision-makers and the main organizations in civil society. ICRAF recently made important choices in this regard in Tanzania and Malawi, and we are re-evaluating the locations in which we establish our physical presence in other countries as well. A related dimension is the balance between international versus national staff. While there can be no firm rules about this, many in ICRAF are increasingly of the opinion that national professionals have access to policy windows and processes not available to international members of staff. This is affecting how and where we allocate senior staff in a number of our country programmes.

Planning element no. 3: agroforestry networks

ICRAF has helped organize and is actively participating in a large number of networks. We know that these networks are of great interest from the standpoint of raising the profile of integrated development and cross-sectoral policy. The African Network for Agroforestry Education has assisted 68 colleges and universities in revising their curricula to incorporate multidisciplinary content and to adopt integrative teaching and learning. At subnational levels, the Consortium for Scaling Up Options for Increased Farm Productivity claims 103 partners in western Kenya, while the Segou Provincial Agroforestry Research and Development Consortium is made up of 30 partners in Mali. These are just three examples from among dozens. Here, the planning questions relate to the scope and reach of the networks. Some of the most fascinating details concern how to identify and incubate 'policy champions' in these networks, and how to build trust and legitimacy into them. Also of interest is when and how to recognize successful versus failed networks, and how to politely but firmly close the door on those which are failing.

Planning element no. 4: participation in policy forums

The planning pertains to choosing which forums to attend and making advance arrangements so that opportunities are not wasted. Whenever possible, ICRAF would like to be invited to help shape agenda, the participants list and important parts of the leading documents. In which scenarios do we command the recognition, respect and influence to be able to do that? Do our delegates correctly understand the composition and tone of the forum, and is our message tailored for it? How are cross-sectoral alliances built before, during and after the forum in order to consolidate and reinforce positions? These are among the universal considerations for policy development, which have to be addressed differently for each new situation.

Specific Policy Aspects

Policy aspects are present in each of the three dimensions of ICRAF's mission: research, development and capacity building. First, the research leads to observations about the relevance, legitimacy and credibility of policies forged by governments, international organizations andnon-overnmental organizations (NGOs). Secondly, ICRAF is a partner in development projects which report to donors and national governments, often with policy questions to be addressed. Thirdly, ICRAF finances considerable capacity building – sometimes on policy questions, or with policy implications. ICRAF's research and development work leads us into engagement or potential engagement with a wide range of cross-sectoral actors. Examples include the following.

Trees and agriculture

ICRAF is scaling up the use of nitrogen-fixing trees to supplement chemical fertilizers on nutrient-poor soils in Africa. ICRAF's research and development work on trees in relation to soil fertility has been a traditional strength. Does national agricultural policy on fertilizer subsidies in countries such as Malawi and Zambia correctly account for the potential of the tree-based options? Here it is not only national agricultural leaders who are the decision-makers, but also individuals in ministries such as finance and trade at a linked level of policy.

Trees and health

About two-thirds of all medicinal plant materials are derived from trees. ICRAF is interested in promoting *Artemisia* for malaria treatment in poor households, and in promoting research and development regarding the way in which tree-based fruits may provide improved vitamins and minerals in the diets of maize-eating households to combat HIV/AIDS. Moreover, ICRAF is embarking on a project to conserve the endangered *Aloe vera* plants in western Kenya. In these examples, the connection is with the agendas and policies of a wide range of health institutions.

Trees and human settlement

Especially in East Africa and South-east Asia, ICRAF is advancing modelling efforts that show the effects of trees on water runoff and infiltration. Agroforests (i.e. scattered trees in cropped landscapes) may be little different from intact forests with respect to their watershed properties for downstream sedimentation, flooding and other external factors. It is important to understand this for ministries of water, as well as for policy-makers in ministries of lands and resettlement who are otherwise intent on relocating peasant smallholders to other areas.

Trees and alien invasive species

A main policy interest for ICRAF is alien invasive tree species such as *Leucaena*, *Prosopis* and others, which can sometimes act as weeds. For example, the Community Museums of Kenya is threatening to sue an international organization for its alleged responsibility in having introduced *Prosopis juliflora* into an area in which the species is aggressively expanding. Efforts to arrest and remove the *Prosopis* (locally called the devil tree!) are difficult and expensive. Here the policy actors range from the World Court of Justice, identified by some Kenyans as the relevant institution for hearing the complaint, to national authorities for environment, agriculture and forestry, as well as local district leaders.

Trees and wildlife

Among the many examples that could be presented is ICRAF's work on mangoes and other fruits (avocados, pawpaws, passion fruit, etc.) to diversify smallholder agriculture, including in semiarid regions in which wildlife is abundant. In the Laikipia District of Kenya, this involves the situation of marauding elephants invading the farms, with wildlife officials alleged to be doing little or nothing to mitigate the situation.

Constraints and Opportunities

ICRAF has never defined itself as a policy-oriented institution. Historically, that was not an element of the institution's design; nor was it particularly prominent in the thinking of our earlier generations of Boards of Trustees and directors. Currently, only a few of our scientists specialize in policy, and only a small share of our programmes and projects directly engage in policy research and development. The main exception is the 'Alternatives to Slash-and-Burn' programme, which bridges forestry and agroforestry, and which applies primarily a policy approach in its operational sphere. Otherwise, ICRAF's influence on policy is mainly indirect. The following reasons may be a part of the explanation for this.

Our reputation, and the reputation of most of our partners, has been built mainly in the biophysical disciplines (soil fertility, tree seed and genetics, carbon measurements, biodiversity indexes, etc.). ICRAF is still learning how to extract policy messages from this blend of predominantly biophysical research and development. That requires time and practice. Even more importantly, we will need to more fully develop a culture that values policy implications. While we find evidence of policy orientation among many of our brightest stars, that quality is not universally shared.

Only four or five individuals in ICRAF can read and speak the language of policy economics. This is a distinct handicap, because economists possess special tools that unlock the mainstream of policy-related work. Typically, the strongest arguments in policy employ economic frameworks, which get to the heart of welfare measurements.

ICRAF has been assertive in taking on a wide range of field projects, and we have liberally defined the scope of agroforestry to explain and legitimize this reach. A side effect of covering a broad range of subject matter is that one is often 'thin' in particular specialties. This has its drawbacks for building a policy agenda, given that we may lack sufficient critical mass to compile policy cases in depth.

Increasingly, many of ICRAF's funding sources favour development over research, on the assumption that there is sufficient knowledge for agroforestry implementation. This assumption is often wrong (Franzel *et al.*, 2002), but it nevertheless applies in any number of practical cases. Most of our development-oriented projects have short time frames, and many projects are not renewable. A composite of many projects, funded over short periods, is not ideal for developing policy insights and capturing them in an agenda.

A slow-moving variable is our composition of scientists and professionals. It is a rule of thumb in the Consultative Group that professional turnover is about 10% annually. Thus, it would take a relatively long period to develop a stronger policy-oriented team, even if we made that a top priority.

While acknowledging these as important constraints, ICRAF's incumbent senior leadership team is aiming to build a greater presence and platform in the policy field. We would cite the following assets and initiatives as being among our strongest opportunities:

- Our headquarters facilities, on a campus setting in one of Africa's most international and increasingly cosmopolitan cities, is an important asset. The campus adjoins the United Nations complex, and we are geographically close to important embassies and consulates. Thus, ICRAF has years of experience in receiving high-level visitors from all over the world. Our campus provides offices for 4 other Consultative Group centres and 11 additional international and national organizations. In other words, our location and facilities provide a superb venue for gathering people from a wide range of countries, organizations, interests and development philosophies.
- Among the organizations hosted on the ICRAF campus is the Millennium Development Goals Centre, reporting to the United Nations Secretary-General and supported by the United Nations Development Group. The Millennium Development Goals professional team provides country-level technical support to several African countries – across disciplines, economic and social sectors, and institutional frameworks (Millennium Development Goals Technical Support Centre, 2004). In particular, ICRAF is a partner with the Millennium Development Goals Centre in western Kenya, in a Millennium Village that combines actions in agroforestry, health, education and other cross-cutting investments in the community. The Millennium Development Goals Centre and its Millennium Villages are a living laboratory of the cross-sectoral policy approach, and ICRAF and its partners are well positioned to learn from it.
- In 2003, ICRAF reorganized itself and added a director's position in Strategic Initiatives. The role of this director is to strengthen ICRAF's alliances and partnerships, not only in Africa but also in world forums. The aim is to be represented at world gatherings on climate change, biodiversity and forests. In Africa, we are aiming to achieve an identity and raise our profile with the Forum for Agricultural Research in Africa, the African Union, and the New Partnership for Africa's Development – among other bodies of importance for policy agendas. Thus, while our past participation at these policy levels was often lacking, ICRAF has shifted its budget and organizational structure to raise our profile in the future.
- In order to better position ICRAF on the policy stage, we are constantly reviewing our work to recast and reconceptualize what we do. For example, a recent project mapped where African hunger intersects the continent's farming systems. From this overlay, we defined different 'hunger hot spots', each of which needs to be characterized for its unique problems–solutions matrix. This refinement provides a new look at the geography of hunger, and in principle opens the door to what should be a spirited dialogue among policymakers in agriculture, economy, environment, water, transport and other sectors.

Conclusions

The preceding section enumerated a number of assets and initiatives that ICRAF can use to its advantage in cross-sectoral policy. Among them are our location and campus; our networks with a range of different partners at strategic and operational levels; and our inherently interdisciplinary and cross-sectoral mission. ICRAF has invested in a director for alliances and partnerships at world forums addressing policy issues; we use and develop science on the ground to afford us credibility and legitimacy. Unlike many sister institutions that work mainly from a central office, ICRAF is literally on the farm and in the landscape, in order to generate real-world learning.

At the same time, we should be explicit about many needed improvements. Our capacity to be effective in cross-sectoral policy from an agroforestry platform needs to grow, as does the capacity of our partners. For scenarios of additional inflows of monetary resources that could be used for this purpose, priorities for capacity building include the following.

University education

The new generation of professionals needs reinforcement of the cross-sectoral approach. If

African universities can break through their disciplinary walls of the past, that could ignite an important spark in cross-sectoral work at an age level that builds for the future (Temu *et al.*, 2003). This is to place an enlightened approach to university education as very high on a list of actions in support of cross-sectoral policy planning.

Policy discovery

Organizations like ours and others that are predominantly engaged in the biophysical sciences could usefully allocate more resources to 'cross-sectoral policy discovery' within our own ranks. The aim is to better interpret ongoing work in terms of its potential significance to a range of policy-makers. Whether this requires new personnel, or whether it can be done by reorganizing existing resources, will vary from case to case. The individuals who take on this responsibility will need to be exceptionally broad in outlook and skilled in formulating policy concepts. Moreover, the agenda built up in this way has to be successfully transmitted to others who can do something with it. This is not a trivial assignment, and to do it well may imply organizational changes and added financial resources.

Cross-sectoral specialization

With or without added resources for policy discovery, ICRAF and others who work in agroforestry should be capable of identifying and finding agreement on where and how to add value to cross-sectoral policy issues. For example, ICRAF has done research to change thinking and alter misplaced misconceptions in selected aspects of watershed management and alien invasive species, to mention but two. On the other hand, more consultation is required with allied actors in a variety of sectors before declaring any sort of comparative advantage. Other organizations, likewise, should make it a priority to explore their strengths and weaknesses in cross-sectoral policy in order to correctly position themselves with respect to the many contributing actors.

Cross-sectoral forums

This is typically a strong point for an agroforestry research and development centre, but ICRAF and our partners should continuously aspire to achieve greater results. Merely convening individuals from different institutions and different backgrounds does not define a cross-sectoral forum. Instead, we should be making a science of the way in which cross-sectoral forums can be made to operate to achieve maximum results. This will turn attention to details of forum design, selection of its participants, options for sharing information, capture and delivery of outputs from the forum, sustainability of follow-up efforts and the like. This adds up to a need for substantial capacity building to be effective.

In summary, ICRAF is among many organizations that believe in evidence-based policy discussions to advance rural development. The goal is to use an agroforestry platform to win credibility with audiences in sectors besides forestry and agriculture. This requires outreach to allies in health, wildlife, human settlement, economic policy and others. On the one hand, our scientists and managers claim to possess many of the assets and initiatives for engaging a range of policy-makers. On the other, we and other organizations like us can greatly benefit from capacity building and institutional realignment to have stronger influence in cross-sectoral policy agendas.

References

Bakengesa, S.S., Otsyina, R. and Ok'ting'ati, A. (2004) Influence of national policies on tree planting and conservation: the case of agroforestry technologies in Shinyanja, Tanzania. In: Rao, M.R. and Kwesiga, F.R. (eds) *Proceedings of the Regional Agroforestry Conference on Agroforestry Impacts on Livelihoods in Southern Africa: Putting Research into Practice*. World Agroforestry Centre (ICRAF), Nairobi, Kenya, pp. 301–306.

Franzel, S., Cooper, P., Denning, G.L. and Eade, D. (eds) (2002) *Development and Agroforestry: Scaling Up the Impacts of Research*. Oxfam GB, Oxford, UK.

Millennium Development Goals Technical Support Centre (2004) *Africa's Green Revolution: A Call to Action*. Proceedings of the July 5th, 2004 High-Level Seminar, Addis Ababa, Ethiopia, convened by the Government of Ethiopia and the United Nations Millennium Project. Nairobi, Kenya.

Temu, A., Mwanje, I. and Mogotsi, K. (2003) *Improving Agriculture and Natural Resources Education in Africa*. World Agroforestry Centre, Nairobi, Kenya.

11 Cross-sectoral Policy-planning Issues in Forest Management: Lessons Learnt from Two Case Studies in Ethiopia

Samuel Gebre-Selassie

Introduction

The government of Ethiopia identifies deforestation, land degradation and diminishing agricultural productivity as key national problems. Increasing population, indiscriminate clearing of forests for cultivation, uncontrolled exploitation of trees for timber, poles and fuelwood as well as overgrazing are the major causes for the alarming rate of deforestation. The government estimated the annual loss of natural forests at 80,000 ha. However, a more realistic figure would be somewhere between 150,000 ha and 200,000 ha (Eyasu, 2003; Tsegaye *et al.*, 2004).

Government-controlled forest conservation initiatives began in Ethiopia in the mid-1970s, after the government had nationalized all rural land. These initiatives have resulted in the establishment of different types of protected areas, such as state-owned forest priority areas, national parks, game reserves, sanctuaries and controlled hunting areas. The government has identified and designated 58 forest priority areas, which include closed forest, woodlands and plantations for nature protection and other forest sites administered by local authorities or communities. Settlement inside the forest priority areas is prohibited, and only the government may harvest forest products. The state has exclusive control over the use, protection and management of these areas, with exclusion of community and local people's rights to use and manage forests (Kubsa *et al.*, 2003).

Between 1976 and 1988, 500 million seedlings were planted on 465,000 ha of communal lands. Community woodlots were introduced during the 1980s through a tree-planting programme organized by the government for catchment protection and production of fuelwood. However, a lack of clear legal rights regarding the ownership, management and utilization of trees has been a major impediment to the sustainable development of community forests. The extent of community woodlots has been estimated as 42,000 ha of peri-urban fuelwood plantations, and approximately 44,000 ha of catchment protection plantations (Eyasu, 2003). Expansion of plantation forestry has been constrained by land shortage and limited government capacity to establish and operate plantations.

The government has proved ineffective in providing the expected level of protection for various reasons, including its poor capacity to provide alternative livelihood for people living around protected forest areas, and its institutional weakness in enforcing established regulations and laws. Individuals have been discouraged due to state ownership of land and tenure insecurity, which makes it impossible or too risky to engage in investment ventures that need more than one decade to pay back.

Since the mid-1990s, several donor-supported projects have been established in different protected areas of Ethiopia to bring about active community involvement, in terms of granting different responsibility and use rights. These projects include the Adaba-Dodola integrated forest management project, the Borana collaborative forest management project, the Chilimo and Bonga joint forest management projects, and Menaggesha-Suba participatory forest management. All of these projects strive to ensure active community engagement through a process-oriented experimental approach (Tesgaye *et al.*, 2004). This chapter discusses two case studies[1] carried out by a team of experts in the southern and northern parts of the country. The studies demonstrate important cross-sectoral planning and capacity-building issues in forest management. Their results should help both the government and the donor community to identify relevant policy instruments and institutional arrangements, with a view to enhancing policy coordination and harmonization.

Case Study I: Adaba-Dodola Integrated Forest Management Project

Purpose

The purpose of this project was to promote and support participatory forest management through community empowerment. It has developed a feasible forest conservation approach locally recognized through the wajib[2] or forest dwellers' association. The association has three main objectives: (i) regulating access – forest dwellers are granted exclusive user rights, with clearly defined and agreed-upon rights and duties; (ii) reducing pressure – non-forest dwellers are encouraged to plant trees for various purposes around their homesteads; and (iii) making trees profitable – opportunities for non-wood income from forests are assessed and implementation of options encouraged (Kubsa *et al.*, 2003).

The Adaba-Dodola forest is situated on the northern slopes of the Bale Mountains at an altitude varying between 2500 and 3500 m above sea level. It represents one of the few surviving afromontane forests and currently covers 53,000 ha, down from 140,000 ha in 1982 (Tsegaye *et al.*, 2004). Stakeholders for the participatory forest management programme were identified and their needs discussed. On the basis of the discussion, forest project areas consisting of different forest blocks were formed. On average, a forest block has a size of 360 ha with a maximum of 30 members – based on an estimated carrying capacity of 12 ha per homestead – of a forest dwellers' group or wajib. The eligibility criteria for becoming a member of the wajib group include ethnic homogeneity among user groups, geographic proximity to a forest area, willingness to participate and length of time lived in the village. Villagers who failed to fulfil these criteria were forced to demolish their houses and leave the forest area. The two other major stakeholders in the project were the forest administration, which consisted of experts hired by a non-governmental organization, and local government institutions providing institutional support and monitoring project activities, together with experts hired for the project.

A forest block allocation contract was prepared and signed by the stakeholders. The contract specifies the rights and the duties of the forest administration and the forest dwellers (Box 11.1). The regional government also endorsed the contract as a legally binding contract. The wajib groups also occasionally permit access to non-members, for example, for the sale of grazing lands and fuelwood collection, based on clearly defined terms. The non-wajib groups were happy with this arrangement. Between June 2000 and September 2002, 19 wajib groups including 507 members in three villages concluded forest block allocation contracts with the forest administration, representing a total forest area of 7526 ha. The process for establishing 20 more wajib groups, representing some 9000 ha of forest area, was completed by mid-2003 (Kubsa *et al.*, 2003).

Results

Due to improved incentive mechanisms and improved control from wajib groups, forest conditions have improved, as witnessed by an

[1]The two studies are: (i) impact of participatory forest management practices in Adaba-Dodola forest priority area of Oromia, Ethiopia (Tsegaye *et al.*, 2004); and (ii) assessment of community-managed forestry sites in South Gonder (Crabtree, 1997).

[2]'Wajib' in the local language stands for *Waldayaa Jiraatoota Bosonaa*, which means forest dwellers' association.

Box 11.1. Rights and duties of wajib groups.

- Settlement in forest blocks that do not exceed the number of agreed homesteads;
- Utilization of forest products for home consumption or sale that do not exceed the annual allowable cut;
- Maintaining a stable tree cover;
- Payment of forest use rent of about US$1/ha for areas not covered by forest, levied by the government in order to encourage the wajib to increase the forest cover;
- Transfer by the government of 40% of the payment to the non-wajib members of the community to share benefits from the forest to other users of this resource;
- Regulating forest use.

increase in natural regeneration and a reduction in the number of donkeys going to the forest for loading forest products. In the area in which a wajib is operational, the forest dwellers have demonstrated their capacity to manage and continuously utilize the forest. The potential of participatory forest management as an alternative to the conventional forest conservation practice has thus been proved, although not yet over a longer period of time or over extended areas. One may conclude that the wajib approach in the context of Adaba-Dodola forestry priority area has improved the capacity and willingness of the local community to actively participate in forest conservation. Moreover, the approach serves as a means to harmonize people and forestry through participatory decision-making and management leading to sustainable forest management.

Despite the positive results of the project itself, the forest conditions in the overall area have not improved, due to two factors inducing a high rate of deforestation. First of all, non-wajib members have speeded up their deforestation activity, the idea being 'let me take my share before the wajib expands into new areas'. Secondly, farmers who were denied access to the forest in their village travelled to other villages for tree cutting where a wajib was not yet established.

Participatory forest management has neither been expanded to, nor replicated in, other areas, for a variety of reasons. Tsegaye *et al.* (2004) identified the following issues as challenges faced by this kind of project:

- Enforcement of rules and regulations has become difficult due to declining food production and high population pressure.
- Members who do not become part of the project should have an equitable share of benefits from forests or other common resources of the village, or should be supported otherwise.
- The creation of new forests has not been considered in the project. The project was based on conservation-oriented intervention.
- Increased population may hamper the successful implementation of participatory forest management if alternative livelihood cannot be promoted in the area.
- Unequal participation among wajib members (users) in forest protection and management activities has negative effects on project performance.
- Weak relationships among a wajib's founding members and more recent members, as well as power struggles for wajib leadership, need to be taken into consideration.
- Various kinds of conflicts, including conflict over resource use (grazing land and forest products) among wajib and non-wajib groups need formal conflict resolution procedures.

Altogether, the success of wajib groups depended on persistent commitment, legal backing, and technical and organizational assistance from different government institutions and bilateral organizations. The findings confirm, as a major point requiring careful consideration, the necessity to incorporate the interests of non-wajib members so that their actions do not counteract the objectives of such projects (Kubsa *et al.*, 2003).

Case Study II: Assessment of Community-managed Forest Sites

Purpose

Crabtree (1997) tried to ascertain what had happened to various forest sites that had been handed over to communities, how the responsibility of forest management had progressed and how benefits were shared among members of communities. The study included 101 representative forestry sites covering 4666 ha or 46% of the total forestry sites in the investigated area. Interviews, group discussions and physical inspection of the forest sites were used to collect the required data. In order to obtain an unbiased opinion reflecting the true view of the people, the questionnaire was prepared in the local language (Amharic). It was designed to be completed by the actual site-managing communities, with a minimum of assistance from development agents working in the area.

The forestry sites had been established between 1977 and 1993 as part of the federal government's policy to hand over to local communities all forestry sites throughout the country that had been established and managed by government agencies. In South Gonder, 218 forestry sites were handed over to communities (peasant associations) by district authorities under the terms of a simple formal letter that requested communities to manage the sites properly and not to expose them to fire or other types of degradation. Communities involved in forest site management were asked about the level of site protection, the level of support provided by the government, cooperation by members of the communities in respecting the regulations set by forest management committees, changes in the forests since their takeover by the communities, benefits or income from the sites or forests, and the suitability of the existing community management system.

Results

The survey results indicated that 73% of the respondents stated that sites are protected by guards, while 27% responded that they were not protected. The communities paid 68% of the guards, with 35% of the amount coming from the proceeds of produce sales. With regard to the level of support from the government, about three-quarters of the respondents (73%) indicated that they received insufficient support. Almost all respondents (97%) demanded governmental support in terms of money for payment of guards (45%), technical advice (30%), training and free seedlings (8% each), tools (4%) and transferring rights to communities for control and use of forests (4%). With regard to the cooperation of community members in respecting the regulations set by forest management committees, about 40% of the respondents said that people respect the regulations, 25% said that the villagers did not respect them; 17% said that some villagers partly violate the regulations and another 17% reported that no regulations had been put in place.

With regard to changes in the condition of the forests since the takeover by the communities, 46% of the community members and members of community forest management committees were of the opinion that the forest cover had deteriorated, and 16% said that the community did not protect the forests. On the other hand, 38% of the respondents said that protection and forest cover were better than before. With regard to benefits or income from the forest sites, 56% of the respondents answered that no benefits of any kind were received; 44% said that some benefits had been obtained, although no income had been distributed to individuals directly.

With regard to the question 'Is the existing community management system suitable for other areas?', 69% felt that the existing management system was unsuitable and 31% considered it suitable. To a related indirect question 'Who should manage those forests not taken over by communities?', 67% answered the government, and 33% said that they should be jointly managed by the community and government. Among the major demands for government participation, payments for guards (88%) were mentioned. Of the respondents 12% said that sites were too large to be managed by the community or committees. Regarding community preferences for forest development, 64% felt that forests should be developed on an individual basis, 31% preferred communal development and only 5% of the respondents were in favour of development by the government. Moreover, the majority (71%) of communities are reluctant to also take over the responsibility for the management of government forest nurseries.

On the basis of these findings, the study made the following six recommendations for improvements in forest management in the investigated project area, as well as for similar interventions in other areas:

- Large plantation and priority environmental protection areas should remain under government management until community forestry management becomes well established and local forestry skills have improved.
- Promotion of joint government and community management practices of forestry sites should be envisaged.
- Participatory strategies need to be developed to ensure that community members have an equal chance to obtain access to forest resources on a competitive basis and in a transparent manner.
- Revenues and other benefits from forest sites should be distributed on the basis of previously agreed criteria.
- The government should continue to provide adequate financial resources for forest guards, given that most forest sites cannot generate adequate income from selling forest products.
- A biannual system of site inspections should be introduced, which will record and disseminate significant information to all concerned bodies. Many of the operational problems could be discussed and solved during the course of these inspections.

Conclusions

Investment in forest management is a long-term activity and venture. More effective policy instruments and institutional arrangements are required for fostering cross-sectoral planning and implementation of forest protection and reforestation programmes. Many actual problems are centred on government ownership of land, which prevents communities and citizens from creating new forests or shifting to sustainable uses of the existing forests. On the other hand, the government is not in a position to manage the forest areas effectively. This has been demonstrated by the government's poor capacity to manage the 58 state forests. There is a gap in terms of policy, institutions and resources. Legislation that provides full responsibility and empowers communities to manage and plant new forests is thus indispensable. Communities need effective support in terms of continuous capacity-building and technical and financial support.

A key policy challenge is finding a cross-sectoral strategy for conservation and sustainable use of forests in today's context of resource demand and competition. In order to be sustainable, consumption and utilization must be stabilized by reducing overutilization. To that end, the following recommendations are made.

Promote collaborative forest management

Forest management could be handled by government, private individuals or communities, or by some collaborative arrangement among these three entities. This requires redefinition of the roles and responsibilities of these three entities in forest management. Experience has shown that when communities are empowered with the responsibility and rights to manage forests, and when they receive the benefits of their engagement, the rate of degradation will be substantially reduced. In many cases, the forest cover improves visibly (Tsegaye *et al.*, 2004).

Provide alternative sources of fuelwood

Alternative energy sources should be found to reduce the excessive dependence on traditional energy sources and rescue Ethiopian forests. Developing and promoting wood-saving technologies to reduce energy loss through the use of improved stoves is important for improving energy efficiency. In the current situation, there is no government institution to look after and guide the future direction of fuel energy development within the country (Bekele, 2005).

Increased productivity of cultivated land

Since most of the land suitable for agriculture is already in production, meeting current and future food requirements will necessitate rapid productivity increases per unit of land to avoid further expansion onto fragile soils. Some options for intensive agriculture include the use of improved crop varieties, application of organic and inorganic fertilizers, conservation-based agricultural practices, irrigation, crop diversification, improved market access, and a variety of other inputs and investments.

Improvement of non-farm employment

Fostering off-farm employment for the purpose of livelihood diversification is inevitable if one wants to reduce degradation of soils and forests. Given the ever-increasing population pressure, development interventions should attempt to organize and promote income-generating activities as a step forward in diversifying income and reducing the current dependence on land resources alone.

References

Bekele, B. (2005) Energy and development in Ethiopia. Paper presented at the Vision 2020 Ethiopia Forum, organized by the Ethiopian Economic Association. Addis Ababa, Ethiopia.

Crabtree, D. (1997) *Assessment of Community-managed Forestry Sites. Integrated Food Security Program in South Gonder*. Amhara National Regional State Bureau of Agriculture and GTZ, Addis Ababa, Ethiopia.

Eyasu, E. (2003) *Environment Role Module. Role of Agriculture in Development. A Study Made for the UNFAO.* Ethiopian Economic Association, Addis Ababa, Ethiopia.

Kubsa, A., Mariame, A., Amante, G., Lipp, H.-J. and Tadesse, T. (2003) Wajib: an alternative forest conservation approach for Ethiopia's forests. Paper submitted to the XII World Forestry Congress, Quebec City, Canada.

Tsegaye, B., Senbeta, F. and Ameha, A. (2004) Impact of participatory forest management practices in Adaba-Dodola Forest Priority Area of Oromia, Ethiopia. *Ethiopian Journal of Natural Resources* 6(1), 89–109.

12 Gender and Off-reserve Forest Management in the High Forest of Ghana

Elizabeth Ardayfio-Schandorf

Introduction

Ghana's 1948 forestry policy became the first formal forestry policy in the country. Its major focus was to maintain the protective purposes of the reserves and apply the concept of productive potential and value on the basis of sustained timber yield. Its preoccupation was with developing scientific systems of management to increase the forest's timber productivity. This policy has had an enormous effect on the forests and people living on the forest fringes. The forest reserves established were managed with local interests relegated to the background, whilst the policy promoted the timber industry and its values to the exclusion of the protective functions of forests and elements of collaboration. This imbalance was to lead to the revision of the policy in 1994.

The implementation of the new policy is being managed through the national forestry programme (1996–2020). In the process, certain policy reforms were approved in 2002 to ensure sustainable management and improved utilization of the country's forest and wildlife resources for optimum benefit to the country and particularly to the resource owners. Of interest in the reforms is the issue of benefit sharing for natural forest timber revenue. Off-reserve forest benefit sharing in favour of landowners has been changed. It changed from 60% for the Forestry Commission to 40% and for the stool landowners from 40% to 60%. Improved incentive structures have also been introduced to ensure sustainable forest management. Social responsibility agreements now enshrined in the law, give forest-fringe communities a financial stake in commercial timber operations, whilst ongoing policy reforms are expected to benefit farmers in the new modified taungya plantations. All these efforts are aimed at eliminating illegal logging and chainsaw operation that still continue (Ministry of Land and Forestry, 1994).

To buttress the protective functions, the collaborative forest management approach has been adopted as an all-inclusive approach in forest management. This inclusiveness seeks to establish partnership with local people and to respond to the subsistence and economic needs of the growing population in sustainable forest management. Stakeholders identified are the scientific and technological communities, women, farmers and forest owners, among others. The success of sustainable collaborative forest management is seen, therefore, to be intricately linked to the active participation of all affected individuals, groups and organizations. Needless to say, all the reforms have been on a general basis, without taking into consideration the gendered interests of resource owners and farmers in the high forest zones.

In Ghana, in principle, all members of the stool and lineage have inherent access rights to land, regardless of sex. Women, however, have never played a traditional role in original land acquisition. As land was acquired under the leadership of stool or lineage heads, women did not play any role in the process. Family headship and stool occupancy have been the prerogative of men. The original role of men in the acquisition of land has given them predominant rights over stool and family landownership. In the farming communities, early marriage is the rule. Related conjugal and domestic obligations also limit the chances of women to acquire land or portions comparably large to those of men. Women in the country have in practice not had equal access to inherent land rights.

This chapter derives from some aspects of research conducted on off-reserve forest management in connection with a 'Socio-economic study of off-reserve tree management in the Goaso Forest District' by the University of Ghana Tropenbos Team between 2003 and 2004.[1] It is intended as a contribution to the understanding of gender relations in off-reserve forest management and its implication for strategic interests in the planning and management of off-reserve forests. It discusses the management of trees on farms and raises issues on equitable access to the participation and benefit sharing of off-reserve forest management in so far as male and female farmers are concerned.

Context, Study Area and Methodology

Context

Various studies have brought to the fore the critical role women play in the management and use of natural resources such as crops and forests, which form an integral part of the rural economy of forest-fringe communities. As far back as 1970, women in Africa were described as farmers par excellence (Boserup, 1970; Vallenga, 1986; Ardayfio-Schandorf, 1994; Benneh *et al.*, 1995). In Africa, women's participation in biodiversity management and the household political economy has been found to vary between various farms, the type of work they do, their access to productive resources such as landownership and the provision of goods and services (Gyasi, 1996). Although non-timber forest products, their production, management and supply are considered to be an integral part of the role of women, these may not apply in all situations in forest communities in Africa. Many women in the high forest are farmers, managing trees in farming systems.

Through current governmental policies, trees and timber cannot be removed without the prior consent of the farmer, whether male or female. This is where the significance of the participation, management and benefits of male and female farmers in managing off-reserve forest resources lies. In the off-reserves, empirical studies were carried out on tree management under a system of clear division of labour between men and women. Men tend to be the landowners, the cash-crop farmers, plantation owners and big time food crop farmers, whilst women are small-scale subsistence farmers, with only a few of them owning lands. Under this system, the issue arises of women's participation and benefits from the management of trees in off-reserves, considering the gender-differential access to the means of production on off-reserves, where most trees are managed for agrodiversity and biodiversity.

Study area

The study area is located in the Goaso Forest District (Brong Ahafo Region) in the high forest zone of Ghana, which has forest characteristics similar to those of other high forests elsewhere in Africa. The 28 communities that formed the basis for the fieldwork are located at the forest reserve fringes. In order to maintain the culture and livelihood of the people living within or close to the forest reserves, there is now worldwide recognition that the livelihood of these people should serve as a major theme in forest management. The recognition of the people's rights to control and manage their own resources is considered a fundamental issue for the maintenance of their culture and for sustainable development in the area. This accounts for one of the major concerns that underlies the purpose of the research conducted in Ghana in the high forest with the involvement of women and men farmers. The off-reserves are characterized by agricultural land use types giving rise to a landscape of agrodiversity and biodiversity. The farmers are engaged

[1]Ardayfio-Schandorf, E., Yankson, P.W.K., Asiedu, A.B., Agyei-Mensat, S. and Attua, M. were the members of the team.

in cash-crop farming for both commercial and subsistence purposes. Although there are other occupations, including administration and commerce, those relating to farming and forestry occupations predominate.

Of all the population, 66.6% of the 26,347 men are farmers, in comparison with 33.4% of the 25,516 women. In these communities, forestry activities and forms of livelihood outside the forest reserves are considered within the agricultural production system. The diversification requires a holistic approach in addressing roles and functions of stakeholders and farmers in effective conservation and management, with a focus on sustainable conservation and management of forest trees in farming systems. One interesting aspect of this forest area is that there are more timber trees in farming areas, where most of the land is under cultivation for cash crops and food crops. Farmers deliberately leave trees on their farms for economic, environmental, spiritual, medicinal, agricultural and livelihood values. The major farming crops consist of cocoa, oil palm and food crops of various kinds.

Methodology

The methodology of the study is based on the distance decay factor in terms of use of resources of the forest reserves. It is meant to demonstrate the relationship between the management and utilization of off-reserve tree resources and the depletion of these resources in the forest-fringe communities in the study area. As women and men in the forest-fringe communities are characterized by varying gender interests and strategic needs, it is expected that their participation in conservation strategies will vary. It is equally expected that benefits derived from ecological services will benefit both sexes once they live in the same forest-fringe communities. Thus, the greater the encroachment is by communities on the reserves, the higher the rate of forest degradation. For this purpose, three distinct buffers were identified and demarcated from the forest reserves in the region. This concept is also expected to demonstrate the land cover types in the buffers and those that are mostly affected by human activities, and how trees could be properly conserved and managed by women and men farmers. In addition, focus-group discussions were undertaken to solicit qualitative data, whilst household surveys from 28 communities generated quantitative data on off-reserve tree resources.

Results

Landownership and tenure arrangements

One of the critical resources essentially needed for sustainable management of forest resources and farming is ownership and control of land. Access to land has been found in many studies in Africa to be a source of grave concern to women in farming systems. In Ghana, women's access to land ranges from 50% in the southern sector to less than 10% in the north. In her work in Ghana, Mikell (1983) has demonstrated the changing fortunes of female cocoa farmers in access to land and other means of production. She pointed out that during the boom for cocoa, which is the major cash crop in the Brong Ahafo Region, women entered cocoa farming in their own right. Of the women she studied 42% were able to acquire their own land and farm their own farms. The remaining 58% obtained land from family members. By contrast, among the males, 70.25% acquired and worked their own land. Beckett (1994) and other scholars, including Arhin (1983), have corroborated these findings. The situation has changed greatly over time with changing fortunes in the economy, especially since independence in 1957, with the adoption of various national economic development policies.

Women's access to land depends on availability, which in recent times is also related to access to credit. In this case, credit has increasingly become an indicator of the extent to which women can operate in their farming enterprises. Lack of capital has in most cases resulted in the inability of women to develop farms on a large scale. Added to this is access to labour, the cost of which is rising. Many women require this resource as farmers in order to operate farms on their own account. The lack of the critical means of production has conspired to limit access of women in the Brong Ahafo Region to the acquisition and ownership of land as a source of economic empowerment in local farming systems.

In spite of the fact that the study area is located in the south, where women's access to resources is supposed to be better, the results indicate that gender allocation of land is unfavourable to women and falls below the south-sector average, even though a greater proportion of the people are farmers. Land for farming in the communities studied could be acquired through purchase, which in this study is considered as self-acquired. It could also be acquired through allocation by the family or the chief, whereby the land is referred to as 'stool land'. Another important means by which farmers acquire land is through share-cropping. By this method, a farmer is granted a plot of land for farming and the owner receives either 50% (abunu) or 33.3% (abusa) of the harvest. The remaining percentage goes to the farmer.

The study also demonstrates that access to land is not uniform between women and men. Out of a total of 399 respondents, self-ownership of land among the females was as low as 19.9%, in comparison with 80.1% for males. Similarly, it was difficult for individual females to enter into personal tenure arrangements, as only 21.7% were able to do so (Table 12.1). The higher access of 25.7% to land for women is obtained through the family. This implies that women's access to land is more favourable when allocated as inherited property or as a gift from family relations such as fathers, husbands or brothers. Patriarchy and its related underpinnings are clearly at play in the access to, and allocation of, land. In all aspects, both landownership and tenure arrangements being practised in the forest area are in favour of men. A direct focus-group discussion held with women's groups underscored the difficult land mechanism of acquisition for females.

Land in the communities is becoming quite scarce, as most lands have been put to cocoa farming. As a result, most farmers – mainly women – who are involved in food production compete for the small parcels of land available. In most cases, the first person to go to the landowner is more likely to acquire the land. It does not depend on sex, but on how fast one gets to the landowner and one's experience in farming. This notwithstanding, women were found to suffer more than their male counterparts in acquiring land for farming. This was because 'most people have the negative feeling about women as not being strong enough to farm on their own' (women's focus-group discussion, 2004).

Apart from this perception, another challenge that women face in acquiring land is a lack of money for outright acquisition through land purchase. In view of this, women reported: 'We, the women in this part of our country, are very poor indeed. We face a lot of challenges in raising enough money to acquire our own land for farming. This is because we have no other income-generating activity than farming, which doesn't earn us much' (women's focus-group discussion, 2004). The sociocultural prejudice and local orientations that introduce certain biases in favour of men in the access to, and tenure arrangements for, land have implications for tree management.

Tree utility and management practices

There are more women farmers generally than men, but most land is owned by men. It follows, then – as is the custom in farming communities elsewhere in the country – that most women are farmers, but not on their own account. They farm in support of their husbands or family. Where they are able to acquire land, they are

Table 12.1. Landownership and tenure arrangement by gender (%).

Landownership (%)			Land tenure arrangement (%)		
Acquisition	Male	Female	Tenure type	Male	Female
Self	80.1	19.9	Self	78.3	21.7
Family	74.3	25.7	Family	76.1	23.9
Stool	84.2	15.8	Share-cropping	79.3	20.7
Share-cropping	77.8	22.2	Lease from chief	75.0	25.0

in a position to manage the trees on their farms. Normally, men tend to be more involved in all the forms of management practices than women. Although women are engaged in a limited manner in these processes, when it comes to pruning, loping and ringing, which require extra energy, women were found to be almost absent from such activities. It is possible for them to engage labour for such management activities, but again lack of finance is an obstacle for them in this respect. In all the various management practices, the engagement of men respondents was far greater than that of women. For pruning as a measure of tree management, a considerable proportion of 82.7% were men, while only 17.3% were women. The management activities in which women are engaged, to some extent, are singling and weeding; these limit their full participation and hence their direct benefits. Through tree management, women have acquired knowledge of trees that impact on their conservation practices. On the basis of indigenous knowledge, trees were classified as ecological, economic, cultural and subsistence. Among them were the uses of trees for the protection of watershed, prevention of soil erosion, soil fertility improvement and provision of shade categorized as ecological. For economic purposes, trees were used for the provision of food, as well as timber for furniture and for construction. Other economic derivatives from tree species were household goods and fuelwood, whilst other trees were useful for sociocultural purposes such as medicine and herbs.

Knowledge about trees and their uses appears to be gendered. Generally, most men are more knowledgeable about trees on farms than women. The men tend to be more knowledgeable about ecological uses than women. All of the men indicated that they use trees to prevent soil erosion; no women mentioned this use. In the case of use for watershed protection, 82% of men utilize trees for this, as against 18% of women (Table 12.2). It was only in knowledge about the use of forest products for furniture that women exhibited a utility knowledge pattern equal to that of men. One area in which women's knowledge on utility far surpassed that of men was in the use of trees for household goods, such as mortar and pestles. All of the women respondents demonstrated knowledge of these uses, with none of the men having similar knowledge. In addition, women demonstrated general knowledge about trees in their farming practices. In their own words, they cited some of the benefits they derive from the presence of trees on their farms: 'Some of the trees help the cocoa to grow. From the beginning the trees protect the cocoa from excessive sunshine. The *Ceiba pentadra* (onyina) tree, for example, helps crops to grow well by providing them with ventilation (air). It also protects the plantain tree from breaking during storms. *Ficus* spp. (dormah) and *Alstonia boonei* (nyamedua) trees provide the land with water, which can be seen around the plant.'

Table 12.2. Tree uses by gender (%).

Tree uses	Male	Female
Prevention of soil erosion	100.0	0
Protecting watershed	81.9	18.1
Providing shade	78.7	21.3
Timber and furniture	50.0	50.0
Fuelwood	82.4	17.6
Charcoal	74.2	25.8
Medicine/herbs	77.6	22.4
Food	61.1	38.9
Household goods	0	100.0

Conservation practices in forest buffer zones

Tree conservation practices are discussed within the concept of collaborative forest management. Collaboration in forestry is defined as 'a working partnership between the local people and the Forestry Department to ensure that management of all forest resources is equitable and more efficient' (Owusu, 1999). The two main goals envisaged here are equity and efficiency. Collaborative forest management is considered as all forms of interaction between the forest-fringe population and the forestry officials that enhance management of the forest and improves the flow of benefits to the local people. It is generally believed by development agencies in the country that active beneficiary participation in forestry projects is a vehicle for ensuring better conservation practices. This is viewed in the forest buffer zones with a gender perspective, to determine the extent of participation and its implications.

Tree conservation practices become more important and imperative during periods of hazards. This is particularly the case in periods of drought, bush fires and in the dry season. It is

Table 12.3. Conservation practices by buffer zone and gender during bush fires (%).

	1st buffer zone		2nd buffer zone		3rd buffer zone	
Conservation practice	Male	Female	Male	Female	Male	Female
Fire prevention	79.5	20.5	82.1	17.9	80.6	19.4
Regular patrols	82.9	19.1	78.6	21.4	88.9	11.1
Plant conservation	n/a	n/a	100.0	0	100.0	0
Compliance with bush fire laws	87.0	13.0	57.1	42.9	87.5	12.5
No conservation	87.5	12.5	73.7	26.3	100.0	0

assumed that the need for conservation will be greater closer to the forest reserve than away from it. The evidence does not seem to provide a firm basis to support this assumption. What is clear is that during periods of bush fire, more plant conservation is undertaken in the second and third buffers, but not in the first buffer. The situation changes in the drought period, during which most plant conservation takes place in the first buffer, with none in the second and third buffers. This is also the area in which all farmers ensure compliance with bush fire laws in order to prevent fire from breaking out in their farms and homes. During the normal dry season, it appears that deliberate conservation practices diminish in all the three buffers, with no plant conservation in any of the buffers. However, with regard to compliance with bush fire laws, all of the male farmers in the three buffers confirmed compliance.

In the identified buffer zones, gender disparities in conservation practices are very clear. Women are less involved in conservation strategies – whether fire prevention, regular patrols, plant conservation or compliance with bush fire laws. In regular patrols, for example, their participation decreases from 19.1% in the first buffer to 11.1% in the third buffer (Table 12.3). With regard to plant conservation, the study shows that women are not involved in the practice in any of the buffers. The results show conservation practices that are generally associated with the roles of men. On the issue of conservation practices during bush fires in the buffers, where conservation generally takes place in the third buffer, the pattern of women's participation shows an interesting trend. Amongst the women, the highest practice recorded was in the second buffer, where 42.9% complied with bush fire laws, in comparison with 13.0% and 12.5% in the first and second buffers, respectively. Bush fires could be disastrous in times of drought. It should therefore be expected that farmers will take extra precautions during such periods to prevent bush fires, and furthermore to prevent them from spreading and completely destroying cultivated crops and valuable trees.

Accordingly, a comparison of Tables 12.4 and 12.5 shows that conservation is practised more in the first buffer during the drought period. This could be attributed to the observation that farmers prefer taking precautionary measures against bush fires during the drought period. The highest level of practice was undertaken by

Table 12.4. Conservation practices by buffer zone and gender during drought periods (%).

	1st buffer zone		2nd buffer zone		3rd buffer zone	
Conservation practice	Male	Female	Male	Female	Male	Female
Fire prevention	78.4	21.6	82.5	17.5	79.4	20.6
Regular patrols	88.6	11.4	92.3	7.7	86.7	13.3
Plant conservation	100.0	0	0	0	0	0
Compliance with bush fire laws	100.0	0	66.7	33.3	0	0
No conservation	86.7	13.3	71.4	28.6	100.0	0

Table 12.5. Conservation practices by buffer zone and gender during dry season (%).

	1st buffer zone		2nd buffer zone		3rd buffer zone	
Conservation practice	Male	Female	Male	Female	Male	Female
Fire prevention	76.9	23.1	76.6	23.4	77.8	22.2
Regular patrols	81.8	18.2	87.0	13.0	95.7	4.3
Plant conservation	0	0	0	0	0	0
Compliance with bush fire laws	71.4	28.6	50.0	50.0	100.0	0
No conservation	92.3	7.7	70.0	30.0	88.9	11.1

men, with all male respondents undertaking plant conservation and complying with bush fire laws. In stark contrast to this, no women in the first buffer undertook any of these two conservation practices, according to the study. The next buffer of protection is the second buffer, in which two-thirds of the men and one-third of the women actually complied with the bush fire laws.

Direct forestry activities, both in the formal and informal sectors, have been identified with men since time immemorial. The hard work and the dangers to which people may be exposed while working in the forest discourage women from working alone in forests far away from the settlement. The fear of wild animals and hostile groups in the forest mean that female family members are encouraged to remain and work nearer home. One of the common requirements for patrolling, as a forestry conservation strategy, is to prevent illegal loggers from cutting trees on farms without the knowledge or permission of the farmers. When this occurs, not only may the farmer lose timber, but valuable trees may also be illegally removed. Existing farms are also damaged, depriving the farmer of good revenue. The trees are thus removed without any compensation. Illegal logging continues in all the buffers and is experienced by both women and men farmers. Some of the destruction caused to the farms and trees by the activities of chainsaw operators and illegal loggers has been severe. This requires some form of compensation to affected farmers. Although some form of compensation was given out to victims whose farms and trees were destroyed, the amount and levels of incentives were mostly inadequate in comparison with the losses incurred. The reasons for failure of compensation to be paid included refusal on the part of the loggers and relatives of chainsaw operators to pay. In other cases, farmers were ignorant about the need to claim compensation, or were absent when the trees were felled, so that it was difficult to identify the perpetrators of the crime. Such experiences, when they occurred, discouraged farmers from conserving trees in the off-reserves. The women stated: 'If we are compensated, every farmer will be encouraged to protect the trees well. But if we know that we will never get anything from the sales of the trees, we will continue to destroy them in our farms' (women's focus-group discussion, 2004). With a local policy that recognizes the landowner as the owner for compensation, farmers – including women on the off-reserves – are now likely to willingly preserve trees on their farms.

Conclusions

Forestry was a male-dominated activity in Africa prior to colonial government rule. It continues to follow the same pattern in the forest-fringe communities and has been worsened by colonial forestry policies. In the off-reserve forests, trees are managed and conserved in farming systems based on gender division of labour, as a result of the patriarchal structures that still characterize African societies. Women throughout the rural areas are actively engaged in farming systems and management practices. One of their major constraints, which is also a matter of general fundamental concern, is legislation that governs land and tree tenure and farmer rights, especially in the rural and high forest areas of Ghana. Tenure reform is crucial in all attempts to improve governance in customary land management institutions. Equally important are the existing incentive mechanisms, which should be further revised to ensure more effective collaborative management and benefits to farmers. Monitoring and stringent enforcement of illegal logging operations, which undermine tree conservation and deprive farmers of their livelihood, should be strictly enforced.

Equity in forest benefit sharing has tended to be top-down, with the Forestry Commission, stool landowners and farmers benefiting most, in that order. Women, who are at the bottom of the hierarchy, receive the least consideration, if any. The traditional system and the family structure, which used to guarantee women's access to land, is breaking down as the result of increasing commercialization, consequently diminishing access to land for women. Corporate efforts in some parts of Africa have assisted women and have succeeded in guaranteeing women access to land and possibly credit and labour. However, to ensure the empowerment of women on similar terms to those of men, national plans with gender mainstreaming perspectives need to focus on ways of increasing women's economic participation and control, so that they can benefit by participating more meaningfully in farming management systems.

References

Ardayfio-Schandorf, E. (1994) *The Status of Women in Ghana, 1985–1994. National Report for the Fourth World Conference on Women*. National Council on Women and Development, Accra, Ghana.

Arhin, K. (1983) Rank and class among the Asante and Fante in the nineteenth century. *Africa*, 33(1), 2–22.

Beckett, W. (1994) *Akokoaso: A Survey of a Gold Coast Village*. Lund Humphries, London.

Benneh, G., Kasanga, R.K. and Amoyaw, D. (1995) Women's access to agricultural land in the household: a case study of three selected districts in Ghana. In: Ardayfio-Schandorf, E. (ed.) *The Changing Family in Ghana*. Ghana Universities Press, Accra, Ghana.

Boserup, E. (1970) *Women's Role in Economic Development*. St. Martin's Press, New York.

Gyasi, E.A. (1996) Land holding and its relationship with biophysical status: case study of tenancy and non-tenancy farming in Ghana. *PLEC News and Views*, 7, 21–25.

Mikell, G. (1983) Filiation, economic crisis, and the status of women in rural Ghana. *Canadian Journal of African Studies*, 18, 195–218.

Ministry of Land and Forestry (1994) *Forest and Wildlife Policy. Republic of Ghana*. Ministry of Lands and Forestry, Accra, Ghana.

Owusu, R.A. (1999) GM technology in the forest sector: a scoping study for the WWF. Available at: www.panda.org/resources/publications/forest/gm-download.doc

Vallenga, D.D. (1986) Matriliny, patriliny and class formation among women cocoa farmers in two rural areas of Ghana. In: Robertson, C. and Berger, I. (eds) *Women and Class in Africa*. Africana Publishing Company, New York, pp. 62–77.

13 Multiple-stakeholder Policy-planning Processes: Lessons from Community-based Projects in Mozambique

Boaventura M. Cau

Introduction

Before 1998, the control and management of forestry and wildlife resources, as well as the formulation and implementation of related cross-sectoral policies, were entirely the responsibility of the state in Mozambique. The national Department of Forestry and Wildlife and its provincial services were the responsible government agencies at central and local level. The district administrations and administrative posts contributed to government natural resource policies at the local level, but planning and implementation measures remained centralized. For instance, decisions linked to the use of forestry and wildlife resources, and to the attribution of concessions and licences for timber cutting, were granted without consultation with the population living in the respective areas (Myers, 1994). Participation by civil society in the formulation of policy was exceptional. This situation corresponded to a period of time when the state was considered as the exclusive entity responsible for safeguarding the interests of the population.[1] The exclusion of stakeholders interested in forestry and wildlife resources from policy-planning and implementation processes brought undesired effects for the sector. The state alone showed an inability to ensure appropriate use and management of forestry and wildlife resources, resulting in numerous infractions and illegal activities (República de Moçambique, 1997a,b).

Since the late 1990s, there has been an effort to revise policies and institutions to accommodate a participatory approach to the management of forestry and wildlife resources, involving multiple stakeholders in the management and sharing of benefits.[2] This process had its origins in the previous decade, with the adoption of a structural adjustment plan, known in Mozambique as the 'programme for economic rehabilitation'. In 1997, a project for supporting community forestry and wildlife management initiatives was launched. It was financed by the governments of Mozambique and the Netherlands, with technical assistance provided by the Food and Agriculture Organization of the United Nations (FAO). The first phase of the project ended in 2002. At present, the project receives support from the forestry and wildlife component of the national investment programme for the agricultural sector. Goba, Monapo and Mecubúri were chosen as pilot

[1]Mozambique became independent from Portuguese colonial rule in June 1975. It introduced a multiparty democracy in 1990, following 15 years of a one-party regime, a centralized planning economy, and civil war between 1977 and 1992.

[2]The forestry and wildlife policy was approved in 1997, its law in 1999 and the accompanying regulations in 2002. However, mechanisms for implementing the forestry and wildlife regulations are still being developed.

Box 13.1. Directives to local forestry and wildlife resource management councils. (From República de Moçambique, 2002, Article 97.)

- Development of actions to ensure the contribution of forestry and wildlife resources for the well-being of local communities;
- Adoption of conflict resolution mechanisms that involve the various stakeholders interested in forestry and wildlife-resource utilization and management;
- Collaboration with state agencies responsible for enforcing improved uses of forestry and wildlife resources;
- Improvement of policy and legislation related to forestry and wildlife resources;
- Initiation of actions to combat illegal fire-making;
- Adherence to the prescriptions of management plans within the geographical area.

areas for the implementation of the project.[3] The project envisaged active involvement of local community members in the planning and implementation of policies aimed at better use and management of resources in these areas. Consequently, multiple stakeholders interested in forestry and wildlife resources were invited to participate actively in the implementation of the forestry and wildlife policy.

This chapter is based on research done by Simon Norfolk, Boaventura Cau and Janete Assulai for the FAO Livelihood Support Programme (Norfolk *et al.*, 2005). It uses cases of community-based projects in Goba, Monapo and Mecubúri to discuss current experiences and needs of multi-stakeholder policy-planning and related cross-sectoral policy issues.

Efforts of the project to promote stakeholders' involvement

Multiple-stakeholder participation in policy planning and implementation was encouraged in the forestry and wildlife law when it established participatory management councils for local forestry and wildlife resources (República de Moçambique, 1999, Article 31). Later, the accompanying regulations provided some directives to local forestry and wildlife resource management councils to take participatory processes into account while developing their activities and integrating stakeholders' interest groups (Box 13.1).

At the community level, the initial stage of the process involving the stakeholders comprised identifying and recognizing the general characteristics of a community, which include: (i) the area's natural resources; (ii) relations between the community and these resources; (iii) the community's organization; and (iv) local survival strategies. Using participatory methods, the project assisted the community to identify their priority development objectives (Mansur, 1999). Equally, it assisted with the realization of a resource inventory within the community area and the elaboration of a management plan for the natural resources. In Monapo, the project sought to assist carpentry, charcoal-making and firewood-collecting groups. In Goba, it worked mainly with the existing charcoal groups and assisted the establishment of new groups, such as for bee-keeping and sewing.

The project helped with the reorganization of the community, with a view to responding better to the development priorities that had been identified. Members of management committees were elected, along with commissions dealing with finance, monitoring and legal issues. The project also assisted with securing greater tenure security by conducting land delimitations in order to obtain formal certificates for using and benefiting from land. It promoted the formation and registration of community associations.

Training was provided for members of the natural resource management committees and for interest groups in improved techniques in sustainable land use. National and international specialists were sought to provide training in various areas, and assistance was obtained from non-governmental organizations (NGOs) that were able to provide

[3]The Goba is located in the southern province of Maputo. Monapo and Mecubúri are located in the northern province of Nampula.

Box 13.2. Necessary characteristics of community organizations.

- *Broad and inclusive*: all citizens have an opportunity to participate, and attention is particularly paid to the participation of less-favoured groups (poorer members, women, etc.).
- *Democratic*: all members have the same opportunity to express their opinion and there are rules in place that make this effective.
- *Autonomous*: the members of an organization take decisions on the basis of their best judgement. Outside stakeholders do not interfere to manipulate decisions to their own advantage.
- *Internally responsible*: there are norms in place for representatives to inform and explain to the members how they have been exercising their functions, and there are rules for the members to sanction or reward the representatives.
- *Accountable*: the representatives are invested with a certain level of authority to represent the members, and the decisions taken in this regard are respected.
- *Sustainable and institutionalized*: the citizens see their organizations as a legitimate means of channelling their voice, rather than expedient necessities in order to obtain benefits.

expertise in areas that had been identified previously. In Monapo, courses in planning, leadership and accounting were organized. Exchange visits were promoted with communities involved in similar initiatives both within and outside the country. Technicians and extension workers were trained and worked subsequently with the community groups in natural resource management.

In Goba, the project encouraged the formation of new institutions. In this way, the 'Ntava Yedzu' (Our Mountain) association was founded. Each suburb of the village had a group of ten people, who jointly formed the association's management council. Of the members of the groups 50% were women. The role of each group was to make the link between the project and the rest of the community. An important result was the participatory elaboration of a management plan for the zone of Goba. Activities that involved members of the community included the diagnosis of problems, the preparation of resource inventories, socio-economic surveys and wildlife-resource inventories, as well as the identification of local problems and sources of income.

Other Efforts to Foster Local-level Stakeholder Involvement

The northern province of Nampula, where Monapo and Mecubúri are located, is implementing a pilot project of district decentralized planning and finance. The project is aimed at supporting local populations in development-planning activities in their areas. The programme involves the government of the district which is multi-sectoral in nature, because it includes district representatives of line ministries, the government of administrative posts, the government of the localities and representatives of local communities. It also receives support from NGOs. Under the district decentralized planning and finance project, initiatives to be financed are based on priorities identified locally by representatives of the population. Six characteristics of community organizations have been identified in order to improve the likelihood of real participation (Box 13.2).

The process of district decentralized planning and finance follows the guidelines developed by the Ministry of Finance and the Ministry of State Administration in 1998. The main activities that are undertaken in the planning process are summarized in Box 13.3.

Through the decentralized planning and finance project, there is greater contact between state civil servants and the local populations. According to Serrano (2002), an interesting aspect reflected on by the members of various local development committees is that these structures have allowed them to establish links with the district government.

Box 13.3. Main activities of district planning and finance process (From: Serrano, 2002).

- Creation of district technical committees to coordinate the formulation of district plans;
- Meetings to launch the planning processes;
- Public meetings regarding planning processes – collecting local views regarding priority issues and solutions;
- Collection of information and statistics on different economic and social sectors;
- Elaboration of cartographic maps – visual aids for planning decisions;
- Summary of problems and potentialities – view of major local constraints and most urgent problems in terms of local priorities;
- Formulation of proposal for a development strategy – identification of strategic interventions;
- Discussions regarding proposed strategies – with feedback from local civil society regarding district government proposals;
- Negotiations – guarantee that proposed actions will receive support from local stakeholders;
- Incorporation of district economic and social plans into the provincial plans – guarantee that Provincial Sector Directorates have taken into account the district plans in sectoral planning at provincial level.

Challenges and Opportunities for Improvement

As forestry resources in Goba and Monapo constitute a basis for subsistence activities not only for the local community itself, but for many individuals from neighbouring community groups, monitoring and prevention of illegal use of resources is a considerable challenge. For example, the groups in Monapo felt they had to compete against people who do not pay for the exploitation licences. The judiciary sector of the country is, however, ineffective. For instance, Johnstone *et al.* (2004) show that many illegal activities from the forestry sector transmitted to the judiciary are not being appropriately addressed. Another recent study on resource conflicts states that the judiciary sector is not adequately involved in the implementation of policies and legislation related to the land, environmental and forestry sectors (Baleira and Tanner, 2004). In view of the limited capacity of community groups in terms of the skills required to engage in multiple-stakeholder policy-planning processes, it is thus essential that the project should support capacity building, with a view to improving the functional literacy of community members.

In Mecubúri, one of the principal difficulties has been in gaining access to markets for the various products. Interest groups dealing with bamboo, bee-keeping, mushrooms, vegetable production, guinea-fowl production and carpentry were therefore formed (Foloma and Zacarias, 2004). In Goba, with the exception of interest groups involved in charcoal production and in information dissemination, the groups that were newly formed have ceased to function. According to the respondents interviewed, bee-keeping proved not to be viable, as most of the flower-producing trees had been cut down before the project began. Similarly, the resources for charcoal had been depleted, and the only remaining resources are within the area defined by the management plan as a reserve area. The group now exploits resources outside of the community. The sewing group, although it has machines, has disbanded because of a lack of market opportunities for its products, and the carpentry group has come to a halt. After the transfer of the project to the national investment programme for the agricultural sector, the lack of information about funds left the community and its structures with a sense of having been abandoned.

In Monapo, where some income-generating activities continue, the lack of equity among community members in sharing benefits is another important challenge. The same can be said for participation in the local decision-making process, where the most powerful groups are also the most visible ones. For example, with regard to the participation of the different strata of the community (rich and poor) in the project decision-making bodies of Monapo (the community councils), Zacarias (2001) revealed that more than 70% of these bodies were also members of the most powerful interest groups. The benefits derived from their activities were distributed amongst themselves, and those who were not part of the interest groups (largely the poorer members of the community) were left out. This raises serious questions concerning the sustainability of these initiatives, since individuals who are excluded also feel that they have some ownership of the resources. Will they be motivated in the same way to use the resources in a sustainable manner when they are not benefiting from any profits generated from their use?

It appears that there is a lack of balance between the legally defined procedures and institutional arrangements, and the human, financial and technical resources available for their execution. Thus, when there is forward-looking and progressive legislation on the table, the potential for implementing it should be evaluated in the light of the prevailing circumstances and capacities for change (Johnstone *et al.*, 2004).

With regard to the process of involving people in decentralized planning, the creation of the various committees is still being done in a top-down manner. The lack of clear and transparent criteria for bottom-up selection of members for posts on district- and administrative-level committees means that it is possible that not all representatives of the community will be invited to participate. Also, those who do participate do not necessarily bring a mandate regarding local preferences from those that they purport to represent. For example, the chiefs or heads of villages – whose presence guarantees a territorial coverage – do not have to hold consultation meetings within their constituencies before arriving at the community council meetings (Serrano, 2002, p. 30).

In spite of the fact that there is a considerable focus on the involvement of people in consultation and planning processes, everything is done at the provincial level when it comes to the execution of the plans. The tenders for the execution of identified activities are all done at provincial level, in such a way that even the monitoring officers come from the province and no local people are involved. This leads to a lack of information, even though we are supposed to be dealing with transparent processes. To justify this exclusion of the local population from the execution of the plans, the state relies on technical arguments related to a lack of local knowledge and capacity. Minorities and marginalized groups do not have ways in which to make their voices heard. The participation of women, for example, in the district community councils and the administrative post community councils has been very low (Matakala and Cavane, 2002).

Serrano (2002, p. 40) observes: 'To participate in the process of planning is good, but not enough. The process of developing district development plans is a long one, during which the interest of the population can vacillate.' His report about participation in the decentralized planning process mentions two things that help community groups to maintain interest in participating: technical assistance and tangible results. The first one has to do with the attitudes of state civil servants or those within NGOs when they are listening to problems and preoccupations and giving advice to the communities. The second one is the need to provide concrete assistance, which is the reason why many participatory planning initiatives have components and funds available for micro-projects that can be completed and have an immediate impact.

It is important for community natural resource management initiatives to create real benefits for all those who feel that they have some ownership and interest over the local resources, so that they have a greater incentive to participate. For example, the members of the interest groups in Monapo are more active because they benefit much more than the rest of the population. The weak participation of community groups in the conception and planning of projects creates a gap in ownership over the initiatives at a later stage, culminating often in their failure and, in the particular case of natural resource management, in the unsustainable use of these resources, putting present and future generations at risk. In addition, the identification of, and assistance to, interest groups proved to be a non-inclusive practice that tends to go against the foundation of collective action that is important for community-level natural resource management. Some community members, who feel equally that they are owners of the resources, are marginalized in favour of members of the interest group. For example, the chief of Cateia, who has participated in meetings and discussions about the initiative since the start of the project in Monapo, feels excluded from any share in the benefits because he is not a member of any of the interest groups.

Conclusions

The Mozambican legal framework in the domain of natural resource conservation and use underlines the participation of multiple stakeholders. However, a favourable legal framework does not necessarily lead to effective participation of the target groups. Lessons learned from the project presented here can be summarized as follows:

- Sometimes the adoption of a discourse on participation at a policy level can conceal important factors regarding a complex social and historical context. It is necessary to build up a greater understanding of this context if we want to analyse the challenges involved in the design and implementation of participatory processes.
- Mozambique has a long history of hierarchical and non-participatory government, which still has a significant impact on the way in which policies – even when they are specifically designed to involve the population – are implemented. Many state civil servants, particularly in the rural areas, continue to have outdated visions and mentalities. Others, who are responsible for the formulation of the actual policies and have hierarchical power to orient government actions, may be resistant to, and fearful of, processes that would devolve this power to lower levels. This means that change is slow.
- There is still a large gap in understanding between community groups and the state which culminates in many top-down decisions and failure to resolve community demands. If community groups are unable to influence policy approaches at higher levels (district and province), they will always encounter difficulties in managing resources in their areas.
- Concepts such as 'communities' and 'villages' can be problematic when they appear in policies without there being a common understanding about who is responsible for implementation. Communities are treated often as if they are homogeneous, with common interests being identified through consensual processes. In reality, however, only the most visible and powerful members of a community may participate in these processes and obtain benefits.
- Transparent norms for taking decisions and divulging information need an effective communication strategy that permits the local population to demand compliance with these norms on the basis of information they have received. A well-defined system is needed to furnish information to stakeholders in order to help them in their implementation of an initiative. However, an effective communication strategy will not necessarily result in increased responsibility if there are no clear norms for decision-making and the divulging of information (Serrano, 2002).

Finally, there is a need for greater training and support for the management and intervention bodies in the area of participatory methodologies and planning. A systematized and institutionalized political network at local level needs to be created in a way that is sustainable and replicable.

References

Baleira, S. and Tanner, C. (2004) *Relatório Final da Pesquisa sobre os Conflitos de Terra, Ambiente, e Florestas e Fauna Bravia*. Projecto Apoio ao Judiciário na Implementação de Novas Leis na Área de Terra, Ambiente, e Florestas e Fauna Bravia, Maputo, Mozambique.

Foloma, M. and Zacarias, A. (2004) *Relatório de Levantamento da Situação Actual dos Grupos de Interesse nas Áreas Piloto Apoiadas pelo Projecto GCP/MOZ/056/NET*. Direcção Nacional de Florestas e Fauna Bravia, Maputo, Mozambique.

Johnstone, R., Cau, B. and Norfolk, S. (2004) *Legislação Florestal em Moçambique: Cumprimento e Impacto em Comunidades Residentes na Floresta*. International Institute for Environment and Development and Terra Firma, Maputo, Mozambique.

Mansur, E. (1999) Irapismu – uma proposta metodológica para o envolvimento das comunidades locais. In: Dias, M.L., Filimão, E. and Mansur, E. (eds) *Comunidades e Maneio dos Recursos Naturais. Memórias da 1a Conferência Nacional sobre Maneio Comunitário dos Recursos Naturais*. União Mundial para a Natureza and Direcção Nacional de Florestas e Fauna Bravia, Maputo, Mozambique, pp. 107–113.

Matakala, P.W. and Cavane, E. (2002) *Identificação de Melhores Práticas na Animação e Participação Comunitária no Processo de Planificação Distrital*. Programa de Planificação e Finanças Descentralizadas, Ministério do Plano e Finanças, Maputo, Mozambique.

Myers, G.W. (1994) *Confusion, Contradiction and Conflict: Land Access in Mozambique in the Post-peace Period. Four Case Studies from Manica, Sofala, Gaza and Inhambane Provinces*. Land Tenure Centre, Maputo, Mozambique.

Norfolk, S., Cau, B. and Assulai, J. (2005) *Analysing Participation in Policy Processes: Experiences in Mozambique*. Document produced for FAO Livelihoods Support Programme, Maputo, Mozambique.

República de Moçambique (1997a) *Lei de Terras. Lei nº 19/1997 de 1 de Outubro*. Imprensa Nacional de Moçambique, Maputo, Mozambique.

República de Moçambique (1997b) *Política e Estratégia de Desenvolvimento de Florestas e Fauna Bravia. Resolução 8/97 de 1 de Abril*. Imprensa Nacional de Moçambique, Maputo, Mozambique.

República de Moçambique (1999) *Lei de Florestas e Fauna Bravia. Lei nº 10/99 de 7 de Julho*. Imprensa Nacional de Moçambique, Maputo, Mozambique.

República de Moçambique (2002) *Regulamento da Lei de Florestas e Fauna Bravia*. Decreto No. 12/2002 de 6 de Junho. Imprensa Nacional de Moçambique, Maputo, Mozambique.

Serrano, R. (2002) *Participação, Transparência e Prestação de Contas de Cima para Baixo na Planificação a Nível Distrital em Moçambique*. Relatório elaborado como contribuição para o desenho do Projecto de Planificação e Finanças Descentralizadas Banco Mundial, Moçambique.

Zacarias, A. (2001) Community-based Management of Miombo Woodlands in Nampula Province, Mozambique: Lessons for Future Community Involvement Programmes. M.Sc. Thesis, University of Reading, Reading, UK.

14 Enhancing Cross-sectoral Linkages in Nigerian Forestry

Sylvester A. Okonofua

Introduction

Nigeria has a total land area of 92.4 million ha, about 10% of which is legally constituted as forest reserves, classified into five broad ecological zones: Sudan/Sahel; Guinea savannah; derived savannah; lowland rainforest; and freshwater/mangrove swamp forest. In addition, by 1998, 269,000 ha of plantations had been established, mostly in forest reserves (Federal Ministry of Environment, 2005). Natural forest types also occur in off-reserve areas, commonly referred to as 'free areas'.

The forests provide socio-economic, environmental, scientific, cultural and educational services. The forestry sector contributes about 2.5% of the national gross domestic product. This value is less than the actual contribution, since activities traded in the informal sectors are not taken into account in the national statistical reporting system. Most of the economic activities in the forestry sector play an important role in the socio-economic development of rural areas. The rural population constitutes over 60% of Nigeria's population, more than 90% of whom depend on the forests for livelihood and domestic energy sources and the majority of whom earn less than US$1/day (National Planning Commission, 2004).

The capacity of the Nigerian forests to provide goods and services sustainably has dwindled over the years. The rate of deforestation and forest land degradation is 3.5% per annum, estimated to amount to a loss of 350,000–400,000 ha every year. The causes include expansion of agricultural frontiers into the forests, infrastructural development, mining, bush burning, felling of trees to meet the revenue targets of some state governments and illegal felling activities. The demand projection for wood requirements in Nigeria between 2000 and 2010 (Table 14.1) indicates a deficit in the demand-and-supply chain for wood for industrial and domestic uses. The situation with non-timber forest products and other goods and services is similar.

A large number of sectoral policies and programmes in Nigeria touch on the forestry sector, directly or indirectly and in different ways. These include agriculture, water resources, education, health, communication, culture, tourism, banking and insurance, energy, technology, trade, marketing, transportation, housing and rural development policies. The cross-cutting issues include land allocation, revenue generation, wood as an industrial raw material, food security, poverty alleviation, environmental services, energy conservation, ecotourism, fiscal policies and political decisions. Examples of cross-sectoral impacts on the forestry sector are numerous. An increase in the price of kerosene and liquefied petroleum gas leads to increased demand for fuelwood, thus putting pressure on the wood deficit situation. Agricultural land expansion leads to large-scale clearing of forest lands. Increases in livestock

Table 14.1. Summary of demand, supply and balance of each log class (2000 and 2010 in '000 m^3). (From: Forestry Management, Evaluation and Coordinating Unit, 1996.)

	2000			2010		
Wood type	Demand	Supply	Balance	Demand	Supply	Balance
Fuelwood	83,521	71,349	−12,172	88,138	63,099	−25,039
Poles	2,183	1,272	−911	2,729	1,153	−1,576
Saw logs	6,378	2,996	−3,382	10,205	2,480	−7,725
Veneer logs	650	136	−514	1,078	114	−964
Pulpwood	410	724	314	760	724	−36
Total	93,142	76,477	−16,665	102,910	67,570	−35,340

holdings in the absence of managed range lands and a lack of intersectoral coordination result in deforestation and degradation of forests through overgrazing, as well as annual forest fires and bush burning. Efforts are therefore being made to achieve a better understanding of the policies of other linkage sectors and to open up dialogue on the need for greater collaboration and intersectoral linkages in addressing forestry issues.

National Development Planning

Nigeria is made up of 36 states, a Federal Capital Territory and 774 local government areas with a federal system of administration. In accordance with the provisions of Section 20 of the 1999 Constitution of the Federal Republic of Nigeria, 'The State shall protect and improve the environment and safeguard the water, air and land, forests and wildlife of Nigeria' (Federal Government of Nigeria, 1999). This provision is one of the fundamental objectives and directive principles of state policy which is non-justiciable.

The federal government does not own, control directly or undertake the management of forests except in the eight national parks. The states and the Federal Capital Territory have direct responsibilities over the ownership and management of the reserves and free areas within their respective jurisdictions, including specific forest policies and legislations. The local governments, on the other hand, are enjoined to participate in forestry development activities. The involvement of other stakeholders, including the communities, depends on the level of integration, devolution and decentralization of authority as enshrined in the forest policies of each state government. Many state governments have accepted the principles of participatory forest management, effective collaboration and decentralization or democratization of forest governance and the need for greater cross-sectoral coordination in forestry development.

The federal government may not directly compel the state governments to implement policies and adopt a specific institutional framework for the forestry sector, but can exert its influence through macroeconomic policy direction, which is within its exclusive legislative power (Federal Ministry of Agricultural and Rural Development, 2003), for example, by directly intervening in forestry development using its resources. Since the Federal Ministry of the Environment was created and the Federal Department of Forestry moved from the Federal Ministry of Agriculture and Natural Resources to the new ministry, 19 out of the 37 states have similarly established Ministries of the Environment and domiciled the Forestry Departments in the new ministries. The federal government can intervene directly in matters of cross-boundary importance and those touching on the implementation of conventions and treaties acceded to by Nigeria.

The national planning process emphasizes cross-sectoral development. The national economic empowerment and development strategy is a fully integrated medium-term action plan for 2004–2007 aimed at achieving long-term objectives of poverty reduction, wealth creation, employment generation and value reorientation from the federal to the grassroots level through well-coordinated and integrated development programmes that are intra- and intersectoral, and involving all stakeholders (National Planning

Commission, 2004). It will in the long run permit greater intersectoral coordination, planning and implementation of programmes. At its conception, the strategy did not take into proper perspective the role of forests in attaining the country's developmental objectives. This was perhaps due to the poor appreciation of the strategic importance of forests in poverty reduction and the limited interaction between players in the forestry sector and other sectors, resulting in a lower ranking of the forestry sector in the government's macroeconomic policies. The process of integrating forestry into the strategy has been promoted and is now being undertaken. In recognition of the role of forestry in sustainable development and environmental integrity, the sector is benefiting from an increased level of federal government funding in 2006 under the millennium development goals programme for afforestation activities in the Sudan/Sahel ecological zone.

The actual government spending on a particular sector can only be accurately analysed by consolidating all spending by the three tiers of government (federal, state and local). The federal government has improved on its allocation to the sector in the last 3 years. The aggregate spending by the states and local governments is low (Federal Ministry of Environment, 2000). The problem lies in fiscal policy within the macroeconomic policy of many states, due to inadequate political will and commitment to the principle of sustainability. The forest revenue system is therefore greatly influenced by political and administrative decisions not supported by empirical data or economic feasibility in relation to the sustainability of the forest resources and the effects on other related or dependent sectors of the economy. This situation needs to be reversed. The private sector and civil society will have to assume an enhanced role and therefore commit more funds to the sustainable development of the forests. The forestry sector is important in delivering immediate to long-term development benefits. As Nigeria moves more into a market-led or private sector-led economy, the role of the government should move towards the provision of a sound regulatory framework and towards a reduction in direct intervention. The sector will then be substantially driven by the states and the private sector, while the role of the federal government will become largely one of coordination and facilitation (National Planning Commission, 2004).

Sectoral Development Planning

At the inception of the scientific management of forests in Nigeria in the early 20th century, forestry activities were mostly based on an outlook focusing only on the wood production sector. Mandates were compartmentalized into sectoral government ministries and agencies. Initiatives on sustainable development following the United Nations Conference on Environment and Development provided the impetus from the 1990s onwards to adopt cross-sectoral approaches to addressing forestry issues in Nigeria. The country benefited from its membership of international organizations and participation in international dialogues on forests such as the Intergovernmental Panel on Forests/Intergovernmental Forum on Forests, the United Nations Forum on Forests and the Committee on Forestry of the United Nations Food and Agriculture Organization (FAO).

A national forestry action plan was formulated between 1989 and 1996. It did not, however, go beyond its finalization in 1996, due partly to the political situation in the country at that time and inadequate counterpart funding. The donor round-table conference to discuss the window of opportunity for local and international funding of actions or activities did, therefore, not take place. A new effort was made with the forestry development programme in 1999 as a comprehensive scheme for massive afforestation in Nigeria over a 4-year period (2000–2003), with a strong emphasis on community participation. This has not been fully implemented, mainly due to budgetary constraints experienced by forestry and other related sectors, inadequate understanding of roles and responsibilities by other stakeholders, insufficient commitment to forest governance and advocacy by the many stakeholders.

In 2000, the President of Nigeria set up an interministerial committee on combating desertification and deforestation, which developed cross-sectoral programmes to address the problems of desertification and deforestation in a holistic way. The scope covered agriculture, energy, petroleum, biotechnology, solid minerals development, land use management, research, monitoring and evaluation, as well as public education, extension, and awareness creation. It was designed to be implemented between 2001 and 2003. The level of implementation of the programmes enunciated has been low, due to factors similar to those experienced in the implementation of the forestry development programme, including:

- Poor response by other sectors to implement related components enunciated through cross-sectoral forest policy dialogue in the face of budgetary constraints in implementing their main sectoral policies;
- Inadequate financial mechanisms to enable stakeholders outside the public sector to undertake their share of responsibilities.

In 2002, the National Forest Programme (NFP) was developed for Nigeria and revised in 2004. Several actions have been started in the full implementation of the NFP, with involvement by multiple stakeholders, including policy review and legal reforms in the forestry sector at the federal level, relating to linkages with other sectors and actors. A new National Forest Policy was produced and approved by the President-in-Council in June 2006, while the draft National Forestry Act has been processed for approval by the appropriate authorities and its eventual promulgation into law by the National Asssembly. Full implementation requires intersectoral cooperation and complementary efforts. Effective mechanisms for this are gradually evolving. One of the strategies for the implementation of the NFP in Nigeria is through community-based forest management. In order to provide baseline information, the NFP Facility hosted by the FAO is funding pilot studies in each of the five ecological zones under contract, through transparent competitive bidding. The process, driven by a multiple-stakeholder committee, will eventually lead to the production of guidelines on community-based forest management.

Based on concern and the need to develop forest resources in a sustainable and cross-sectoral way, two of the major presidential initiatives in 2005 were:

- A national committee on transforming *Azadirachta indica* (neem) into wealth was established, with the objective of harnessing the potential for neem tree growing and industrial utilization in Nigeria for economic growth, environmental amelioration and poverty reduction.
- A forum on forests was held on 15 December 2005 by the President, which examined broad national forestry issues with multiple stakeholders.

There are several non-governmental organizations (NGOs) that are involved in cross-sectoral initiatives. However, many of them lack the ability to generate funds as well as the capacity to implement plans, with considerable dependence on government and donor agencies or institutions. Many community-based organizations exist informally at the grass-roots level, responsible for communally driven activities, while some others have institutionalized. However, many of them are struggling with problems of weak leadership, capital and capacity.

Enhancing Cross-sectoral Linkages

The traditional institution for planning and implementing government policies and strategies for development is the ministry or government agency that has mandate for a specific sector. The recognition of cross-linkages between forestry and other sectors has led to increased recognition of multiple stakeholders' participation in forest policy planning and implementation towards achieving sustainable management of the dwindling forest resources. The process involves different ministries and agencies, academia, NGOs, community-based organizations, the private sector, indigenous communities and donors.

There are existing institutional arrangements that engender cross-sectoral planning (directly and indirectly) in the forestry sector. The National Council on Environment is the statutory organ for policy formulation with a national outlook, and is made up of the Commissioners in charge of Environment in the 36 states of the Federation and representatives of the Minister of the Federal Capital Territory, with the Minister of Environment as the chairperson. It provides a forum for agreeing on effective mechanisms for coordinating state and federal government action plans, as well as monitoring and evaluation. At its meetings, policies, strategies, issues and developments in the environment sector are considered. The Council's decisions and recommendations are brought to the attention of the federal and state governments for implementation. There is a report-back mechanism at the subsequent meetings on levels of performance and adherence to decisions of the National

Council on Environment. The National Forestry Development Committee is the technical committee of the National Council on Environment, and consists of members drawn from the forestry departments in the states and the Federal Capital Territory, relevant ministries, research institutes, the universities, NGOs, community-based organizations and the private sector, with the Federal Director of Forestry as the chairperson. It initiates policy development and has an oversight responsibility over forestry activities nationally.

The Federal Ministry of the Environment is responsible for national environmental policy development and technical support for the states. It has eight departments: Finance and Supplies; Personnel Management; Forestry; Drought and Desertification Amelioration; Flood, Erosion and Coastal Zone Management; Environmental Assessment; Environmental Health and Pollution; Planning, Research and Statistics. Interdepartmental cooperation is strengthened and harmonized through regular meetings of the committee of directors in the ministry and ad hoc committees to address cross-sectoral issues. In order to relate effectively with the states and grassroots communities, the Federal Ministry of the Environment has field offices in all the states of the federation and the Federal Capital Territory. However, due to funding constraints, there are weak linkages between these offices and the state forestry departments, which need to be reviewed and strengthened. The Federal Department of Forestry is the leading department responsible for initiating and formulating forestry policies of national significance as well as fostering forestry and institutional development. It has field offices in all the 36 states of the federation and the Federal Capital Territory, Abuja. Every state has a forestry department responsible for forest policy, planning, management and administration.

Some national councils with relevance to forestry matters include the National Council for Water Resources and the National Council on Agriculture, responsible for the development of policies, strategic plans and programmes in the agricultural sector. The former is responsible for formulating and coordinating water resource policies and advising federal and state governments. In the implementation of the responsibilities of sectors that touch on forestry, the Federal Department of Forestry as the national lead department is consulted and represented in committees or similar bodies in which issues relevant to forestry are to be considered.

Other national institutions also affect forestry:

- The National Planning Commission is responsible for national planning and intersectoral coordination. Its functions include the preparation of 3-year rolling plans and annual plans, as well as joint monitoring and missions to assess the field situation. The sectoral minister collaborates with the National Planning Commission and the respective state government ministries to develop national sectoral strategies and plans.
- The National Bureau of Statistics is responsible for national statistical reporting, including collection, analysis, dissemination and storage of socio-economic data.
- The Federal Ministry of Finance is responsible for setting development and recurrent ceilings for sectors, sector releases and accountability guidelines.
- The Federal Ministry of Solid Minerals Development is responsible for solid minerals development, including developing coal briquette as a potential alternative to fuelwood.
- The Federal Ministry of Petroleum Resources is responsible for energy policies on kerosene and liquefied petroleum gas, which are alternatives affecting the demand for fuelwood. Increases in price have in the past resulted in a rising demand for fuelwood.

Several restraining or enhancing factors influence the effectiveness of cross-sectoral linkages in the forestry sector. Some of the restraining factors are:

- Inadequate mechanisms for managing forestry information and securing political will, as well as inappropriate statistical tools and inadequate economic and environmental data for evaluating the forestry sector within the national accounting system;
- Slow pace of the implementation of the NFP countrywide by multiple stakeholders in order to reverse the loss of forests and attain sustainable forest management;

- Serious imperfections in the forest revenue system;
- Improperly harmonized plans and policies, increasing uncertainty and making it difficult to develop appropriate strategies for cross-sectoral implementation;
- Inadequate institutional capacity to address cross-sectoral issues and mandate protection and development within the various sectors;
- Inadequate cooperation and capacity to address cross-sectoral issues.

Among the enhancing factors one can point out:

- Greater awareness of the value of forest goods and services, backed up with statistics and documentary evidence based on appropriately integrated environmental and economic reporting systems, so as to increase the general perception of forestry outputs within the public and private sector;
- A lead agency must therefore be responsible for marketing the sector's values and attributes;
- Development of appropriate mechanisms for economic valuation of forest goods and services and assessment of the impact of forests on the activities of other sectors;
- Promotion of supportive policies and a legal framework to enhance the status and functionality of the forests in the provision of goods and services;
- Fostering effective multiple stakeholders and intrasectoral coordination, participatory monitoring and evaluation, by developing an enabling policy and a legal environment for the respective lead agencies;
- Promotion of an integrated ecosystem and landscape management approach in forestry development;
- Development of effective response mechanisms within the forestry sector to policy changes in related sectors;
- Cross-sectoral capacity building for planning and implementation.

Conclusions

1. Planning and implementation in the forestry sector in Nigeria until the late 1980s were led by government agencies and ministries, with a top-down approach based on the forestry sector mandate. A paradigm shift in the forestry sector began with the involvement of stakeholders through consultations and participation, and the process is still intensifying.

2. The cross-sectoral linkages of various processes in the forestry sector with other sectors have become increasingly clearer and defined through various programmes. These include the NFP and the national economic empowerment and development strategy, which is a national holistic development strategy that brings out intersectoral linkages and provides a framework for priority setting and implementation aimed at addressing poverty reduction, amongst other issues.

3. Although the forestry sector did not receive a special mention in the implementation of the national economic empowerment and development strategy and poverty reduction strategy programmes, the need for it to have an enhanced role in these programmes is receiving more attention from government and other stakeholders.

4. The forestry sector has also benefited from the country's membership of international organizations and participation in international dialogues on forests such as the Intergovernmental Panel on Forests/Intergovernmental Forum on Forests, the United Nations Forum on Forests and the Committee on Forestry of the United Nations Food and Agriculture Organization.

5. Coordination and harmonization of cross-sectoral issues are receiving central-stage consideration in the forestry sector in Nigeria, though they are still in an embryonic phase. Some measure of success has been achieved at the planning level in addressing issues cross-sectorally through processes such as the forestry development programme and the national forest programme, as well as the institutional arrangements already in place.

6. The forestry sector has approached many cross-sectoral issues through a multiple-stakeholder process. Stakeholder-driven processes now have greater potential than hitherto in the sustainable management of the forests. However, there is still wide scope for improvement in the planning and implementation process, as well as the need to remove or minimize the limitations imposed by the compartmentalization of mandates along sectoral lines.

7. The genuine way forward includes effective information and communication exchanges between sectoral players, transparency, accountability and fuller integration and coordination of cross-sectoral activities by the lead stakeholders' agencies and government ministries or bodies.

References

Federal Government of Nigeria (1999) *The Constitution of the Federal Republic of Nigeria 1999*. Federal Government Press, Lagos, Nigeria.

Federal Ministry of Agricultural and Rural Development (2003) A review of rural and agricultural sector institutions. Draft main report and annexes, Vol. 1.

Federal Ministry of Environment (2000) Report of the interministerial committee on combating desertification and deforestation submitted to Mr. President. Abuja, Nigeria.

Federal Ministry of Environment (2005) The status of the forest resources and environment in Nigeria. Paper presented by the Honourable Minister of Environment, Dr Iyorchia Ayu, at the Presidential Forum on Forestry on 15 December 2005, Abuja, Nigeria.

Forestry Management, Evaluation and Coordinating Unit (1996). National forestry action programme. Vol. II, main report. Federal Department of Forestry, Federal Ministry of Agriculture and Natural Resources, Abuja, Nigeria.

National Planning Commission (2004) Meeting everyone's needs – National Economic Empowerment and Development Strategy. Abuja, Nigeria.

15 External Policy Impacts on Miombo Forest Development in Tanzania

Gerald C. Monela and Jumanne M. Abdallah

Introduction

In Tanzania, forests and woodlands are estimated to occupy 33.5 million ha of the land area. They consist of high closed forests, closed and open Miombo woodlands, and coastal mangroves (Table 15.1). Out of the 33.5 million ha, 12.5 million ha are set aside and gazetted as production and protection forest and woodland reserves, of which 11.9 million ha are under the central government's Forest and Bee-Keeping Division and 0.6 million ha are under village councils (local governments). This means that about 21 million ha of forests and woodlands are unreserved forest lands under private management by farmers, which have come to be known as 'forests on general lands'. Extensive Miombo woodlands are unique forest ecosystems available in this huge forest resources endowment, and are potentially a very useful frontier for economic development of the country.

Pressure on natural resources has progressively escalated and ecological degradation has become evident, especially in arid and semiarid areas of the country. Deforestation is estimated to be taking place at an average rate of 91,000 ha/year, and it mainly occurs in the unreserved forest lands. The main reasons for deforestation are clearing for agriculture, overgrazing, wildfires, charcoal burning and overexploitation of wood resources. Also, the sudden and large-scale inflow of refugees into the border districts has created major ecological, social and environmental problems. Environmental problems associated with refugee pressure are of several kinds, but the foremost among them has been rapid deforestation and resource depletion in a widening belt around refugee camps. Deforestation is very high in old camps such as Katumba, Ulyankulu and Mishamo that have been in existence for more than 25 years now.

The government of Tanzania has attempted to curb the problem of deforestation by promoting village and community forestry, aimed at producing sufficient amounts of forest products and services to meet both local demands and promote the forest contribution to global environmental conservation. Despite these efforts, environmental degradation is continuing at a fast rate. This is partly due to unsustainable land uses such as shifting agriculture, but more so because of the relationship between environmental degradation and poverty in attempts to satisfy basic needs. Other problems include water pollution, fragmentation of habitats, biodiversity loss and soil erosion. The consequences have been a general decline of forest products and services such as fuelwood and water catchment values (Ministry of Natural Resources and Tourism, 1994, 2001; National Environment Management Council, 1994).

Generally, markets have failed to capture the externalities associated with forestry, such as biodiversity conservation, climate change mitigation, needs of local forest users, soil and water

 Cross-sectoral Policy Developments in Forestry (Dubé and Schmithüsen)

Table 15.1. Tanzanian forests and woodlands. (From Ministry of Natural Resources and Tourism, 1998.)

Forest land type, use and legal status	Area (1000 ha)
Type	
Closed forest	1.1
Miombo woodland	32.3
Mangrove	0.1
Total	33.5
Use	
Production forest and woodland	23.8
Protection forest and woodland	9.7
Total	33.5
Legal status	
Forest and woodland reserve	12.5
Forest and woodland within national parks, etc.	2.0
Unreserved forest and woodland	19.0
Total	33.5

protection, and recreational and aesthetic values. In addition, forests may suffer from unclear tenure arrangements if markets are enabled to function freely. Furthermore, equity issues abound in connection with the way in which forests can contribute to poverty reduction. Ironically, intervention by government, justified by these concerns, can sometimes worsen the situation if policies are inappropriate or corruption exists. None the less, it is this framework of competing policies and stakeholders in which forestry is placed and which has motivated this study focusing on the plight of forestry and the causes of forest decline.

The objective of this case study[1] is to shed light on the influences of external policies on forestry development in Tanzania. The specific objectives are to identify the main cross-sectoral policy issues, report any successful policy instruments in dealing with these policy issues, and make recommendations to increase positive, and reduce negative, impacts on Miombo forest development.

[1]The study involved interviewing representatives of the key ministries, institutions, donor agencies, development projects and non-governmental organizations. It also involved collecting and studying background documentation with relevance to the task at hand. The people met were selected on the basis of their involvement in policy and/or planning in areas related to development and natural resource management.

Macroeconomic Policy Framework

Towards a market-oriented policy regime

From the 1960s to the 1980s, Tanzania pursued policies for economic growth and development that were based on restrictions, controls and direct state investment for commercial activities and management in all service and economic sectors. In the late 1970s and early 1980s, the economy and social services sectors faced major crisis, including a general decline in gross domestic product (GDP) and per capita income, fiscal deficit and deterioration of the terms of trade and social services. Most of the internal factors that contributed to the economic crisis can be connected to the wrong choice of development policies and strategies, and misappropriation of domestic and external resources.

Since 1986, economic policies in Tanzania have changed towards a more market-oriented policy regime. Macroeconomic indicators such as the country's GDP have been increasing for the last decade. For example, GDP growth at constant 1992 prices rose by 4.5% in 1999, compared with the growth rate of 3.6% recorded in 1997 (Bank of Tanzania, 1997). GDP growth has been higher than population growth during this period. The large decline in inflation, from close to 30% in 1995 to around 5% in early 2002, is a sign of strong improvement in macroeconomic policies (Table 15.2).

The integration of the national economy into the world market has led to an intensification of agriculture in both commercial and smallholder farms in order to meet market demands for export crops. Such intensification has led to accelerated conversion of woodland areas to crops and pasturelands, and increased demand for fuel for tobacco curing. Today, globalization is a multifaceted phenomenon involving a number of interrelated economic, technological, social, political and cultural changes that together have intensified an interdependence in which everyone who wants to exercise control over the future has to do so with others locally, nationally and globally (Karlsson, 2001).

Before the introduction of the macroeconomic and policy reform, most publicly owned forest industries in the country had virtually come to a standstill for various reasons, mainly economic. Similarly, the management of government-owned

Table 15.2. Key economic indicators. (From Tanzania Bureau of Statistics, 2000.)

Economic indicator	1995	1996	1997	1999	2000
GDP (million US$)	4622.5	5676.2	6785.4	8226.0	
GDP per capita (US$)	168.6	200.6	233.0	284.5	
GDP growth (%)	3.6	4.5	3.5	4.6	
GDP growth/capita (ratio)	0.7	1.6	0.7	1.2	7.9
Inflation (%)	28.4	21	16.1	12.8	7.9
Balance of payments (million US$)	–382.1	–231.2	–633.4	–636.7	–639.6
Population (million)	27.5	28.3	29.1	30.00	30.9

forest plantations was very poor. In the wake of recent policy reforms, the drive has been to privatize most of these enterprises. Positive results have become evident in most of the privatized enterprises. The present macroeconomic policies in the country favour private investment. Since 1993, the government has been taking a variety of measures to encourage such investment. These measures include adopting the Investment Act, privatizing public enterprises and establishing the Tanzania Investment Centre.

Macroeconomic policy changes have had fundamental impacts on the forestry sector, which has gradually been forced to face international competition and to increase its efficiency both in productive and conservation activities. Important parts of the previously fully controlled forest sector are being privatized through various means, ranging from outright sales to handing over management roles to local government and villages. With regard to productive activities such as forestry, the government is currently limiting its share to infrastructure investments, policy formulation and sectoral planning, collection and dissemination of information on prices and markets, public awareness campaigns, and carrying out basic and applied researches and general extension. This implies that the role of the government is being confined to providing an enabling environment for the expansion of production, trade and investment by the private sector.

Forestry sector contribution to the national economy

With regard to the forestry sector's contribution to the national GDP, there are no reliable data at present. Most estimates place the sector's contribution close to 3.3% of GDP (including some wildlife-related services). It is from the forest and woodlands that people's needs for fuelwood, timber, building materials, fruits, mushrooms, fodder, medicines, honey, beeswax, gum arabic and many other products are met. In both the rural and urban areas, wood-based energy consumption is estimated to account for about 92% of the total energy consumed in the country. However, the value of fuelwood, with an estimated per capita consumption of 1 m^3 roundwood per annum, is not recorded. This alone amounts to more than 30 million cubic metres per year, or 30 billion shillings per year when valued at 1000 shillings per cubic metre, the present royalty rate. Unfortunately, this royalty is hardly collectable in most of the rural areas. Besides, not taking into account the value of forest products that are traded informally, the GDP calculations also do not take into account the positive influences of forests on agricultural production. The official GDP figures used therefore do not reflect the true economic importance of the forest sector in the national economy.

The forest sector contributes 10% of official foreign-exchange earnings, or 11% of total merchandized exports. It provides 730,000 person-years of employment. At the national level, the value of forests is estimated at US$750/ha on the basis of royalties collected, exports and tourist earnings. At global level, the value of the Tanzanian forests is estimated at US$1500/ha on the basis of the value of recycling and fixing of carbon dioxide. There are also other indirect uses of forests and woodlands, such as for catchment areas, prevention of soil erosion, habitat for flora and fauna biodiversity, and climate regulation, which are not captured by conventional national accounting methods, thus greatly understating their contribution to the national economy. As a result of this, the forestry sector is not given the high priority it deserves in national development plans.

Cross-sectoral Policy Issues

Decentralization and devolution to local government

The government's present administrative and political decentralization policy has resulted in two main features – first, significant changes in the role of the government at ministerial and regional levels; and secondly, political devolution and decentralization of functions and finances to the local authorities.

At the ministerial level, the policy includes the government's withdrawal from direct production of goods and services, so that the roles and functions of ministries are mainly policy formulation, monitoring and evaluation, and regulatory roles. There are ongoing studies to elaborate on the establishment of executive agencies that will take up management responsibilities for the forests under the central government. The functions of the central government are to be restricted to facilitating local government authorities in their responsibilities to provide services.

At the regional level, the main feature of the new set-up is the formation of a team of sector experts who now form the regional secretariats. They are to provide back-up for local government authorities and other stakeholders. Their role basically is advisory to the Regional Consultative Council on their technical specialties. They also provide technical support to the local authorities in their respective sectors. The regional secretariats are expected to develop cross-sectoral collaboration and coordination to improve national resource utilization and increase efficiency. The practical implication of this for the Ministry of Natural Resources and Tourism is to establish, through the regional secretariat, some mechanism for each of its sub-sectors (forestry, bee-keeping, fisheries and wildlife) to monitor and obtain feedback from local governments on their performance. The ministry has to have measurable indicators to monitor that performance. Equally importantly, through this monitoring process, the ministry's role is to have a national picture of sub-sector performances and to provide mechanisms for adjusting any imbalances across districts.

Under the policy of decentralization and devolution to local government, the promulgation of new policy and the launch of the implementation of local government reform programme in Tanzania have introduced issues and problems in the coordination between ministries, sectors and district councils (Box 15.1). There are four major problems: (i) the parastatal sector reform programme does not appear to address adequately the reorientation to sectoral ministries needed to be consistent with the decentralization programme; (ii) staff in the forestry sector need to be reoriented to embrace decentralization and the roles and mandate of the district councils; (iii) the role and status of regional secretariats, which potentially can facilitate effective coordination, remain ambiguous; and (iv) the capacities of the district councils across the country remain rather weak and uneven.

Formal policy coordination

Forest management is affected by a number of functions and services in other related sectors.

Box 15.1. Major coordination issues related to decentralization and devolution.

- Weak links between the central government and the local governments and undefined rights and responsibilities;
- Poor institutional set-up and unclear mandates between the Ministry of Natural Resources and Tourism (MNRT) and the Ministry of Regional Administration and Local Government (MRALG);
- Redefining power relations between the local people, local governments and central government.

Major sectoral policies that have a bearing on the forest sector include land, agriculture and livestock, minerals, environment, wildlife, bee-keeping, energy, water and health. This necessitates formal cross-sectoral coordination and emphasis on harmonizing the respective policies and instruments to ensure sustainable land productivity, including enhancing soil fertility, hydrological balance and conservation of biological diversity (Table 15.3).

The key policy objective of the 1995 *national land policy* is to ensure equitable distribution of land and equal access to it by all citizens, especially smallholder peasants and herdsmen, and ensuring its optimal use. The 1999 Land Act and Village Land Act state that all land in mainland Tanzania shall continue to be public land and remain vested in the President as trustee for, and on behalf of, all citizens of Tanzania. The acts provide for the use and occupation of land through the system of rights of occupancy. The state grants rights of occupancy and tolerates customary rights. According to these acts, land in Tanzania falls under three categories: general land, administered by the Commissioner of Lands; reserved land, administered by the statutory bodies (e.g. the Forest and Bee-Keeping Division); and village land, administered by the village councils. Forests on general lands have been under constant pressure for conversion to other competing land uses, such as agriculture (shifting cultivation), livestock grazing, settlement and industrial development. They suffer from repeated forest fires because of unclear ownership, lack of security of tenure and unclear formal user rights.

Table 15.3. Important cross-sectoral policy issues and their implications.

Issue	Implications
• Land tenure, ownership and use conflicts; land use planning for forest development	• Demarcation of forest land to facilitate operations; participatory and collaborative land use planning
• Agriculture-caused deforestation; non-harmonized extension service	• Improved agricultural production through coordinated extension service and integrated land use; enhanced food security
• Environmental conservation with respect to mineral sector development	• More environment-friendly technologies applied; reduction of conflicts related to land use mining activities
• Environmental degradation of land, water and vegetation, including desertification; conserving and enhancing biological diversity of unique ecosystems; restoring degraded areas	• Reduced crop yields and increased distances to fetch firewood and water; integration of environmental consideration in all land and forest development activities; closer coordination and collaboration with environmental agenda at local, national, regional and global level
• Wildlife conflicting interests in land use and fragmented institutional framework	• Improved coordination in sustainable management of overlapping forest and/or wildlife biological diversity; harmonizing conflicts in wildlife and forest land uses
• Ensuring an environmentally sustainable energy supply; promoting and disseminating affordable energy technologies	• Expanded supply of fuelwood; availability of alternative sources of energy supply; reduction of forest loss for fuelwood through sustainable utilization
• Conflicting interests in water resource management	• Participatory management of watershed through reduction of conflicts
• Health-related research on the usefulness of plant species and collaborative management of useful species sites	• Increasing utilization of plant species in medicine and biodiversity conservation

There has been little incentive for investment and sustainable management of these forests, leading to continued degradation.

The Village Land Act contains *statements on land tenure,* which are fundamental in fostering forest management. This is very important because almost all forest resources, with the exception of reserved forests, lie on village lands. Conflict in management and utilization of forest resources has often occurred in areas adjacent to village land where, in most cases, tenure rights are insecure. An instrument that is considered to be relevant for sustainable forest management on general lands is the establishment of village forest reserves with properly surveyed boundaries and management plans. Forest areas that have catchment, biodiversity and other amenity values can be managed under joint forest management arrangements between village communities and central government or local authorities. Currently, the formation of village forest reserves must have legal backing from the forest acts approved by the government in 2002.

Agricultural extension services are a crucial cross-sectoral matter linking *agriculture policy* with forest development. Experience has revealed that conflicting extension messages are given by the extension agents on matters related to land use, especially on agriculture and afforestation programmes. Harmonization of the extension messages is required in order to avoid confusing the target population. Conflicting extension messages also have the effect of degrading the confidence people have in extension agents and cause some sectors to be perceived as more important than others. This causes negative impacts on forest resource management, as well as on the investment environment in the forest sector. Another important issue is the common perception in the agricultural sector that views forests, particularly on public lands, as a land reserve for agricultural expansion. From the point of view of forestry, increasing agricultural productivity and efficiency would be a preferable agricultural development policy instead of merely expanding the area under cultivation. Recent studies (Monela *et al.*, 2000) indicate that forest-related incomes represent a large share, sometimes more than agriculture, of rural household incomes. Small-scale farming should be seen by the agricultural administration as a multiple production activity that also relies on forest-related income.

The *minerals policy* aims at promoting private sector participation in the exploitation, mining, processing and marketing of minerals. It aims to strengthen institutional capacity in the mining sector. According to the Mining Act, the forest ordinance cannot prevent exploration rights or the issue of an exclusive prospecting licence. In fact, the Mining Act allows prospecting for minerals anywhere, including in forest reserves. Establishing an obligatory consultative process between the mining and forestry authorities before prospecting licences in forest reserves are issued could mitigate conflicts of interest.

The major goal of the *1997 adopted environment policy* relating to institutional and human resources is to achieve sustainable development for the maximum long-term welfare of the present and future generations of Tanzanians. The main emphasis is to promote cross-sectoral matters related to management guidelines, environmental impact studies, and criteria and indicators for environmental assessment.

The aim of the *1998 wildlife policy* is to promote involvement of the local communities in wildlife conservation in and outside protected area networks, and to create an enabling environment for international cooperation in wildlife conservation. The strategies include the transfer of wildlife management areas to local communities, and the sharing of benefits and responsibilities. The objectives and strategies regarding local community involvement and international cooperation are similar to those in the forestry sector. Lessons could be learnt by the forestry sector from experiences in the wildlife sector.

The *bee-keeping policy* calls for strengthening of strategic planning and development of bee-keeping in harmony with forestry and other sectors. Bee-keeping is often combined with the collection of forest produce. It has been reported to have an important role in the economy of small-scale farming households, contributing to about 50% of household income in bee-keeping reserve areas (Monela *et al.*, 2000).

Policy Instruments for Enhanced Miombo Forest Development

In 1998, the government approved the National Forestry Policy. This concerns the design and

Box 15.2. National Forest Policy goal and objectives.

- The overall *goal* of the National Forest Policy is to enhance the contribution of the forest sector to the sustainable development of Tanzania, and conservation and management of her natural resources for the benefit of present and future generations.
- The *forest sector objectives* on the basis of the overall goal are as follows:
  - Ensuring a sustainable supply of forest products and services by maintaining sufficient forest area under effective management;
  - Increasing employment and foreign-exchange earnings through the development of sustainable forest-based industries and trade;
  - Ensuring ecosystem stability through conservation of forest biodiversity, water catchments and soil fertility;
  - Enhancing the national capacity to manage and develop the forest sector through collaboration with other stakeholders.

implementation of plans and programmes in the forestry sector, the increase of forestry's contributions to social and economic advancement, and the operationalization of commitments and obligations derived from international agreements and intergovernmental processes.[2] The new policy recognizes cross-sectoral linkages between forestry and other sectors, and takes into account macroeconomic and social policy developments related to the country's natural resource base, such as land, environment, water, energy and agriculture (Box 15.2). A new forest law was approved in 2002. Together with the National Forestry Programme of 2001, it provides for new and appropriate instruments, including the establishment of a national forestry advisory committee, the elaboration of management plans for national and local authorities, the promotion of collaborative forest practices and joint management agreements, the introduction of environmental impact assessment, and the elaboration of criteria and indicators for the sustainable use of resources. The Forest Act recognizes related legislation, such as the Land Act, the Village Land Act and other laws and regulations supporting natural resources conservation and environment protection.

[2]These include the Convention on Biological Diversity; the United Nations Convention on Combating Desertification and Drought, and its protocol; the United Nations Framework Convention on Climate Change and the Kyoto Protocol; the United Nations Forum on Forests; and the Convention for the Protection, Management and Development of the Marine and Coastal Environment of the Eastern African region and related protocols.

Participation and decentralization

The focus on participatory management and decentralization is a radical divergence from the earlier forest policy and legislation, which insisted on preservation and control under centralized government authority. Participation of all stakeholders in forest management and conservation is now encouraged through joint management agreements, promotion of the private sector ownership of plantations, reduction of bureaucracy in the acquisition process of forest land, and the attribution of permits and licences for investment in afforestation. The intention of the central government in local government reform is to create local institutions at grassroots level for community participation, in order to foster rapid social and economic development. Local people are being granted the privilege of worship on the forest land, hunting rights and collection of forest products. More incentive schemes are being designed to give them additional benefits, such as collection of revenues that accrue from the forest resource, thereby improving their livelihood.

The devolution of power goes together with promotion of gender balance and addresses inequality in all fields. It was noted that present customary practices as well social and civil laws have led to gender inequalities, which in turn have a negative impact on the goal of sustainable resource management. There is a need to integrate gender issues better into policies, programmes and projects.

Infrastructure, investment and market development

The development of viable road networks is essential for promoting rural development and forestry production. Investments in the supply of energy and water, as well as in telecommunications, are central in order to stimulate local and foreign investment and create employment-generating activities in all sectors, including forestry. Following the recent macroeconomic reforms, most trade and market restrictions have been removed. This has encouraged increased production of non-traditional crops, such as flowers and other horticultural crops, and has led to a dramatic increase in timber and non-traditional forestry product exports. According to Dani (2000), no developing economy can progress within a protected wall. The need is to combine the opportunities offered by the market economy and to foster public institutions that stimulate domestic entrepreneurs. This is what is happening in Tanzania and in the forestry sector in particular at present.

Improved health and livelihood

The government intends to promote health by reducing mortality, improving nutrition and strengthening access to health services and safe water. In pursuit of these objectives, the government is making substantial progress towards the immunization of children, improving the availability of drugs, establishing revolving funds, conducting active HIV/AIDS awareness campaigns and establishing a national health insurance fund. Ill health due to poor nutrition and disease outbreaks has severely affected the entire workforce in all sectors, including forestry.

Criteria and indicators for sustainable forest management

Tanzania has already embarked on developing its own national criteria and indicators for sustainable forest management, but this work is still in its initial phases. Criteria and indicators are operational tools for monitoring and measuring trends in forestry development over time at national and subnational levels. They can also be used as tools for promoting sustainable management at various levels, serve as an early warning system and help in identifying policy gaps and threats in sustainable forest management. They also allow national, subnational and global comparability, providing a relevant basis for forest certification criteria.

Conclusions

Many of the problems currently surrounding the utilization and management of Miombo woodlands are intimately bound with wider issues of land access and control. The apparent land use problems are a reflection of the lack of coordination among different sectors, with each sector fostering its own interests without due consideration of the impact. Community-based forest management is still in the pilot phase, although the new forest policy has opted for it. The present systems of decentralization between the central government and local government reflect weaknesses of overlapping of roles and functions. Local authorities are financially weak because they lack reliable, cost-effective and buoyant sources of tax and non-tax revenue. They are therefore more concerned with the extraction of forest products to raise revenue at the expense of forests.

There is a need to continue with the review of policies and rules governing land use and tenure, which are still complex, ambiguous and arbitrary, particularly with regard to the legal status of customary rights. A clarification of tenure rights is seen as a fundamental step towards improved and sustainable use and management of land resources. There is also a need to define the legal basis on which village surveying, demarcation and titling can take place in order to ensure recognized systems for resolving disputes over forest land and access to it. Clarifying individual and communal tenure rights must take account of the interests of women as heads of the household, while the interests of pastoralists in gaining seasonal access to grazing land must be safeguarded.

Finally, monitoring systems for the implementation of the national forestry programme are to be set up, with assessment of clearly defined criteria and indicators for sustainable forest management. Provisions for regular meetings with key decision-makers at local level are required in order to disseminate information and promote district-level planning and implementation.

References

Bank of Tanzania (1997) *Economic and Operational Report for the Year Ended 30th June, 1997*. Bank of Tanzania, Dar es Salaam.

Dani, R. (2000) Development strategies for the 21st century. Paper presented at the conference on "Developing Economies in the 21st Century", Institute for Developing Economies, Japan External Trade Organization, 26–27 January 2000, Chiba, Japan.

Karlsson, M. (2001) Opening address: development strategies for the 21st century. In: Boris, P.N.S. (ed.) *Annual World Bank Conference on Development Economics 2000*. World Bank, Washington, DC, pp. 1–3.

Ministry of Natural Resources and Tourism (1994) *National Environmental Action Plan: A First Step*. Division of Environment, Ministry of Natural Resources and Tourism, Dar es Salaam.

Ministry of Natural Resources and Tourism (1998) *National Forest Policy*. Ministry of Natural Resources and Tourism, Dar es Salaam.

Ministry of Natural Resources and Tourism (2001) *National Forest Programme in Tanzania, 2001–2010*. Forestry and Bee-Keeping Division, Ministry of Natural Resources and Tourism, Dar es Salaam.

Monela, G.C., Kajembe, G.C., Kaoneka, A.R.S. and Kowero, G. (2000) Household livelihood strategies in the Miombo woodlands of Tanzania. *Tanzania Journal of Forestry and Nature Conservation*, 73, 17–33.

National Environment Management Council (NEMC) (1994) *Tanzania National Conservation Strategy for Sustainable Development*. National Environment Management Council, Dar es Salaam.

Tanzania Bureau of Statistics (2000) *Yearbook of Statistics*. Tanzania Bureau of Statistics, Dar es Salaam.

16 China Environmental and Economic Accounts for Forestry: Case Studies of Jiaohe County and Fengman District in Jilin Province

Guangcui Dai and Dong Huang

Introduction

The system of national accounts (SNA) does not currently account for the full contribution of forest resources to the national economy, or for the depletion or degradation of forest resources. More specifically, non-market forest products and environmental services are not recorded in the gross domestic product (GDP). In addition, forests provide intermediate inputs to other sectors such as livestock grazing or tourism, but the value of these services is often wrongly attributed to these sectors. Also, ecological services such as watershed protection and carbon storage may not be represented at all (FAO, 2004).

This chapter presents the results of the pilot application[1] of environmental and economic accounts for forestry (EEAF) in Jiaohe county and Fengman district, respectively, located in Jilin province in China. Forest asset and flow accounts are presented both in physical and monetary terms. The results show that forest resources contribute to local, provincial and national economies providing timber products, non-timber products, forest recreation and environmental services such as carbon storage, biodiversity, habitat preservation or water resource protection. They also improve the livelihood of local farmers. Specific policy instruments and institutional arrangements are recommended to improve sustainable forest management.

Jilin province is located in a temperate zone in the north-east of China. In 2003, its population was 26,586,000 inhabitants, and the GDP per capita was US$1167.30.[2] The total land area is equal to 187,400 km^2, of which the forest area accounts for 42.5%, making it one of the most important state-owned forest regions of China. Forests play a unique ecological role in watershed protection, landscape conservation, fresh water supply for downstream users, hydropower development, agriculture, fisheries and tourism. Fengman district, located in the southern part of Jilin City in Jilin province, is mostly a mountainous area (70%) with water surfaces (20%) and some arable land (10%). It has a population of 190,200 inhabitants, more than 44% of whom live in rural areas. Fengman district is rich in forest resources, most of which are distributed in the confluent area of the Songhua Lake and the Songhua River, and along the banks of the source of the Songhua River. By contrast, Jiaohe county, located in the south-eastern part of Jilin City, is a mountainous area (70%) with water surfaces (10%) and arable land (20%). It has a population of 460,000 inhabitants, more than 63% of whom live in rural areas, with 70,353 households (Table 16.1).

[1]The application study was sponsored by the Food and Agriculture Organization of the United Nations (FAO) and was based on the FAO Forestry Policy and Institutions Service Working Paper: 'Manual for environmental and economic accounts for forestry: a tool for cross-sectoral policy analysis' (FAO, 2004).

[2]US$1 = RMB 8 yuan.

Table 16.1. Socio-economic aspects, 2003.

Items	Jilin province			Total China
	Province	Jiaohe	Fengman	
Population (thousand inhabitants)	26,586.5	460	190.2	1,292,270
Proportion of rural population (%)	55.0	63.0	44.2	59.47
Net income of rural residents (US$)	316.3	431.6	367.5	327.8
Disposable income of urban residents (US$)	875.6	689.4	962.5	1059.0
GDP per capita (US$)	1167.3	1334.9	2142.5	1171.4
Gross output of forestry (million US$)	1610.7	35.3	2.9	73254.1

The application of EEAF in the Fengman district and Jiaohe county consisted of: (i) forest asset accounts for forest lands and standing timber; (ii) flow accounts for timber, non-timber forest products, forest recreation and environmental services; and (iii) a rural household survey to assess the degree of dependence of farmers' livelihoods on forest resources. The distribution of forest benefits by sector and level was then estimated. Most of the basic data for the accounts were provided by the Department of Forestry in Jilin province, the Forest Bureaus of Jiaohe county and Fengman district, and the township forestry stations. The rural household information was obtained using a questionnaire and personal interviews with 120 households chosen randomly in eight villages during two case studies. The parameters for environmental services were taken from domestic academic research and consultant reports.

Forest Asset Accounts

Forest lands

Forest lands are divided into two categories: available for wood supply and not available for wood supply (State Forestry Administration, 2005). Afforestation, deforestation, changes in classification and other changes were the accounting variables monitored during the reporting period 1999–2003. The price of forest lands was determined in accordance with the compensation criterion of forest land levies by local governments. In 2003, the area of forest lands in the Fengman district amounted to 58,201 ha, of which the area available for wood supply increased by 1.7% and the area not available for wood supply decreased by 0.3%. On the other hand, the area of forest land in Jiaohe county was 285,875 ha, of which the area available for wood supply decreased by 3.4% and the area not available for wood supply increased by 13.4%. In 2003, the value of forest lands in Fengman district was US$155.5 million, representing a drop of 0.2% in comparison with the value in the opening period in 1999. The value of forest lands in Jiaohe county was US$623.1 million, representing an increase of 0.95% compared with the value in the opening period in 1999 (Table 16.2).

Standing timber

The standing timber is divided into two categories, corresponding to forest lands available for wood supply and those not available for wood supply. To calculate the monetary value, young forests and middle-aged forests are accounted by area and reforestation cost, while the near-mature forests, mature forests and overmature forests are valued by stock volume with timber market price (Luo and Chen, 2002). In 2003, the standing timber in Fengman district totalled 6.56 million cubic metres, representing an increase of 3.8% in comparison with 1999. The volume of young and middle-aged forests decreased by 6.7%, while the volume of near-mature, mature and overmature forests increased by 36.3%. By contrast, the standing timber in Jiaohe county totalled 22.49 million cubic metres, representing a drop of 7.5% compared with the total in 1999. The volume of young and middle-aged forests and of near-mature, mature and overmature forests decreased by 9.3% and 4.4%, respectively. During the reporting period, the volume

Table 16.2. Forest asset accounts (*forest lands*). (From Forestry Bureaus of Fengman district and Jiaohe county, 1999–2003.)

	Fengman		Jiaohe	
	Area (ha)	Value (million US$)	Area (ha)	Value (million US$)
1999, opening period	58071.0	155.7	288944.0	617.2
Afforestation	406.0	0.7	8562.0	23.3
Deforestation	–383.1	–0.2	–5953.5	–12.4
Changes in classification	0.0	–0.2	0.0	5.0
Others	107.1	–0.5	–5677.5	–10.0
2003, closing period	58201.0	155.5	285875.0	623.1

of standing timber that is not available for wood supply increased both in Fengman district and in Jiaohe county. The value of standing timber in Fengman district in 2003 was US$127.5 million, a rise of 23.1% over that in 1999, and the value of standing timber in Jiaohe county was US$602.0 million, a marginal increase of 0.2% in comparison with that in 1999 (Table 16.3).

Flow Accounts

Timber, non-timber forest products and forest recreation

Forest products comprise use values such as: (i) timber, including logging both for commercial and farmers' households' own consumption and fuelwood; (ii) non-timber forest products (NTFPs); and (iii) forest recreation. The statistics for NTFPs and fuelwood for farmers' own consumption are underestimated due to the lack of reliable data. The direct economic value of forest products in Fengman district in 2004 was US$2.5 million, of which the contributions of timber, NTFPs and forest recreation represented 59.8%, 17.1% and 23.1%, respectively. On the other hand, the total output value of forest products of Jiaohe county in 2004 was US$25 million, of which timber, NTFPs and forest recreation accounted for 27.5%, 53.0% and 19.5%, respectively (Table 16.4).

Environmental services

Four categories of environmental services were appraised: (i) water source storage (or water-

Table 16.3. Forest asset accounts (*standing timber*). (From Forestry Bureaus in Fengman district and Jiaohe county, 1999–2003.)

		Fengman district		Jiaohe county	
Year	Age group	Volume (1000 m^3)	Value (million US$)	Volume (1000 m^3)	Value (million US$)
1999	Juvenile and middle-aged forests	4,779.4	36.2	15,512.3	163.2
	Near-mature, mature and overmature forest	1,544.3	67.5	88,09.9	437.9
	Subtotal	6,323.7	103.6	24,322.2	601.0
2003	Juvenile and middle-aged forests	4,457.3	33.3	14,062.9	165.2
	Near-mature, mature and overmature forest	2,104.5	94.2	8,424.8	436.8
	Subtotal	6,561.8	127.5	22,487.7	602.0

Table 16.4. Flow accounts, 2004 (*timber, non-timber forest products and forest recreation*). (From Forestry Bureaus of Fengman district and Jiaohe county, 2004.)

Forest products	Volume		Value (million US$)	
	Fengman	Jiaohe	Fengman	Jiaohe
1. Timber forest products	15.3	102.7	1.49	6.88
Commercial timber (1000 m³)	14.3	77.0	1.43	5.79
Fuelwood for farmers (1000 m³)		23.9		1.01
Timber for farmers' own use (1000 m³)	1.0	1.8	0.06	0.08
2. Non-timber forest products			0.43	13.24
Plant product or raw material (1000 kg)	200.0	3293.1	0.18	10.76
Nuts (1000 kg)	200.0	901.9	0.01	0.43
Edible mushroom and plants (1000 kg)		2129.5	0.00	9.90
Medicinal herbs (1000 kg)		261.6	0.00	0.35
Flowers (container)	20000	4000	0.16	0.00
Tea and fruits (1000 kg)		124.9	0.00	0.08
Animal products or raw materials (1000 kg)	465.1	1625.2	0.25	2.48
Wild animal breeding (1000 kg)	465.1	1624.6	0.25	2.43
Fodder and livestock grazing (1000 kg)		0.6	0.00	0.05
3. Forest recreation			0.58	4.88
Total			2.49	24.99

shed protection); (ii) water and soil conservation; (iii) carbon storage; and (iv) conservation of biodiversity. The valuation of carbon storage was established on the basis of existing carbon taxes, and the value of water and soil protection and water resource storage was estimated on the basis of the cost of the damage prevention programme (Hou and Zhangying, 2005). The value of habitat and/or biodiversity preservation was estimated by combining the opportunity cost and the average annual income, adjusted with a development coefficient (Zhang, 2004).

Table 16.5 shows that the total value of forest environmental services in Fengman district in 2004 was US$8.4 million, of which the value of water and soil protection represented 32.4% of the total; the value of watershed protection accounted for 59.7% of the total; and the values of carbon storage and biodiversity conservation were 1.4% and 6.5% of the total, respectively. The total value of forest environmental services in Jiaohe county was US$39.6 million, of which the value of water and soil protection represented 34.4% of the total; the value of watershed protection accounted for 58.9% of the total; and the values of carbon storage and biodiversity conservation were 1.4% and 5.3% of the total, respectively.

Beneficiaries of Forest Resources and Contribution to Rural Livelihood

The total annual forest output of Fengman district in 2004 was US$10.9 (2.5 + 8.4) million and that of Jiaohe county was US$64.6 (25.0 + 39.6) million. The current SNA includes the direct economic use value (timber, part of NTFPs and forest recreation), for which only a percentage of the total value of forest products was recorded as the output of the forestry sector, while forest environmental services were not accounted for at all. Also, some forest services, as intermediate inputs, were wrongly attributed to other sectors.

Table 16.6 shows that the total annual forest output mainly benefited non-forestry sectors, which include hydropower and irrigation, agriculture, urban water supply, tourism and animal husbandry. In Fengman district, for example, the largest beneficiary was hydropower and irrigation, which received 42.7% of the total. In Jiaohe county, 20.4% of the total annual forest output went to hydropower and irrigation, followed by agriculture (12.7%) and urban water supply (10.6%).

If the total annual forest output is broken down in terms of the level of the beneficiaries, it is found that the local community level is the

Table 16.5. Flow accounts, 2004 (*environmental services*).

Classification	Item	Fengman Value (million US$)	Fengman Percentage	Jiaohe Value (million US$)	Jiaohe Percentage
Carbon storage	Annual carbon storage	0.1	1.4	0.5	1.4
Biodiversity protection	Indirect output of biodiversity protection	0.6	6.5	2.1	5.3
	Subtotal	5.0	59.7	23.3	58.9
Watershed protection	Value of flood control and water storage	2.8		12.7	
	Value of runoff regulation	1.7		8.5	
	Value of water purification	0.5		2.1	
	Subtotal	2.7	32.4	13.6	34.4
Water and soil protection	Value of reducing soil disuse	0.8		4.0	
	Value of soil improvement	1.8		9.2	
	Value of silt reduction	0.1		0.4	
Total		8.4	100	39.6	100

main beneficiary. For example, in Jiaohe county, 59.8% of the total annual forest output benefits the local community, 36.1% benefits the province and 4.1% the national or global level; while in Fengman district, the proportions are 47.9%, 46.1% and 6.0%, respectively (Table 16.7).

The findings from the survey of 120 rural households showed that the extent of dependence of farmers' livelihood on forestry varies according to the richness of forest resources and the distance from the city. In Jiaohe county, for example, fuelwood is the main energy source for cooking

Table 16.6. Percentage distribution of forest output by beneficiary sector, 2004.

	Agriculture		Animal husbandry		Forestry		Hydropower and irrigation		Water supply		Tourism		Others	
Forest output	FM	JH	FM	JH	FM	JH	FM	JH	FM	JH	FM	JH	FM	JH
Timber					13.6	10.6								
NTFPs	1.5	0.7	2.3	3.8	0.1	16.0								
Forest recreation											5.3	7.6		
Carbon storage													1.0	0.8
Watershed protection	7.0	5.8					26.0	19.7	13.1	10.6				
Water and soil protection	7.5	6.2			0.9	14.2	16.7	0.7						
Biodiversity protection													5.0	3.3
Total	16.0	12.7	2.3	3.8	14.6	40.8	42.7	20.4	13.1	10.6	5.3	7.6	6.0	4.1

FH = Fengman district; JH = Jiaohe county.

Table 16.7. Percentage distribution of forest output by beneficiary level, 2004.

	Local/county		Provincial		National/global	
	Fengman	Jiaohe	Fengman	Jiaohe	Fengman	Jiaohe
Timber	13.6	10.6				
NTFPs	3.9	20.5				
Forest recreation	5.3	7.6				
Forest environment services	25.1	21.1	46.1	36.1	6.0	4.1
Total	47.9	59.8	46.1	36.1	6.0	4.1

and heating. About one-third of the total annual gross income of households comes from timber and trees as wages, fuelwood or collected NTFPs. In Fengman district, on the other hand, only 5% of the total annual gross income of households comes from timber and trees. Nevertheless, they play an important role in rural livelihood, particularly during cold winters, when households mostly rely on fuelwood for heating.

Conclusions and Recommendations

The pilot application of the EEAF in Fengman district and Jiaohe county shows that the total annual forest output is grossly underestimated in the SNA. The main beneficiaries of sustainable forest management are non-forestry sectors and people at local and provincial level. This result should lead to an enhanced awareness of the contribution of forest resources to local, provincial and national economies and should improve cross-sectoral policy decision-making aimed at achieving sustainable forest management. It also shows that the environmental and economic accounts for forestry:

1. *Constitute a useful tool for policy analysis and coordination.* In addition to the commercial timber forest products, it includes non-market and non-timber forest products and forest recreation, and environmental services values, which are not included in the SNA. It shows the total environmental, social and economic contribution of forests to the national and local economies. The accounts show the value of forest assets, the total annual forest output value and its beneficiaries, and they can provide strong support for decision-makers at the national and local levels in improving cross-sectoral policy coordination with the aim of achieving sustainable forest management.

2. *Provide a rational framework for increasing investments in rehabilitating and protecting forest ecosystems.* The study shows that the value of environmental protection services represents as much as two-thirds of the total annual forest output. The central government has established a forest ecological benefit/service compensation fund for the national key ecological forests (State Forestry Administration, 2004). However, few local governments have established a forest ecological benefit/service compensation fund for local ecological forests, although this is stipulated in the Forest Law. In fact, the actual demand for funds exceeds the amount invested for rehabilitating and protecting forest ecosystems, because compensation standards are still too low. Government at national and local levels should therefore increase the related financial support.

3. *Establish a valuation basis on which payments for environmental services can be improved.* The study shows that the main beneficiaries of forest outputs are non-forestry sectors such as agriculture, hydropower and irrigation, urban water supply and tourism. According to the Forest Law (State Council, 2000), 'the beneficiary of forest environmental services should pay'. It is therefore recommended that a system of payments should be established for environmental services, building on the current national forest ecological benefit/service compensation fund. It could work out like this: (i) the agriculture sector is charged a proportion of inputs for cropland rehabilitation/protection; (ii) the water conservation sector is charged a proportion of the investment for water conservation construction works; (iii) the urban water supply sector levies a resource tax on water consumption; (iv) a levy or forest fee is charged for forest recreation and medicinal plant collection and transformation; and (v) a proportion of the cost of access to water hydropower is charged to the electricity fee and/or price.

4. *Are an incentive to improve current statistical system.* The *China Forestry Statistical Yearbook* only records data concerning forestry construction and economic operation and lacks cross-sectoral information, particularly with reference to rural wood energy and NTFPs, and on the value of forest environmental services.

5. *Require capacity building and institutional strengthening for their development and application.* First, the environmental and economic accounting for forestry is a huge and complex system. It needs large amounts of data, professional skills and a capacity for multidisciplinary policy analysis. It is therefore necessary to adopt standard environmental and economic accounts for forestry. Secondly, the construction of accounts at the national or provincial level should coincide with forest resource inventories. Thirdly, it is essential to build up the capacity of departments such as the National Statistics Bureau and the State Forestry Administration. Finally, the dissemination of concrete results obtained from environmental and economic forestry accounting to government departments, academic institutions and universities needs to be improved.

References

FAO (Food and Agriculture Organization of the United Nations) (2004) *Manual for Environmental and Economic Accounts for Forestry: A Tool for Cross-Sectoral Policy Analysis*. Glenn-Marie Lange, Forestry Policy and Institutions Service Working Paper. FAO, Rome.

Hou, Y. and Zhangying, C.K. (2005) *Account for Forest Resource*. China Science Technology Publishing House, Beijing, p. 233.

Luo, J. and Chen, P. (2002) *Asset Evaluation of Forest Resource*. China Forestry Publishing House, Beijing, pp. 61–88.

State Council (2000) *Regulation for the Implementing of the Forestry Law of the People's Republic of China*. China Forestry Publishing House, Beijing, p. 23.

State Forestry Administration (2004) *Implement Forest Ecological Benefit Compensation Fund System in China*. China Forestry Development Report. China Forestry Publishing House, Beijing.

State Forestry Administration (2005) China Forest Resources Assessment Report for FRA 2005. Research report submitted to FAO.

Zhang, Y. (2004) *Theory and Method on Green GDP Accounts*. China Forestry Publishing House, Beijing, p. 194.

17 Non-timber Forest Product Linkages in the Tribal Economy of the Western Ghats Region of Karnataka, India[1]

Mogenahalli R. Girish and Mamatha Girish

Introduction

The Kanara forest circle was chosen deliberately to study the linkages between non-timber forest products (NTFPs) and other sectors in the tribal economy of the Western Ghats region of Karnataka state in India. It has the largest area of forest cover, yields a variety of NTFPs and is truly representative of the Western Ghats region. On the basis of a preliminary survey, it was found that the Yellapur forest division is rich in a variety of NTFPs and that forest communities depend on them not only for their subsistence, but also for earning their cash income. Among the different forest dwelling communities in the division, the degree of dependency of the *Siddi* community on NTFPs for their livelihood is significant. Yellapur forest division has six forest ranges, namely Idgundi, Manchikeri, Mundgod, Katur, Kirwatti and Yellapur. Two of these ranges, namely Mundgod and Yellapur, were selected, as they had a relatively larger tribal population. Detailed data were collected by personally interviewing, with the help of a pre-tested questionnaire, 40 tribal households and 20 non-tribal households to study the linkages between the NTFPs and other sectors of the tribal economy. Secondary data on the crop area, population and livestock were collected from various government offices (Tahsildhar offices and the veterinary hospitals in Mundgod and Yellapur). This chapter presents and discusses the results of the application of a social accounting matrix (SAM) to show and discuss these linkages.

Methodology

The SAM combines diverse sets of data on all aspects of an economy, such as production, consumption, savings and investment, income generation and distribution, transfers and external trade and income flows, and presents them as a set of consistent accounts in the form of a square matrix. Each row contains receipts accruing to that account, and the corresponding column shows how that account's total receipts are spent on (or distributed to) other accounts. There are two basic rules for understanding this type of matrix: (i) for every row, there is a corresponding column, and the system is complete only if the corresponding row and column totals are identical; and (ii) every entry is a receipt when read in its row context and expenditure from the point of view of its column. The description of SAMs as single-entry accounts derives from this rule (Pyatt and Roe, 1977).

A village economy is characterized by a relatively simple set of production accounts but relatively complex labour allocation patterns. For each production activity, the rows contain payments received by the activity for the commodities that it produces (and sells to the commodity

[1]Based on Girish (1998).

accounts). The corresponding column account breaks up the value of the total output into the value of intermediaries, payments to factors, profits accruing to the owners of the activity (in this case, households), etc. The commodity row accounts give the components of total demand, intermediate use, consumption demand by household group and exports. The commodity column accounts show what part of each commodity's total supply comes from each production activity and imports. The factor accounts are of great importance, because they show how factor incomes are generated and distributed to households and other institutions. The household and other institutional accounts show the sources of each institution's income (along the row) and the objects of expenditure (down the column), such as consumption, savings, etc. The capital row account shows each household group's savings, and the column account breaks up total investment (which equals total savings). Finally, the rest of the world row account shows payments made by the village to the rest of the world. The column account shows payments received by the village from outside the village (Subramanian and Sadoulet, 1990).

The input–output (I–O) matrix for the village economy (Table 17.1, entry 1,1) consists of four sectors: NTFPs, agriculture, livestock and trade. The NTFP sector is divided into fodder, food and non-food sub-sectors. The trade sector refers to the retailing (sales and purchases) of goods produced within the village. The village economy is likely to have a large import component (5,1), since the economy is a subsistence one. As the transition from a subsistence economy to a market economy unfolds, an increasing share of village production will tend to be exported (1,5). The village production activities result in income payments to labour (2a,1) and capital (2b,1). Payments to capital include imputed returns to capital when no explicit payment takes place. Both types of payment are contained in the capital factor account. Separate entries are included for hired and non-hired (family) labour services, or labour value added, in order to provide a sharper focus on inter-household farm-labour linkages in village production activities. Together, accounts 1 and 2 represent the flow of commodities across product markets and of factors across factor markets within the village economy (Adelman *et al.*, 1988).

The village value-added and wage income are channelled into four village household institutions and the forest department. The household institutions are defined by the size of the holdings. The institution accounts (3a–e) provide a detailed breakdown of payments for labour services supplied by village households to employers both inside and outside the village. Total payments to households for labour services in village production from the two labour value-added accounts are represented by entries (3a–d, 2). The household accounts, in turn, channel household income into the final village (i.e. domestic) consumption demand for village products – entries (1a–d, 3) – and savings. The household savings are represented by the entry (4,3).

The endogenous accounts include production sectors, institutions/households and factors of production. Accounts such as 'rest of the world' are treated as exogenous accounts. The endogenous accounts are balanced by construction, having equal row and column sum vectors. When there is only one exogenous account, the total of the leakages or outflows out of the village is equal to the sum of the initial injections into the economy. However, when there are several exogenous accounts, this does not hold for each account individually, but only in the aggregate.

The data collected from the sample households were averaged out and then blown up for the population in each institution by multiplying the corresponding average by the total number of households in that category for the region. SAM analysis was performed, and the production multiplier matrix and household income multiplier matrix were derived from the Leontief inverse matrix.

Results and Discussion

The SAM for the region is presented in Table 17.2. From the table, it can be seen that the fodder sector of NTFPs supplied fodder worth Rs 10,210,312 to the livestock sector (Rs 10,210,312 implies one crore, two lakhs, ten thousand three hundred twelve rupees). The non-food sector supplied forest litter worth Rs 693,842 and Rs 245,700 to paddy and areca nut crops, respectively. The paddy sector of agriculture supplied dry fodder worth Rs 3,744,542 to livestock.

Table 17.1. Outline of the social accounting matrix (SAM).

Expenditures Receipts	1 Activities	2 Factors	3 Institutions	4 Capital	5 ROTW	6 Total
Activities	Village input–output table		Consumption	Investment	Exports	Total sales
a. NTFPs						
i. Fodder						
ii. Food						
iii. Non-food						
b. Agriculture						
c. Livestock						
d. Trade						
Factors	Value added in village production					Total labour and capital value added
a. Labour						
i. F. labour						
ii. H. labour						
b. Capital						
Institutions	Payments to households for labour and capital services used in production	Payments to households for labour services outside the village			Wages earned from outside the village	Total household income Total FD receipts
a. Landless						
b. Marginal						
c. Small						
d. Large						
e. FD						
Capital			Household savings			Total savings
ROTW	Imports					Total imports from ROTW
Total	Total sales	Total payments to labour and capital	Total institutional expenditures	Total capital investment	Total exports to ROTW	Total receipts/ expenditures

NTFPs = non-timber forest products; F. labour = family labour; H. labour = hired labour; FD = Forest Department; ROTW = rest of the world.

Table 17.2. Social accounting matrix (SAM) for the region (amounts in rupees).

	Fodder	Food	Non-food	Paddy	Cotton	Areca nut	Fertilizer	PPC	Livestock	Concentrate	F. labour	H. labour
Fodder	0	0	0	0	0	0	0	0	10,210,312	0	0	0
Food	0	0	0	0	0	0	0	0	0	0	0	0
Non-food	0	0	9,344,792	693,842	0	245,700	0	0	0	0	0	0
Paddy	0	0	0	3,422,326	0	0	0	0	3,744,542	0	0	0
Cotton	0	0	0	0	193,800	0	0	0	0	0	0	0
Areca nut	0	0	0	0	0	0	0	0	0	0	0	0
Fertilizer	0	0	0	3,956,423	765,000	135,000	0	0	0	0	0	0
PPC	0	0	0	0	969,000	351,001	0	0	0	0	0	0
Livestock	0	0	0	8,019,848	989,400	221,401	0	0	0	0	0	0
Concentrate	0	0	0	0	0	0	0	0	2,659,675	0	0	0
F. labour	2,859,327	30,344,380	43,015,835	16,041,464	2,210,083	226,193	0	0	25,733,942	0	0	0
H. labour	0	0	0	48,124,392	3,315,124	4,297,668	0	0	0	0	0	0
Profit	7,350,985	–731,347	8,446,267	13,230,985	3,005,390	45,947,641	0	0	–18,756,923	0	0	0
Landless	0	0	0	0	0	0	0	0	0	0	19,726,486	36,229,170
Marginal	0	0	0	0	0	0	0	0	0	0	26,301,981	11,147,437
Small	0	0	0	0	0	0	0	0	0	0	46,028,467	8,360,577
Large	0	0	0	0	0	0	0	0	0	0	39,452,973	0
FD	0	0	0	0	0	0	0	0	0	0	0	0
Savings	0	0	0	0	0	0	0	0	0	0	0	0
Trade	0	0	0	0	0	0	2,428,211	330,000	0	1,994,756	0	0
ROTW	0	0	0	0	0	0	2,428,212	990,001	0	664,919	0	0
Total	10,210,312	29,613,033	60,806,894	93,489,280	11,447,797	51,424,604	4,856,423	1,320,001	23,591,548	2,659,675	131,509,907	55,737,184

	Profit	Landless	Marginal	Small	Large	FD	Savings	Trade	ROTW	Total
Fodder	0	0	0	0	0	0	0	0	0	10,210,312
Food	0	2,333,755	1,249,631	697,086	1,421,280	0	0	3,228,023	20,683,258	29,613,033
Non-food	0	14,471,467	5,475,124	7,831,094	5,960,304	0	0	0	16,784,571	60,806,894
Paddy	0	0	9,756,747	12,655,970	10,683,440	0	0	21,290,462	31,935,793	93,489,280
Cotton	0	0	0	0	0	0	0	1,125,400	10,128,597	11,447,797
Areca nut	0	0	284,762	369,380	311,809	0	0	7,568,798	42,889,855	51,424,604
Fertilizer	0	0	0	0	0	0	0	0	0	4,856,423
PPC	0	0	0	0	0	0	0	0	0	1,320,001
Livestock	0	442,638	3,098,467	4,094,403	3,430,445	0	0	1,647,473	1,647,473	23,591,548
Concentrate	0	0	0	0	0	0	0	0	0	2,659,675
F. labour	0	0	0	0	0	11,078,683	0	0	0	131,509,907
H. labour	0	0	0	0	0	0	0	0	0	55,737,184
Profit	0	0	0	0	0	0	0	0	0	58,492,998
Landless	584,930	0	0	0	0	0	0	0	0	56,540,586
Marginal	5,264,370	0	0	0	0	0	0	0	0	42,713,788
Small	8,773,950	0	0	0	0	0	0	0	0	63,162,994
Large	43,869,748	0	0	0	0	0	0	0	0	83,322,721
FD	0	7,536,296	1,998,496	1,543,891	0	0	0	0	0	11,078,683
Savings	0	−16,061,049	4,810,299	5,998,578	7,398,404	0	0	0	0	2,146,232
Trade	0	47,817,479	14,436,236	23,978,074	37,881,927	0	0	0	0	128,866,683
ROTW	0	0	1,604,026	5,994,518	16,235,112	0	2,146,232	94,006,527	0	124,069,547
Total	58,492,998	56,540,586	42,713,788	63,162,994	83,322,721	11,078,683	2,146,232	128,866,683	124,069,547	

Non-timber forest products (NTFPs) consist of fodder, food and non-food sub-sectors; agriculture consists of paddy, cotton and areca nut sub-sectors. FD = Forest Department; F. labour = family labour; H. labour = hired labour; PPC = plant protection chemicals; ROTW = rest of the world.

The livestock sector supplied farmyard manure (FYM) worth Rs 8,019,848, Rs 989,400 and Rs 221,401 to paddy, cotton and areca nut crops, respectively. Among the different institutions, the landless category showed dissavings, while the other three categories – marginal, small and large households – showed savings. Out of the total exports from the region, agriculture contributed about 69%, followed by NTFPs at 30% and the livestock sector at 1%.

The production multiplier matrix for the forest-based tribal economy is presented in Table 17.3. It reveals that the livestock sector had the highest production multiplier, at 3.76. This implies that for every one-rupee increase in final demand in the livestock sector, the total production would increase by Rs 3.76. In the case of agriculture, paddy had the highest multiplier, at 3.37. For every one-rupee increase in final demand for paddy, the total production would increase by Rs 3.37. For the other two agricultural sub-sectors – cotton and areca nut – the multiplier values are 3.07 and 2.90, respectively. In the case of NTFPs, the non-food sub-sector had the highest multiplier of 3.21, implying that for every one-rupee increase in final demand, their total production would increase by Rs 3.21. The other two sub-sectors of NTFPs – food and fodder – had multipliers of 3.07 and 2.91, respectively. The trade sector had the lowest multiplier, at 1.87.

Table 17.4 presents the household income multiplier matrix for the forest-based tribal economy. It reveals that the food sub-sector of NTFPs had the highest income multiplier, at 1.90. This implies that for every rupee of investment, the income from the sector would increase by Rs 1.90. Out of the expected increase in income, 31.62% will go to small households, 27.82% to large households, 20.94% to landless households and 19.62% to marginal households. Non-food and fodder sub-sectors had multipliers of 1.87 and 1.79. In both of these sub-sectors, large households were expected to derive the maximum share of the increased income. In the case of agriculture, paddy had the highest multiplier, at 1.84. For every one-rupee of investment, the income would increase by Rs 1.84. Of the expected increase in income, 34.38% will go to landless households, 24.23% to small households, 22.70% to large households and 18.69% to marginal households. Paddy was followed by areca nut and cotton, with multiplier values of 1.77 and 1.61, respectively. In both of these sub-sectors, large households were expected to obtain the maximum percentage of the increased income. In the case of the livestock sector, every rupee of investment would increase the income by Rs 1.77. Small households were expected to derive the maximum share of this increased income. The trade sector had the lowest multiplier, at 0.49.

Table 17.3. Production multiplier matrix for the forest-based tribal economy.

Sectors		NTFPs			Agriculture				
		Fodder	Food	Non-food	Paddy	Cotton	Areca nut	Livestock	Trade
NTFPs	Fodder	1.06	0.06	0.06	0.09	0.09	0.06	0.50	0.03
	Food	0.06	1.07	0.07	0.07	0.06	0.06	0.07	0.04
	Non-food	0.26	0.31	1.48	0.36	0.28	0.26	0.33	0.09
Agriculture	Paddy	0.44	0.47	0.46	1.49	0.41	0.43	0.62	0.29
	Cotton	0.01	0.01	0.01	0.01	1.03	0.01	0.01	0.01
	Areca nut	0.06	0.06	0.06	0.07	0.06	1.06	0.07	0.08
Livestock		0.13	0.14	0.14	0.22	0.21	0.13	1.15	0.06
Trade		0.89	0.95	0.93	1.06	0.93	0.89	1.01	1.27
Total		2.91	3.07	3.21	3.37	3.07	2.90	3.76	1.87

NTFPs = non-timber forest products.

Table 17.4. Household income multiplier matrix for the forest-based tribal economy.

Sectors/ Institutions	NTFPs			Agriculture				
	Fodder	Food	Non-food	Paddy	Cotton	Areca nut	Livestock	Trade
Landless	0.27	0.40	0.37	0.63	0.45	0.28	0.48	0.14
Marginal	0.27	0.37	0.35	0.34	0.29	0.25	0.39	0.09
Small	0.42	0.60	0.55	0.45	0.40	0.36	0.63	0.12
Large	0.83	0.53	0.60	0.42	0.47	0.88	0.27	0.14
Total	1.79	1.90	1.87	1.84	1.61	1.77	1.77	0.49

NTFPs = non-timber forest products.

Conclusion

SAM analysis revealed that the livestock sector had the highest production multiplier, implying that any investment would generate relatively large returns in this sector in comparison with other sectors. Investment in the agriculture sector (especially in the case of paddy crop) was found to benefit largely the landless households. Hence, efforts are called for in the direction of increasing investment in these two sectors, which will ensure a more equitable distribution of income in the forest-based tribal economy.

References

Adelman, I., Taylor, E.E. and Vogel, S. (1988) Life in a Mexican village: a SAM perspective. *Journal of Development Studies* 25(1), 5–24.

Girish, M.R. (1998) Role of non-timber forest products in the tribal economy: an economic study in Western Ghats region of Karnataka. Ph.D., thesis, University of Agricultural Sciences, Bangalore, India.

Pyatt, G. and Roe, A. (1977) *Social Accounting for Development Planning, with Special Reference to Sri Lanka.* Cambridge University Press, Cambridge.

Subramanian, S. and Sadoulet, E. (1990) The transmission of production fluctuations and technical change in a village economy: a social accounting matrix approach. *Economic Development and Cultural Change,* 39(1), 131–173.

18 Road Construction and Population Effects on Forest Conversion in Rubber Villages in Sumatra, Indonesia

Motoe Miyamoto

Introduction

This chapter investigates factors accelerating forest clearance in Indonesia. The main factor behind forest loss has been agricultural expansion. Cash tree crop cultivation in particular has had a greater impact than subsistence-oriented shifting cultivation (Chomitz and Griffiths, 1996). The main tree crops of Indonesia, that is, those whose total planted area exceeds 1 million ha, are rubber, coconut, oil palm and coffee. Most tree crops are cultivated mainly by smallholders, not by large-scale companies. For example, in 2001, 87% of rubber, 98% of coconut and 95% of coffee were planted by smallholders. Oil palm was an exception for which large-scale companies accounted for 63% (BPS, 2003).

Why has agricultural land so rapidly expanded into the Indonesian forests? Population growth, land rent increase, land tenure insecurity, improvement of agricultural technology, poverty and other possible reasons have been suggested and discussed (Duraiappah, 1998; Angelsen and Kaimowitz, 2001; Barbier and Burgess, 2001). Population growth has long been considered to be the primary factor behind agricultural expansion into forests. Recent studies have questioned this because the effect of population on forest loss is not clear (Angelsen and Kaimowitz, 1999). Although the effect of the total population may be ambiguous, the effect of the agricultural population is undoubtedly very important. Not only the agricultural population within a certain limited site, such as a village or province, but also that in the larger area extending perhaps far beyond that site, such as the entire country, must be considered since people will move in order to acquire jobs and/or land. Java, the Indonesian island neighbouring Sumatra, has an excessive agricultural population and, consequently, many Javanese migrate to Sumatra, Kalimantan and neighbouring countries. On the other hand, land rent, defined as the profit to the owner of the land, has been argued to be an important factor (von Thunen, 1875; Angelsen, 1995; Serneels and Lambin, 2001). This approach holds that economic factors, such as more roads and higher prices for agricultural products raise land profitability and lead to the conversion of forest to agriculture.

Study Area and Methodology

Study area

The chapter examines the effects of agricultural population and road construction on forest conversion to rubber, based on household surveys conducted in four rural rubber villages in Sumatra, Indonesia. The study villages, Pulau, Bunga, Air Ulu and Sungai

Table 18.1. Key variables for selected villages in Sumatra, Indonesia.

Study village/variable	Air Ulu	Sungai Indah	Pulau	Bunga
Total area of village (km^2)	178	43	72	75
Population	1,354	539	3,635	1,768
Altitude at village centre (m)	70	105	50	95
Road to village	Unpaved	Unpaved	Paved	Paved
Distance to processing centre in Jambi City (km)	174	288	279	291

Indah (fictitious names), are located in the west central part of Jambi Province. Sungai Indah village lies at elevations from 100 to 200 m near the foot of mountains (Table 18.1). The other three villages are situated in lowlands 100 m or less above sea level. All the villages are located along tributaries of the Batang Hari River, which is the longest in Sumatra. Natural vegetation consists of lowland rain forest. The climate is characterized by abundant rainfall (about 2,500 mm annually), high humidity and high temperatures. The inhabitants of the four villages belong to the Melayu ethnic group.

Accessing the villages required travelling more than 100 km on a national road west from Jambi City and then taking one or several branch roads. In Pulau and Bunga, roads were constructed early and the roads running to the villages were paved. Roads to Sungai Indah and Air Ulu were constructed late and were still unpaved at the time of the survey.

Village land use data based on the community surveys are shown in Fig. 18.1. In the two villages to which paved roads ran (i.e. Pulau and Bunga), the forests had completely disappeared. Rubber fields accounted for about 90% of the total village area, while upland rice fields and bush fallow comprised about 10%. By contrast, in the two villages to which

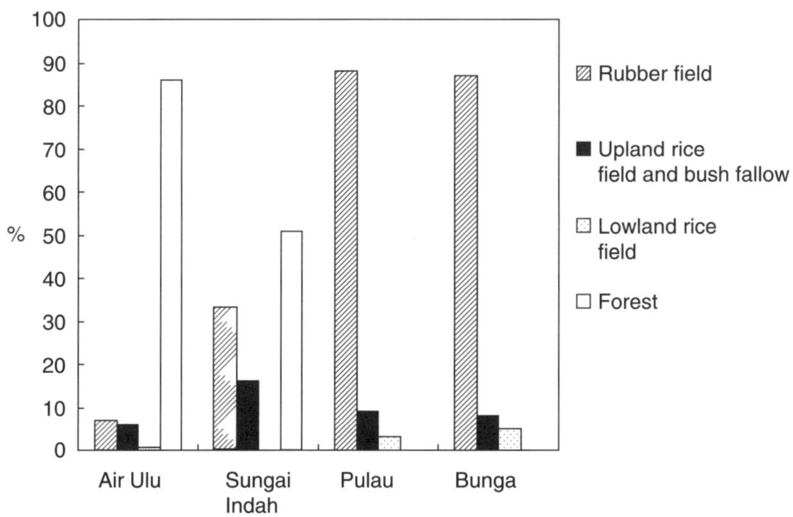

Fig. 18.1. Land use of total village area.

only unpaved roads ran (i.e. Sungai Indah and Air Ulu), forest accounted for more than 50% of the area, while rubber fields made up 33% or less.

Swidden-based rubber cultivation was the dominant economic activity in each village. This farming system is a combination of shifting cultivation and traditional rubber (*Hevea brasiliensis*) planting. After clearing and burning forest, dryland rice and other food crops are planted during the first 1 or 2 years. Rubber trees are planted in the empty spaces between the rice and other crops in the first year. The plots are then left alone for about 10 years until the rubber trees begin to produce latex. Many other kinds of wild trees also grow in such smallholder rubber fields. The villagers make use of these non-rubber trees for fuel wood, fruit and rattan. This kind of rubber cultivation, called 'jungle rubber' or 'rubber agroforest', is common in Indonesia, especially in Sumatra and Kalimantan (Gouyon *et al.*, 1993). Mature rubber trees are productive for 30–40 years. Rubber can be tapped all year long, although productivity decreases from July to September when the leaves have fallen. The villagers tap rubber trees for 4–5 days every week. Once or twice a week they collect the latex and coagulate it. The solid product is called a 'slab', which they sell to the rubber collectors in the village once or twice a month.

One means for the villagers to acquire land for agriculture is to clear primary forest. According to customary law, any villager can clear primary forest and the person who does so can thus acquire tenure. Such customary law is common in Sumatra (Marsden and Bastin 1986, Suyanto and Otsuka 2001). In Air Ulu and Sungai Indah, an average of more than 70% of the rubber fields was acquired by forest-clearing. In Pulau and Bunga, although forests had already been converted to agriculture, rubber fields acquired by forest-clearing still accounted for 33–41%. Thus, forest-clearing was considered to be the main method of land acquirement in all the study villages.

Methodology

Household surveys were conducted in four villages situated in Jambi Province, Sumatra. Jambi is one of Sumatra's main rubber-producing provinces. The four villages were selected so as to meet the following three common requirements: (i) the villagers were not migrants but local people; (ii) rubber was the main source of income; and (iii) the villages had been settled for more than 100 years. They were also selected so that they differed as to the time roads were built related to the transportation of rubber from the village to Jambi City, where the rubber processing centres were located.

Field work was conducted from 1998 to 2000 as follows. First, to collect information on socio-economic conditions at the study sites, a community survey was performed with the key persons of each village including the village headman. Second, surveys were conducted of 160 households in the four villages, i.e. 40 randomly selected households in each village. The interviews inquired about land use, land acquisition methods and household income. Third, four or five rubber collectors were interviewed in each village about rubber transportation. There were a few households which owned more than 100 ha of rubber fields in Pulau and Bunga villages. Such households comprised less than 1% of the total number of households in each village. More than 90% of the study households in each village owned rubber fields amounting to 10 ha or less. Since the data for these larger land-holding households could be considered as far outside the norm, we excluded households having more than 100 ha of rubber fields from our random sampling. However, we interviewed all the larger land-holding households about the area of their owned rubber fields.

Results and Discussion

Road construction

In Jambi Province, many roads have been constructed since the 1970s as vital infrastructure, based on the economic policies of the Suharto era (1967–1998). Two arterial roads in the study site were built in 1974 and 1978, and following their completion, branch roads were constructed and lengthened. In Air Ulu and Sungai Indah, the roads between each village and Jambi City were constructed and lengthened after these arterial roads were built. The final unpaved road reaching each village was completed in 1992 in Sungai Indah and in 1997 in Air Ulu. On the other hand, the roads reaching Pulau and Bunga were constructed in the 1950s, but remained in poor condition until they were widened and paved beginning in the 1970s.

This road construction profoundly affected rubber transportation from the study villages to the processing centres in Jambi City by changing it from via river to via road, and especially to via paved road. Rubber transportation from the villages to Jambi City shifted from via river to via mostly paved road in 1974 in Bunga, 1978 in Pulau, 1992 in Sungai Indah and 1997 in Air Ulu.

Figure 18.2 shows how transportation cost changed after road construction, based on community surveys and interviews with rubber collectors. Until the early 1970s, the rubber produced in all the study villages had to be transported by river to Jambi City. This was both expensive and time-consuming. By contrast, road transportation was inexpensive and quick. The road construction process involved at least two steps: first, unpaved roads were constructed and, second, these roads were paved. The construction of unpaved roads shortened transportation time, but often increased the cost due to bad road conditions. However, paving greatly reduced transportation cost. After the transportation route changed to via mostly paved road, transportation by road became cheaper than by river.

The road construction improving access to the villages significantly improved rubber transportation and, as a result, increased both the frequency of transport and the quantity sold. The rubber collectors' profits increased and therefore the prices they paid the farmers for their latex also increased. In all the study villages, the farm gate price rose after the transportation route changed to mostly via paved road. Rubber profitability increased for the farmers and rubber land rent rose. Thus, road construction clearly accelerated forest conversion to rubber in the villages.

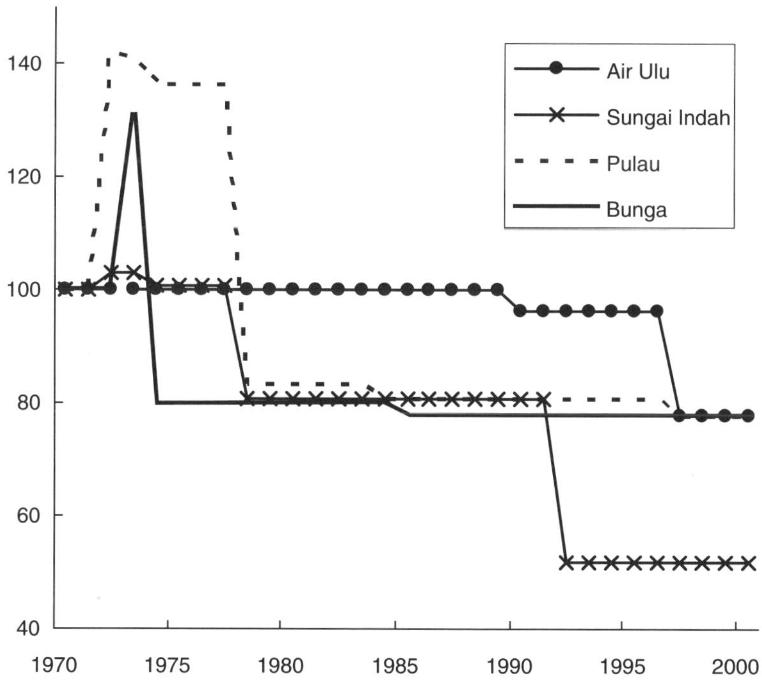

Fig. 18.2. Cost of rubber transportation change. (The cost for transportation via river in 1970 is considered as 100 in each village.)

Agricultural population

The population in the study villages was considered to be agricultural, because farming families comprised all the study households and most members over the age of 15 performed agricultural work. The average household income of each village is shown in Fig. 18.3. This income includes both production for sale and for direct consumption (at local market prices). In all four villages, agricultural income accounted for more than 70% of total income, thus occupying an overwhelming share. Rubber was the main source of this accounting for 35–60%. Other agricultural income included rice and other annual crops from lowland and upland rice fields. Off-farm income accounted for only around 15% of the total in all the villages and included income earned through business (e.g. retail sale of miscellaneous goods, carpentry, sewing, farm product-trading, rubber-collecting and mini-bus operation) and labour (e.g. farmworkers, log carriers, rubber carriers, teachers and village officials). Additional income, shown in Fig. 18.3 as 'other', was acquired by sale of assets, such as gold, buffalos and rubber fields, or by money sent from children.

The manpower for rubber production in the villages was provided not only by the villagers but also by migrant workers from outside the villages, mainly Javanese. In particular, the large rubber holders owning more than 20 ha of rubber fields, of which there were about 10–20 households in each village, employed more Javanese than villagers as their sharecroppers. Most Javanese sharecroppers working in Pulau, Bunga and Sungai Indah came on their own accord from Java Tengah Province, while the Javanese sharecroppers in Air Ulu were mainly settlers in the transmigration site near the village. Java island has an excessive agricultural population and a great number of landless farmers. Consequently, many poor farmers leave Java to seek jobs and/or land. The Javanese sharecroppers did not settle in the study villages. Husbands and wives lived in temporary sheds on the rubber fields they rented while their children remained in Java. They saved their money and returned to Java once a year or less. Such Javanese migrant workers move on if there is a more profitable opportunity in some other place and, in fact, many of them left the study villages to mine gold after the monetary crisis in 1997.

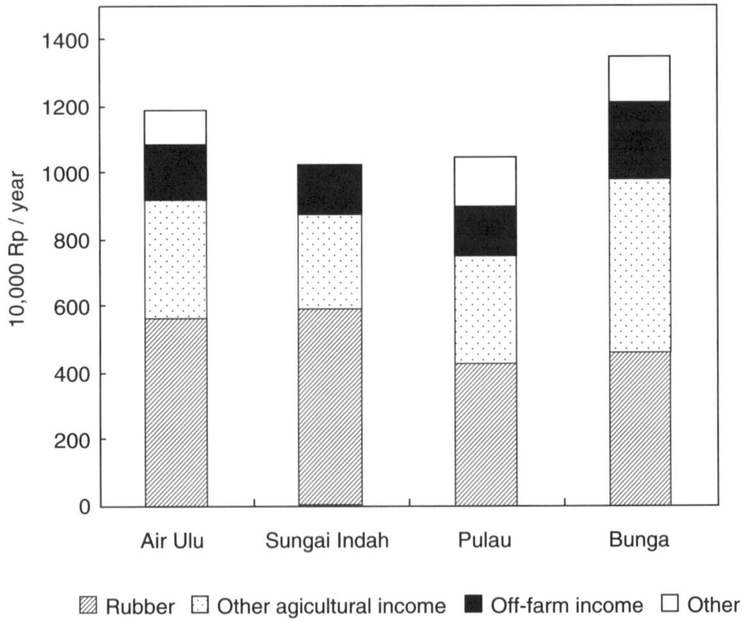

Fig. 18.3. Average household income of the study villages.

The large rubber holders in all the villages tended to hire Javanese rather than villagers as sharecroppers for several reasons. First, the Javanese were hard-working and their productivity was higher than that of the villagers. Second, they would work for the same landowner longer than the villagers. Finally, they caused less trouble for the landowners, especially regarding debts. Sometimes a villager-sharecropper who owed a landowner a lot of money left without paying and became a sharecropper of another landowner. The Javanese did not create such trouble. The Javanese themselves, whose only source of income was rubber-sharecropping, also preferred to work in the rubber fields of the large rubber holders, because they could not produce a large amount of latex in the smallholders' rubber fields. Thus, Javanese sharecropped in largeholders' rubber fields, while the villagers sharecropped in those of smallholders. In this manner, rubber production in the study villages was supported by abundant labour originating from both inside and outside the villages.

Forest conversion

The extent of forest conversion to rubber in the four villages is shown in Table 18.2. Data for the extent of the remaining forest or the year the forest disappeared were obtained by the community surveys. The area of forest converted to rubber for each village was estimated from the village population and the area of rubber fields held by the 40 study households and by all of the rubber holders who possessed more than 100 ha. Air Ulu and Sungai Indah, where the population density was low and rubber transportation via mostly paved road started in the 1990s, still had plenty of forest. The area of converted forest was relatively small, i.e. 1000–1200 ha. On the other hand, in Pulau and Bunga, where the population density was high and rubber transportation via mostly paved road began in the 1970s, 2400–2900 ha of forest were converted to rubber, resulting in the disappearance of all forest by 1991 and 1985, respectively.

This clearly demonstrates that more forest was converted in the villages where the population density was higher and roads were constructed earlier. This conclusion is consistent with regression results examining the significance of various factors on forest conversion to rubber in the same four villages, which suggested that road construction and population density were significant determinants (Miyamoto, 2006).

Conclusions

In Indonesia, agricultural expansion has been the main factor behind forest loss, especially that of cash tree crops, such as rubber and oil palm. This study examined why forest conversion to rubber has accelerated, based on household surveys conducted in four rural rubber villages in Sumatra. It has demonstrated that road construction studied in the rural areas accelerated the conversion of forest to rubber by reducing both the cost and time of rubber transportation and raising land rent. This notwithstanding, roads are necessary to improve the standard of living of rural people. Their infrastructure must by all means continue to improve. In order to reduce forest loss it is necessary to focus on excessive agricultural population. The population density in the study villages was not particularly high, but because it consisted mainly

Table 18.2. Population, road construction and forest conversion to rubber in the study villages.

Study village/variable	Air Ulu	Sungai Indah	Pulau	Bunga
Population density (persons/km^2)	8	13	50	24
Year rubber transportation to Jambi City changed from river to mostly paved road	1997	1992	1978	1974
Extent of forest	Large	Large	Lost in 1991	Lost in 1985
Area of forest converted to rubber (ha)	1,033	1,229	2,973	2,411
Forest conversion to rubber	Low	Middle	High	High

of low-income agricultural households it was enough to increase forest conversion. This was exacerbated by the abundant supply of labour from Java, a densely populated area with a huge number of poor people among which are many landless farmers. Since people migrate in order to acquire jobs and/or land, the mobile agricultural population over a large area must be duly considered to both explain forest conversion and develop a strategy for its reduction.

References

Angelsen, A. (1995) Shifting cultivation and 'deforestation': a study from Indonesia. *World Development* 23(10), 1713–1729.

Angelsen, A. and Kaimowitz, D. (1999) Rethinking the causes of deforestation: lessons from economic models. *The World Bank Research Observer* 14(1), 73–98.

Angelsen, A. and Kaimowitz, D. (2001) *Agricultural Technologies and Tropical Deforestation*. CAB International, Wallingford, UK.

Barbier, E.B. and Burgess, J.C. (2001) Tropical deforestation, tenure insecurity, and unsustainability. *Forest Science* 47(4), 497–509.

BPS (2003) *Statistical Year Book of Indonesia*. Badan Pusat Statistik, Jakarta, Indonesia.

Chomitz, K.M. and Griffiths, C. (1996) *Deforestation, Shifting Cultivation, and Tree Crops in Indonesia: Nationwide Patterns of Smallholder Agriculture at the Forest Frontier, Research Project on Social and Environmental Consequences of Growth-Oriented Policies*. Working Paper Vol. 4. World Bank, Washington, DC.

Duraiappah, A.K. (1998) Poverty and environmental degradation: a review and analysis of the nexus. *World Development* 26(12), 2169–2179.

Gouyon, A., De Foresta H. and Levang, P. (1993) Does jungle rubber deserve its name? An analysis of rubber agroforestry systems in Southeast Sumatra. *Agroforestry Systems* 22, 181–206.

Marsden, W. and Bastin, J. (1986) *The History of Sumatra*. Oxford University Press, Singapore.

Miyamoto, M. (2006) Forest conversion to rubber around Sumatran villages in Indonesia: comparing the impacts of road construction, Transmigration projects and population. *Forest Policy and Economics* 9(1), 1–12.

Serneels, S. and Lambin, E.F. (2001) Proximate causes of land-use change in Narok District, Kenya: a spatial statistical model. *Agricultural Ecosystem Environment* 85(1–3), 65–81.

Suyanto, S. and Otsuka, K. (2001) From deforestation to development of agroforest in customary land tenure areas of Sumatra. *Asian Economic Journal* 15, 1.

von Thunen, J.H. (1875) *Der isolierte Staat in Beziehung auf Landwirtschaft und Nationaloekonomie*. Wiegandt, Hempel and Parey, Berlin.

19 Community Forest Management in the Terai Region of Nepal: Contribution to the Local and National Economy

Maheshwar Dhakal and Misa Masuda

Introduction

Nepal's overall economy is based on subsistence farming, mainly supported by the agricultural production of the Terai region. Human resource development indicators show that the country has a poor economic situation and that more than 38% of the population are still living below the absolute poverty line (NPC, 2002; UNDP, 2004). The current political crisis and the poor peace and security situation are making people's lives more vulnerable and critical. They are also having negative impacts on the conservation and management of forest resources (FAO, 2005). Government records show that forest resources cover 29% of the land (DFRS, 1999). These play an important social, economic and environmental role and are crucial for the livelihoods of poor and forest-dependent people (World Bank, 2001; Gurung, 2003).

In the past, the Terai region was covered by dense sal forests (*Shorea robusta*), and human pressure on the forests was low. Today, despite various efforts by the government to keep the people in the mid-hills, poor farmers are migrating to this region to escape from subsistence farming and acute mid-hills poverty and to obtain a more secure livelihood. The population density (330 persons/km^2 in 2001) is constantly increasing (CBS, 2004). The migration has had negative consequences on the conservation and management of forest resources, in terms of deforestation and encroachment by the migrants. As a result, the Terai forests declined at a rate of 0.8% annually between 1991 and 2001 (Table 19.1).

The government of Nepal has been implementing community forestry programmes since the 1970s. In the mid-hills, such programmes have already had a positive impact on people's livelihoods. Government action focused until recently on plantation sites and natural forest areas, where disputes are less frequent and consensus among the people can be achieved more easily. On the other hand, in the southern flatland of the Terai region, community forestry is still in its infancy. This chapter examines some basic relationships between people and forest resources in a case study of the Dhuseri community forest in the Terai region.

Forest Management and Policy

Forest management

The national forests are state-owned, and the government has the sole authority for their management and utilization. Except for a few small areas of tree lands, private forests are almost non-existent. The Department of Forests coordinates the activities of the Regional Forest Directorates. Subordinated units are the District Forest Offices, the Ilaka Forest Offices and Range Posts at the

Table 19.1. Forest and population changes in the Terai region. (From DFRS, 1999; CBS, 2003, 2004; DOF, 2005a; MPE, 1998.)

Region	Total area (ha)	Forest area (ha)		Population (millions)		Population density (per km²)	
		1991	2001	1991	2001	1991	2001
Terai	3,351,294	1,158,545	1,149,494	8.3	11.2	244	330
Change in the parameter		−9,051 ha		+2.9		+86	

field level. As many as 7125 persons are engaged in managing the forests of Nepal (DOF, 2005a,b). There are five types of forest management, namely: community forests, government-managed forests, protection forests, leasehold forests and religious forests (HMG, 1993). Community and government-managed forests are the most important ones in the local and national economy.

In the Terai region, the supply of forest products is served by the District Forest Products Supply Board. The District Forest Office and its subordinate organizations are responsible for felling, logging and transportation work and directly sell forest products to the people, collecting a minimum fee or royalty. In the urban areas, the Timber Corporation of Nepal oversees the supply of forest products. It buys the timber from the government at a minimum royalty rate and sells it with a certain profit. Its activities are primarily focused on the Terai region in order to supply forest products to the main cities, such as Pokhara and Kathmandu. However, the lack of fiscal transparency and accountability and the monopolistic control by the government are major underlying causes of dwindling forest resources in the Terai region (Bhattarai *et al.*, 2002).

Table 19.2 shows that the forests of the Terai region contribute more than three-quarters of the total annual forest revenue. It also shows the amount of revenue collected by the government agency in the region. The revenues mainly come from sales of timber, firewood and medicinal plants. In addition, large quantities of forest products are consumed by rural households on a daily basis and are not included in the national statistics.

Forest policy

Before 1950, population pressure on forest resources was low and there was no need for a specific national policy for managing forest resources. Most interventions were guided by the local feudal system. The *jamindar* (the largest landholder in the village) and the *mukhiya* (government

Table 19.2. The Terai region's share in total forest revenue and government agency revenue collection (1 US$ = 70.10 NRs). (From DOF, 1996; 1997; 1998; 1999; 2000; 2001; 2002; 2003.)

Fiscal year	Forest in Nepal revenue	Contribution of Terai regions in total revenue collection				Terai region (%)
		DFO	DFPSB	TCN	Total	
1996	319,617,317	139,127,097	22,281,974	106,767,494	268,176,565	83.91
1997	242,768,866	82,768,073	21,462,253	82,704,643	186,934,969	77.01
1998	253,743,403	129,588,237	22,346,000	48,341,245	200,275,482	78.92
1999	348,580,795	286,525,460	6,705,072	35,538,724	328,769,256	94.32
2000	323,215,482	238,349,467	391,943	75,926,189	314,667,599	97.35
2001	458,442,177	300,599,076	28,049,342	19,138,005	347,786,423	75.86
2002	554,886,607	335,292,389	24,622,960	79,393,405	439,308,754	79.17
2003	611,437,095	308,718,173	18,704,898	117,944,803	445,367,874	72.84

DFO = District Forest Office; DFPSB = District Forest Products Supply Board; TCN = Timber Corporation of Nepal.

tax-collector in the village) were the key persons in charge of local resource management. The feudal system collapsed in 1950 (Fisher *et al.*, 2000; Gautam, 2006), and forest ownership became an important issue in the mid-1950s.

The Private Forest Nationalization Act of 1957 was misinterpreted and many forest areas were destroyed after its enactment. Despite this negative result, the Forest Act of 1962 and in particular the Forest Act of 1967 gave even more power to government officials. In 1978, almost 20 years after the adoption of the Forest Nationalization Act, a participatory forest management approach was introduced, taking the form of forests protected by the *Panchayat* (the local village council), which later became the community forestry programme (Karki *et al.*, 1994; Fisher *et al.*, 2000; Gurung, 2003; Gautam, 2006). Initially, the purpose of the programme was to conserve forests and to guarantee a sustained supply of forest products to local communities. Empirical studies have found that it was also important for improving people's livelihood (Kanel, 2004). In effect, it had become increasingly clear that government efforts alone could not effectively address forest problems without being supported by local people (Gilmour and Fisher, 1991; Fisher *et al.*, 2000). Today, Nepal is recognized as one of the world's pioneering countries in adopting community forestry practices (World Bank, 2001).

The 25-year master plan for the forestry sector, developed in 1989, promoted the community forestry programme through decentralization of central government functions. The new Forest Act of 1993 and Regulations of 1995 confirmed and institutionalized the concept of community forestry. They put the emphasis on decentralization, devolution and good governance, with a view to alleviating poverty and enhancing opportunities to earn a livelihood (Kanel, 2004). The revised forest policy of 2000 introduced a collaborative forest management concept into the Terai region. Despite the positive aspects of these changes, there are still many shortcomings in the laws, as well as overlapping and contradictory decision-making by different government agencies.

Figure 19.1 shows that the community forestry programme has been constantly expanding since 1998. However, the rate of its expansion has been declining in recent years. In 2005, there was a total of 13,791 community forests established and managed by local communities (DOF, 2005b). The community forestry programme currently covers 27% of forest land and 37% of total households in the country (DOF, 2005b). There is still a large proportion of the population that is not part of the community forestry programme. In the Terai region, there are 1920 community forests, comprising 364,537 households and 286,618 ha of forest land (DOF, 2005a).

Case Study of Dhuseri Community Forest

The Dhuseri community forest was established in 1996 and is under the jurisdiction of Nawalparasi

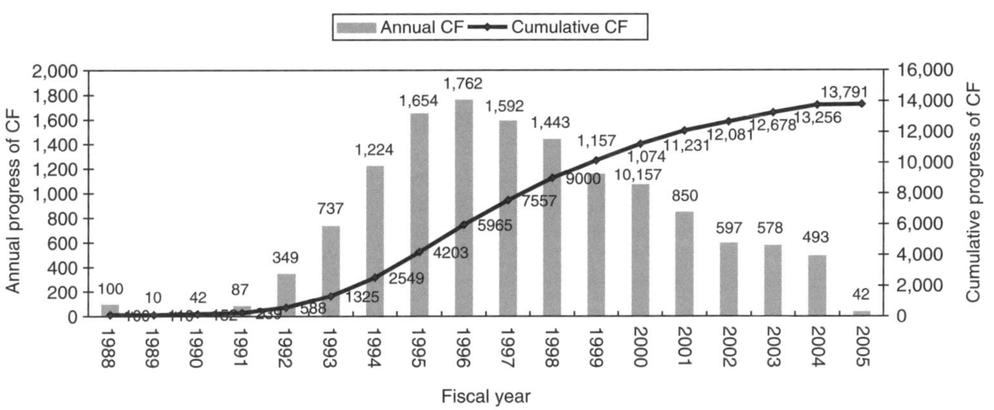

Fig. 19.1. Annual and cumulative number of community forests (CF) in Nepal. (From DOF, 2005b.)

Table 19.3. Social attributes affecting forest product consumption. (From Field Survey, 2005.)

Social attribute	Minimum	Maximum	Mean	SE	SD
Family size	2.00	9.00	5.68	0.24	1.92
Land size (ha)	0.01	2.70	0.53	0.07	0.55
Livestock unit	0.00	7.20	1.92	0.18	1.42
Education (years)	0.00	10.00	5.02	0.32	2.60

1 livestock unit = 1 buffalo; 1 livestock unit = 0.8 cattle; 1 livestock unit = 5 sheep/goats, as adopted by Malla *et al.* (2003).

district. The case study analysed the relevant contributions of forest resources to improving local livelihoods. A stratified random sampling was carried out to collect the primary information. The sample was divided into four strata in two Village Development Committees. In addition to the household survey, interviews were carried out with government officials and senior villagers. Additional secondary information was obtained from government documents, community reports and forest records.

The results of the inquiry show that rural households have different sources of off-farm income and that the rate of forest product consumption depends on family size, land-holding size and the number of livestock units possessed (Table 19.3). During field observations and formal and informal discussions, it was found that educated people have a better awareness of forest resource conservation, but that there is no significant difference in forest resource consumption in comparison with less educated people. Migration was found to be the most important factor affecting the forest management process. Of a total of 66 households sampled, 80% had migrated from the mid-hills. A newly migrated household needs 1650 kg/year of firewood, 1785 kg/year of fodder and grasses and 0.14 m^3/year of timber (Table 19.4).

Forest products

The study focused on mainly three types of product – firewood, timber and fodder/grasses – that are commonly consumed by the people in the community forestry area (Table 19.4). In addition to these dominant forest products, local people use various types of non-timber forest products (NTFPs) and some medicinal plants as well. To collect and distribute forest products, two practices are used. The first is self-collection by individual households; and the second is the collection and selling of forest products by forest user group committees. The second approach makes people's lives easier, but it was observed that economically better-off people benefit from it more than poor people. Additional studies are needed to examine social justice and equity aspects in more detail.

Altogether, the case study revealed that rural development was not initially an important concern in community forestry. It has only become an important issue in the Terai region in recent years. The Dhuseri community forest is progressively increasing its budget allocation for various rural development and forest management activities (Fig. 19.2). However, the majority of the community forest budget continues to

Table 19.4. Annual consumption of forest products. (From Field Survey, 2005.)

Forest product	Minimum	Maximum	Mean	SE	SD
Firewood (kg/year)	678.00	2,345.00	1,650.98	50.11	407.06
Fodder (kg/year)	400.00	3,000.00	1,785.20	69.87	567.64
Timber (m^3/year)	0.02	0.80	0.14	0.01	0.11

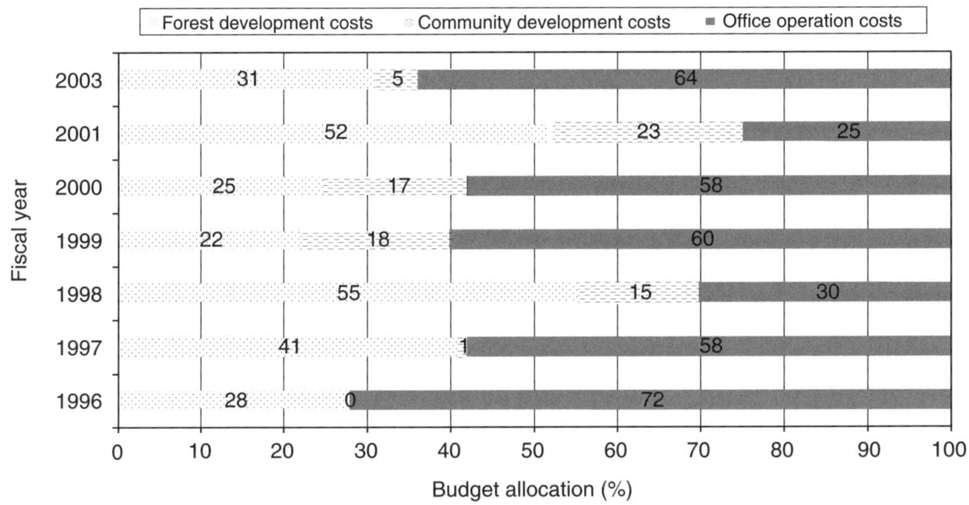

Fig. 19.2. Allocation of community forest income. (Dhuseri Community Forest User Group Records, 2005.)

be spent on office operation and costs of forest management activities, such as harvesting timber and firewood. Capacity building in more effective management practices for forest committee members could help reduce these costs. On the other hand, the activities currently being engaged in create many skilled and non-skilled employment opportunities, enhancing the rural livelihood of poor and unemployed people. Similarly, the building of schools and drinking water facilities directly supports, for instance, school students and poor women in the community.

Conclusions

Forests contribute a wide range of social, ecological and economic benefits to the rural people of Nepal. The contributions to the local economy in terms of direct consumption of forest products and rural development support is crucial for offering more livelihood opportunities to poor and forest-dependent people in the Terai region. At present, it is difficult to predict whether community forest practices can be made sustainable. If the population increase (including migration) continues at the same level as today, current community forestry management alone will probably not be sufficient to address the growing demand for forest products in the long run. The Dhuseri case study reveals that the relationship between people and forest resources is governed by social and natural attributes, and as such that it varies from place to place. For example, in the mid-hills (Adhikari, 2004), the quantitative and qualitative analysis suggests that the consumption of forest products depends on socio-economic variables, such as family size, livestock units, land holding size and off-farm income.

The study shows that forests support both the national and local economy. The government collects forest revenues from the sale of forest products, which have a direct link to the national economy. It confirms that local people are collecting various types of forest products on a daily basis and indicates that better-off households are benefiting more than poor households. The current sale and distribution system of forest products by forest committees also supports the better-off households. On the whole, the study concludes that the establishment of community forests contributes to improving forest management, community development and more productive land use. It stresses the importance of involving all local user groups in order to strengthen participatory forestry management practices and to ascertain new management options that are acceptable to all stakeholders.

References

Adhikari, B. (2004) Household characteristics and forest dependency: evidence from common property forest management in Nepal. *Ecological Economics* 48(2), 245–257.

Bhattarai, K., Conway, D. and Shrestha, N.R. (2002) The vacillating of forestry policy in Nepal: historically manipulated, internally mismanaged. *International Development and Planning Review* 24(3), 315–338.

CBS (2003) *Population Census 2001: National Report*. Central Bureau of Statistics (CBS), Kathmandu, Nepal.

CBS (2004) *Statistical Pocket Book, Nepal 2004*. Central Bureau of Statistics (CBS), Kathmandu, Nepal.

DFRS (1999) *Forest Resources of Nepal (1987–1998)*. Department of Forest Research and Survey (DFRS), Kathmandu, Nepal.

Dhuseri Community Forest User Group Records (2005) *Dhuseri Community Forest User Group*. Nawalparasi, Nepal.

DOF (1996) *Annual Report, 1995/1996*. Department of Forests (DOF), Kathmandu, Nepal.

DOF (1997) *Annual Report, 1996/1997*. Department of Forests (DOF), Kathmandu, Nepal.

DOF (1998) *Annual Report, 1997/1998*. Department of Forests (DOF), Kathmandu, Nepal.

DOF (1999) *Annual Report, 1998/1999*. Department of Forests (DOF), Kathmandu, Nepal.

DOF (2000) *Annual Report, 1999/2000*. Department of Forests (DOF), Kathmandu, Nepal.

DOF (2001) *Annual Report, 2000/2001*. Department of Forests (DOF), Kathmandu, Nepal.

DOF (2002) *Annual Report, 2001/2002*. Department of Forests (DOF), Kathmandu, Nepal.

DOF (2003) *Annual Report, 2002/2003*. Department of Forests (DOF), Kathmandu, Nepal.

DOF (2005a) *Forest Cover Change Analysis of Terai Districts: 1990/1991–2000/2001*. Department of Forests (DOF), Kathmandu, Nepal.

DOF (2005b) *Records of Community Forestry Division*. Department of Forests (DOF), Kathmandu, Nepal.

FAO (Food and Agriculture Organization of the United Nations) (2005) *The State of the World's Forests, 2005*. FAO, Rome, Italy. Available at: http://www.fao.org/documents/show_cdr.asp?url_file=/docrep/007/y5574e/y5574e00.htm

Fisher, R.J., Durst P.B., Enters, T. and Victor M. (2000) *Decentralization and Devolution of Forest Management in Asia and the Pacific: Report of an International Seminar*. RECOFTC Report No. 18, Regional Community Forestry Training Center (RECOFTC) for Asia and the Pacific, Bangkok, Thailand. Available at: http://www.recoftc.org/site/index.php?id=234

Gautam, K.H. (2006) Forestry, politicians and power – perspectives from Nepal's forest policy. *Forest Policy and Economics* 8(2), 175–182.

Gilmour, D.A. and Fisher, R.J. (1991) *Villagers, Forests and Foresters: the Philosophy, Process and Practice of Community Forestry in Nepal*. Sahayogi Press, Kathmandu, Nepal.

Gurung, S. (2003) Impacts of policies and legislation on Nepal's forests and its future. *Journal of Forest Economics* 49(2), 1–10.

HMG (1993) *Forests Act 1993*. His Majesty's Government of Nepal (HMG), Kathmandu, Nepal.

Kanel, K.R. (2004) *Twenty-Five Years of Community Forestry: Contribution to Millennium Development Goals. Proceedings of the 4th National Workshop on Community Forestry*. Community Forestry Division, Department of Forests, Kathmandu, Nepal.

Karki, M.B., Karki, J.B.S. and Karki, N. (1994) *Sustainable Management of Common Forest Resources: an Evaluation of Selected Forest User Groups in Western Nepal*. International Center for Integrated Mountain Development, Kathmandu, Nepal.

Malla, Y.B., Neupane H.R. and Branney, P.J. (2003) Why are not poor people benefiting more from community forestry? *Journal of Forests and Livelihood* 3(1), 78–92.

MPE (1998) *State of the Environment, Nepal*. Ministry of Population and Environment (MPE), Kathmandu, Nepal.

NPC (2002) *Tenth Five-Year Plan, 2002–2007*. National Planning Commission (NPC), Kathmandu, Nepal.

UNDP (2004) *Human Development Report, 2004*. United Nations Development Programme, New York.

World Bank (2001) *Community Forestry in Nepal*. Technical paper no. 217. World Bank, Washington, DC.

20 Cross-sector Linkages in Northern Thailand Mountain Development

Mingsarn Kaosa-ard and Jeff Rutherford

Introduction

This chapter[1] addresses the institutional web of pressures and influences that are at work in the continuing transformation of patterns of livelihood and land use in the rugged and multicultural terrain of the uplands in northern Thailand. Forestry, agriculture, national security and tourism have combined in northern Thailand, often under the rubric of highland development, to facilitate the integration of remote and semisubsistence-based mountain villages into a rapidly changing regional, national and global economic milieu. The chapter argues – looking at both formal and informal cross-sectoral linkages – that state agencies' policy rigidity neglects potential avenues toward sustainable development in the north.

The chapter first provides an overview of the institutional setting and key sectoral issues, focusing on the pressures and subsequent changes in upland livelihoods that have so thoroughly transformed the socio-economic landscape of northern Thailand. It then discusses the response of various stakeholders, including an increasingly assertive civil society. Finally, recommendations are made to further improve the overall socio-economic situation of the northern communities.

[1]Based on Mingsarn and Rutherford (2002).

Institutional Setting

The highlands of northern Thailand are populated by a number of ethnic groups known officially and unofficially as hill tribes. In 1998, the Tribal Research Institute reported that the population of the nine officially recognized hill tribes was 752,728 people (Asian Development Bank, 2000, p. 17). About 92% of the highland minorities are found in the north, where they make up 5.8% of the regional population (ICRAF, 2001, p. 10).

Ethnically, almost half of the highland minorities are Karen (46.8% or 352,295 people). The second largest group is the Hmong, with 15.8% (118,779) of the ethnic highland population, followed by the Lahu with 11.2% (84,414). The smallest group is the Khamu, with 1.4% or 10,476 people. The highland population is not limited to minority groups, however, with lowland Thais and other Thai ethnic groups, such as the Shan and Lue, swelling the total highland population to nearly 940,000 people.

For the people of northern Thailand, regardless of ethnicity or the altitude zone they traditionally occupy, one of the primary resources has been the forested mountains and foothills. Both Thai and ethnic minority highlanders have practiced non-irrigated farming for centuries in the upland slopes using methods variously referred to as slash and burn, bush or forest fallow, swidden fields or shifting cultivation (Kunstadter *et al.*, 1978, p. 3).[2] Prior to the 1960s, Thai forest policy was concerned primarily with managing timber extraction and governed by the Forest Act of 1941. Although a number of conservation-oriented laws were passed in the 1990s, they are highly centralized in nature and some of them follow western models. For example, the National Park Act is only concerned with nature and neglects the fact that the livelihoods of many rural communities depend on non-timber forest products (NTFPs).

The nature of natural resource management in Thailand is fragmented. By this it is meant that the bureaucracy is essentially vertically oriented within each sector, with few horizontal linkages. For instance, there are now three departments that look after forests: the Royal Forest Department; the National Parks, Wildlife and Plant Conservation Department and the Marine and Coastal Resources Department. The highly specialized and sectoral nature of management has rendered an integrated resource management approach, which is much needed for ecosystem management, all the more difficult.

Issues

Deforestation

In 1960, more than half of the kingdom was cloaked in forest. A generation later, when the government banned logging in 1989 and cancelled logging concessions in natural forests, half of the forests were gone. Yet, the logging ban did not end the haemorrhage of natural capital. From 1990 to 1995, Thailand lost 12% of its remaining forest cover (FAO, 1999).

Three broad and interrelated types of deforestation occurred in northern Thailand. First, conversion of forests by the lowland population in two successive phases, initially into farmland and then for urban–commercial purposes; secondly, logging, which ceased legally in 1989 but continues to persist illegally and thirdly, farmers in the forest or the land use practices of highland minorities (ICRAF, 2001).

Agriculture

The driving interest in highland agricultural development was the push to end opium-growing by highland minorities. The king's directive in the 1970s that no opium fields be destroyed until crop replacement efforts had been attempted was crucial for charting a course of development that could otherwise have been far more violent (Renard, 2001). Today, while some highland production activities, from cabbages to barley, ginger and some fruit crops, are conducted through private channels, a range of Royal Project centres specializing in fruits, vegetables, flowers or ornamentals have come together under the umbrella of the Royal Project Foundation. Importantly, the only state agency with the expertise and technical means to promote temperate fruits and vegetables is the Royal Project. The expertise of the Agricultural Extension Office and other agencies is still limited to tropical agriculture. Today, most of the foreign opium replacement projects have been phased out or absorbed by Thai government agencies. Issues of importance are whether sedentary agriculture in the highlands will prove sustainable amid competition with imported temperate fruits and vegetables, following the conclusion of the free trade agreement between Thailand and China.

Drugs

In the last decade or more, Thailand has been transformed from a major opium-producing country to a net importer of opium. The 'Golden Triangle' of yesterday, with highlanders growing opium for Western consumption via a complex distribution system comprising horse caravans, no longer exists. In the northern highlands, where the poppy was grown for generations, nowadays heroin and amphetamines from Myanmar curse communities with a myriad of problems – addiction, acquired immunodeficiency syndrome (AIDS), crime and violence.

[2]Swidden cultivation is a practice in which vegetation felled in patches of forest during the dry season is burned before the onset of the rainy season to open the site and release nutrients. The cleared fields are cultivated and harvested for one or more years, and then left to lie fallow for varying periods to allow secondary forest to regrow (Fox, 2000).

Tourism

The tourism industry is the top earner of foreign currency in Thailand. Its growth has been dramatic in recent decades. The industry affects the mountains of the north and their people in four main ways. First, many tourists who visit the north engage in 'hill-tribe trekking', which involves visiting and often staying overnight in highland villages after walking through the forest or riding elephants to reach them (which today can be done much more conveniently, though less dramatically, by road). Secondly, selling handicrafts to tourists is an important source of income for poorer families. This is done either in the village, at markets along major mountain roads, or in tourist areas of cities like Chiang Mai and Chiang Rai. Thirdly, in the 1990s developers scrambled for land to build upscale tourist resorts in the mountains to serve both the boom in foreign tourism and the demand of urban Thais for an escape from the increasingly congested cities. A fourth area of importance is tourism in national parks. Trekking, along with roads, electrification and communications technology, bring the outside world to the village. The resulting commodification of customs – villagers will sometimes put on shows and dances, for a fee – and hunger for material objects and the cash to purchase them, is decried by many (Michaud, 1997), but trekking should be seen as just one of many forces of change in the uplands. In 2001, the Ministry of Natural Resources and Environment piloted a programme of people's participation in natural resources management in six national parks. Yet several studies have shown that community-based tourism in Thailand has generated much less cash than the initial expectations and worsens the income distribution pattern (Mingsarn *et al.*, 2004).

Tenure insecurity

The combined impact of forestry and highland development projects has wrought dramatic changes. It is interesting to note that while state enclosure of forests has limited the amount of land available to upland farmers, the development of roads and integration with lowland markets has given the remaining land an increased economic value. The impact is most obvious in the case of pioneer swiddeners, who no longer have the option of moving settlements to new forest areas after the soil fertility in the old settlements has been depleted. The increase of state reserved forests has also limited the options of forest-fallow swiddeners, such as the Karen and Lua. Kanok and Benjavan (1994) note that one reason many Karen and Lua villages have existed for hundreds of years and maintained small populations was because they historically could respond to population pressures or community conflicts by fragmenting and establishing new settlements. This option no longer exists.

The last two decades – most dramatically since 1990 – have seen an expansion of strict regulation of land previously only nominally enclosed by the state. This has the greatest impact on the ethnic minorities largely populating the upper tributary watersheds. It is estimated that in the upper watersheds and the conservation areas, there are around 8–15 million people in about 150,000 villages with predominantly insecure settlement and use rights for the land they are living on (Lohmann, 1993; Vandergreest, 1993; Neef, 2000). Such tenure insecurity is acute for the majority of upland villages of ethnic minorities. It is estimated that 68% of village clusters (a total of 2490) are not officially recognized by the government. Moreover, 85% of these village clusters are located in the forest reserve areas, blocking some public and private development efforts (Asian Development Bank, 2000, p. 20).

Another important factor is that Thailand does not recognize communal ownership or control of land (Benchaphun, 1994; Anan, 2000; Tongroj, 2002). The implications of this for government land policy are significant. No matter how extensive forms of communal land management may be, they are largely ignored by policy-makers, rarely studied by academics and generally overlooked by rural development practitioners. Nearly all tend to accept as given the government's classification of land as either private or public, with no alternatives existing or possible. Tongroj's 'conservative estimate' is that at least 3 million ha of Thailand's area are under various types of common or communal property regimes. These areas are often not under the exclusive control of one village, but are shared among several neighbouring communities (Tongroj, 2002). This situation of insecure land tenure is perhaps most extreme in the north, which possesses a large share of Thailand's forest resources. The residences and farmland of these people are technically illegal under state law.

Citizenship

In the 1950s, when modern highland policies were first being formulated, most people in remote villages failed to register. The integration of the hills with the valleys was still a few years off and the benefits of citizenship – and competing claims on traditional farmland – were as yet unclear issues (Renard, 2001). Eventually in 2000, the Department of Local Administration of the Ministry of Interior revised its methodology for granting citizenship, giving considerable authority to the district officer. Since then, hundreds of thousands of hill people have become eligible for citizenship, though the process is still ongoing. On 8 January 2002, Prime Minister Thaksin approved the granting of Thai citizenship for 200,000 so-called hill-tribe people, including former Chinese nationalist soldiers. Remnants of the Kuomingtang (KMT) army were instrumental in the Thai government's anti-insurgency campaigns in the 1970s and 1980s, as well as acting as an informal border force along the frontier with Myanmar (Wassana, 2002). The latest estimate of the National Security Council is that of the 800,000 upland minorities in northern Thailand, nearly 500,000 have received citizenship while the status of the remainder must still be established (Thawin, P., Bangkok, 2003, personal communication).

Urban migration

Several studies have indicated urban migration as an adjustment to state forest enclosure. In one case study of a village in a national park in northern Thailand, one out of every three young men and one out of every five young women migrated to town to look for jobs following the tightening of the National Park Act (Anan and Mingsarn, 1995). A more recent study in Chiang Mai and Chiang Rai revealed that between 1996 and 2001, ethnic out-migration rose rapidly from 2500 to almost 5000 (Kwanchewan *et al.*, 2002). The study concluded that while some migrants had succeeded, mainly as traders and also as non-governmental organization (NGO) workers or government officials, life was grim for the majority.

Land use diversification

The expansion and commercialization of agriculture has followed both from opium replacement efforts in the highlands and from expansion of lowland commercial agriculture up hill slopes from the valley bottoms. A number of fundamental changes have occurred (Kunstadter and Tippawan, 2001, pp. 49, 55). First, there was a substantial drop in opium production, with government pressure being given as the overwhelming reason. The farming of traditional crops dropped, with several communities explaining that this was because no seeds were left – an important loss of genetic material of domestic cultivars. Secondly, shifting cultivation is still practised. In cases in which an 8- to 10-year fallow period is allowed, the absence of purchased inputs and reasonable rice yields produces a profitability level that is quite competitive among upland rice practices, with relatively few negative environmental externalities. In other cases, where fallow periods dropped to 3 or 4 years, they required inputs of fertilizers and pesticides and became vulnerable to soil erosion. Thirdly, the new cash crops can be categorized into four major groups in order of decreasing profitability; these are vegetables, field crops, paddy rice and upland rice. A broad range of production methods and household strategies has been identified, with considerable variation in profitability and environmental impact. Fourthly, wage labour has become common. Finally, the diet of the communities changed as well, with villagers eating many introduced cash crops, although these might have been in part crops that could not be sold.

Land speculation, too, has put pressure on the uplands. The concept of land as a basic production factor has given way to the idea of land as a financial asset, with skyrocketing prices for land marking the advent of speculation and its impacts on the rural sector. The price of land rose spectacularly near areas of urban–industrial zone expansion, road corridors or mineral deposits, as well as scenic piedmont areas considered prime zones for resorts and golf courses (Thomas, 1996). Land speculation has problematized land reform thinking and efforts, as it has become increasingly difficult to distinguish between legitimate claimants to long-degraded forest land with arable potential and stand-ins for powerful interests seeking to expand their landholdings. 'Legitimate' targets for land reform are swayed by younger generations urging parents to consider land as an asset to be liquidated when the price is right. Such factors mean that previous beliefs that ownership or usufruct rights would lead to productive investments by smallholders may have been shaken.

Negative effects of chemicals and other costs

'The attempts to suppress opium cultivation, decrease swidden agriculture and restrict access to forest land for farming have been largely successful in the northern highlands, but intensive permanent field cultivation in the highlands has had environmental, ecological, economic and human costs' (Kunstadter and Tippawan, 2001). These costs included loss of cultivars; introduction of weeds and plant pests associated with exotic crops; soil degradation and erosion associated with permanent field cultivation; widespread environmental pollution with pesticides; detrimental health effects of pesticides and large-scale migration of relatively uneducated and unskilled highlanders to urban centres, where exploitation is common and sometimes horrific.

While there have been few studies on the environmental and health effects of chemical agriculture, findings by Kunstadter *et al.* (2001) indicate that the health problems are extensive. Participants in this study named 120 different varieties of chemicals they used. In this study 60% of the rural Hmong and 70% of the urban Hmong participants mentioned health problems in response to an open-ended question about negative consequences of use of pesticides. Even higher proportions (74% of rural and 77% of urban Hmong) named specific health problems when asked directly about health consequences for themselves, and most also thought there were health problems for their children associated with pesticides (Kunstadter *et al.*, 2001, p. 9).

The high prices of inputs have resulted in widespread indebtedness and increasing landlessness, as farmers hedge future crop output for pre-planting loans for fertilizer and pesticides (Thomas, D., Chiang Mai, 2003, personal communication). In one village in Mae Chaem valley studied by Sanitsuda, only four out of 200 families were free of debt (Sanitsuda, 1994). Many farmers end up as tenants farming land worked by their ancestors – or they end up clearing forest land for a season or two of marginal returns. The use of chemical inputs has also become very controversial, as lowland farmers worry – and agitate – about degraded water quality. That lowland farmers have also rapidly increased the use of chemical pesticides is 'far less visible in the political arena'.

Conflicts

Conflicts over natural resources are mounting in northern Thailand. During the dry season, conflicts can be found in literally every catchment in the region. Conflicts over resources have led to road blockades by lowland farmers, demonstrations by upland minorities at government offices and even mob action – allegedly with official sanction – against property of ethnic minorities.

These conflicts can be grouped into two broad categories: conflicts over forest between people and the state, and conflicts between people, generally over water usage. Conflicts over forests involve state policies and actions limiting local access and increasingly assertive villagers demanding rights to land and natural resources. Near the border, these conflicts often involve security forces and have sometimes resulted in eviction and even deportation of highland villagers (Eudney, 1989; Kammerer, 1989; Panadda, B., Chiang Mai, 2003, personal communication).

The second form of conflict is more closely related to the agrarian transformation of the region, especially the rapid increase in commercial agriculture. Lowland Thais and their urban NGO supporters hold demonstrations and demand the relocation of upland water users, who they say are destroying the forest in their quest for more land for cash cropping. Although the expansion of commercial agriculture in the highlands and midlands, especially in the dry season, is clearly a quantitative change from the recent past, rapid expansion in the lowlands is also evident. For instance, Mingsarn *et al.* (2001) found that in one 50,000-rai watershed in Nan Province, Hmong farmers planted 15,000 rai of lychee in the midlands in the last decade (the rai, the traditional unit of land area in Thailand, is equivalent to 0.16 ha). Meanwhile, the lowland area planted with longan trees in Chomthong District increased from 5000 rai in 1975 to 31,500 rai in 1997 (Watershed, 1998). Expansion of lychees and longans (and oranges in Fang District of Chiang Mai) is a general phenomenon in the north. Both types of trees – grown in the uplands and the lowlands, respectively – demand a great deal of water in the very season in which water is most scarce.

These factors are all leading toward a crisis in water demand, though official attention and lowland propaganda focus exclusively on water supply, with upland minorities targeted as the villains for clearing forests. Despite some scholarly attention to what Andrew Walker calls 'ecological orthodoxy' – such as the idea that forests make rain – sustained research and conflict resolution measures are needed to study the relationships between forests, water and the rapid expansion of dry season agriculture (Alford, 1992; Enters, 1995; McKinnon, 1997; Mingsarn, 2000).

Response

Social mobilization

Although the state is, without question, the dominant actor when it comes to control over resources and livelihoods in the mountainous north, in recent years it has been increasingly balanced by a vibrant civil society. Both middle-class and elite-based NGOs and grassroot people's organizations have joined the fray and are increasingly challenging the role of the state and powerful economic actors in the control and use of key resources, such as land, forests and water.

Civil society emerged over the course of the last several decades in distinct periods. First, there was a period of institutional formation in the late 1960s and 1970s, when the leading figures of the nascent Thai NGO movement began to form organizations and develop their ideology.[3] The late 1960s was a time of 'growing disillusionment with government economic growth policy, While economic growth had boomed, inequalities were perceived to increase and rural people experienced increasing dislocation as differences between urban and rural life increased' (Prudhisan and Maneerat, 1997, p. 197). This formative period of the Thai NGO movement was initially part and parcel of the nascent democracy movement that resulted in the 1973 uprising and ousting of the military government.

Secondly, the mid-1970s was a period of increasing ideological differentiation, confrontational tactics and violent reaction. Politically sophisticated students began to guide the democracy movement leftward and some elements of the movement began to link up with the Communist Party of Thailand (Pasuk and Baker, 1995, p. 304). The state reaction was ultimately violent and many students fled to the hills to join the communists. It was an important time for the NGO movement, in the sense that many radical students criticized the movement as reformist (Prudhisan and Maneerat, 1997, p. 198). Thirdly, the 1980s saw a gradual opening of political space, first by business and later by civil society organizations. Fourthly, the NGO movement increasingly began to adopt environmental themes in the late 1980s and early 1990s as resource depletion and urban pollution helped raise consciousness. This corresponded with a growing shift in understanding within the global South: where previously the nexus of poverty and population growth were seen to be the prime causal agents in environmental degradation, the relationship was increasingly seen to be the opposite. That is, environmental degradation – seen as a result of economic development – was causing poverty and rural dislocation.

[3]Some leading figures in the early years were Dr. Puey Ungphakorn, a respected economic advisor who set up two rural development organizations; Niphon Thianwihan, a Catholic priest active in the north of Thailand, especially with the Karen (Chatthip, 1991, p. 118); Bamrung Bunpanya, one of the first Thai development workers to begin developing the community culture perspective (ibid., p. 120) and Prawet Wasi, who has been called 'the moral torchlight of Thailand's prodemocracy movement' (Eng, 1997, p. 187).

In Thailand, some of the key environmental themes were dam-building, which displaced rural villagers and produced deleterious effects on agriculture; eucalyptus plantations, which required large-scale evictions of grassroots actors in the forests; and a 1988 disaster in which torrential rains in the logged-out hills of a southern province caused mudslides and a 'sea of logs' to wash over several villages, killing hundreds (Prudhisan and Maneerat, 1997, pp. 202–203). In response to these stimuli, development NGOs began to see 'outside' pressures (from state and powerful business interests) on natural resources as a formidable threat to rural society and adopted a three-pronged strategy that continued grassroots development work but included alliances with academics and policy advocacy (Prudhisan and Maneerat, 1997, p. 203). Some of the key players in the north are Terra and the Northern Development Foundation. Also of real significance are advocacy groups for highland minorities. Notable among these is the Inter Mountain People Education and Culture Association. This group links people of diverse ethnic groups in a coordinated effort to address issues of land rights and citizenship through dialogue with the state.

The final and current phase has seen an increasingly prominent role for grassroots actors, which often stands them in opposition to the state, provincial business interests and even the middle class. The mid-1990s have seen many mass rallies by people's organizations demanding that their grievances be addressed. The most well-known have been actions by the Assembly of the Poor, which has on different occasions led marches to Bangkok (Prudhisan and Maneerat, 1997, p. 207), but in the north, groups such as the Northern Farmers Network and intra-watershed networks like the Mae Wang Watershed Network have been most active in organizing minority peoples, especially in demands for the return of resource management rights to local communities (Pratuang, 1996, p. 146).

The development of large-scale rural action of this sort – for instance, mass protests at the provincial hall in Chiang Mai over issues, such as community forestry and citizenship – has driven the NGO movement into uncharted waters and has strained middle-class support for grass-roots actors. On the other hand, the Thaksin government has shown itself to be more even-handed than the preceding regime, which occasionally resorted to violence to end peaceful protests (Pinkaew, L., Chiang Mai, 2003, personal communication).

Public participation

In terms of mountain development, an integrated community approach was promoted in the 1990s as a means to stop deforestation and to improve the standard of living of the people who depend on the forests. Within the Royal Forest Department, at least at field level there is a growing understanding that in order to succeed in watershed management, the focus must begin with the community's ability and desire to participate because only then will plans for land use be realized (Mingsarn, 2000). Although community resource management has been an important focus for development work of NGOs, some promising moves in a similar direction are also the subject of new government initiatives. A notable example is the Sam Mun Highland Development Project, under the auspices of the Royal Forest Department. This watershed management scheme emphasizes participatory approaches to conflict resolution among the communities involved (Hirsch, 1996, p. 27). However, people's participation in the management of forest resources has not been adopted as a policy by high-ranking administrators in the Royal Forest Department.

Public sector reform, which started to focus on public participation in 2004, aimed to create a paradigm shift among public agencies to people-focus development and to establish public or popular participation as a mandate. This move has driven the Royal Forest Department to seriously consider the promotion and transfer of community forests. For the Department of National Parks, initial activities in compliance with the people-participation mandate have been in the areas of community-based tourism, where resource-use conflicts between the state and the park residents are considered to be minimal (Mingsarn *et al.*, 2004).

Community forestry

Thailand has a rich, although mostly unwritten history of local participation in the management of natural resources (Chatchawan and Lohmann, 1991; Uraiwan, 1995; Anan, 2000). The existence of rights-in-use arrangements for many upland Karen communities, for example, reveal an understanding of the importance of forest conservation that largely mirrors modern ecological knowledge. The term generally used to describe such local arrangements for forest conservation and use is 'community forestry'. In Thailand, an initial survey by the Action Research Project on Community Forestry by the Social Research Institute of Chiang Mai University reported approximately 150 cases of locally initiated community-protected forests (Anan, 2000, p. 183).

Since the public service reform, which highlighted the importance of public participation, and the emergence of the populist party Thai Rak Thai, the Royal Forest Department has given serious consideration to community forests as its main responsibility. The Royal Forest Department staff at field level have expressed the need for a Community Forest Bill so that the authority of the community forest committee can be officially recognized. At the time of writing, the Bill is in the Senate and its future is still unclear.

Conclusions

Three broad areas have to be addressed in order to meet the environmental and social challenges of the mountainous north. First, the problem of tenure insecurity has to be addressed. Secondly, the human security of local communities has to be incorporated into national security policies. Finally, multi-actor cooperation in empirical assessments of the environmental and social sustainability of mountain livelihoods and development plans needs to be promoted. The decade-long fight over the Community Forestry Bill is evidence of resistance to institutional flexibility and innovation in state-enclosed areas. What is needed is government flexibility and policies that encourage and promote sustainable systems that balance resource protection and both subsistence and commercial crop production. Integrated rural development programmes that incorporate protected forests, organic farming, multicropping, agroforesty, ecotourism and cottage industries, for instance, are severely constrained by government policies. While many communities are taking great pains to demonstrate their commitment to sustainable agriculture, their efforts go unheeded by the authorities, which deny their right to exist in their present locations. Policies that link rights of tenure with sustainable practices should be encouraged. It is critical that these rights should be issued to communities, not individuals, and that steps are taken to ensure that speculation and land alienation should not be allowed to proliferate.

In addition, human security has to be incorporated into official development and national security thinking and action. Instead of viewing ethnic minorities as threats and problems that have to be solved through law enforcement, the government needs to consider the human damage inflicted on communities by drug-trafficking, environmental degradation, AIDS-human immunodeficiency virus (HIV) and other social problems. Seeking participation and input from community members, rather than treating them like passive actors or obstacles to development, will make it possible to gain a more sophisticated idea of the threats to communities, whether internal or cross-border. As a key part of this effort, social welfare efforts should be modified so that public services are not linked to assimilation of diverse cultures. The Thai Constitution guarantees protection of these freedoms, and it is crucial that state authorities act in accordance with it.

Finally, it is recommended that research into sustainable development – building local capacity for adaptive management of natural resources – be promoted as collaborative enterprises among various actors: villagers, academics, authorities and NGOs. In general, the authors are in agreement with the World Agroforestry Centre (ICRAF) that 'What is urgently needed is a widely acceptable and accessible set of criteria, indicators and measurement tools that are based on appropriate calibrations with science and local knowledge for empirical assessment and monitoring of watershed and related environmental services.' This will only be meaningful if it includes the participation, input and understanding of local actors. Collaborative efforts of this type, linked to land tenure security and concern for human security, offer a way out of trends currently leading toward greater social conflict and environmental degradation.

References

Alford, D. (1992) Streamflow and sediment transport from mountain watersheds of the Chao Phraya basin, northern Thailand: a reconnaissance study. *Mountain Research and Development* 12, 3.

Anan, G. (2000) *Local Control of Land and Forest: Cultural Dimensions of Resource Management in Northern Thailand* (in Thai). Monograph Series, no. 1. Regional Center for Sustainable Development, Chiang Mai, Thailand.

Anan, G. and Mingsarn, K.-S. (1995) *Evolution of Land Access in Forest Area: A Case Study of the Northern Thailand*. Monograph no. 13. Thailand Research Development Institute, Bangkok.

Asian Development Bank (2000) *Health and Education Needs of Ethnic Minorities in the Greater Mekong Subregion*. Thailand Country Report, June 2000. Asian Development Bank, Manila, Phillipines.

Benchaphun, S. (1994) *Property Rights, Institutions and Access to Resources: Impacts on Sustainable Intensification in Northern Thailand*. Agricultural Economics Report, no. 41. Multiple Cropping Center, Chiang Mai University, Chiang Mai, Thailand.

Chatchawan, T. and Lohmann, L. (1991) The Muang Fai irrigation system of Northern Thailand. *The Ecologist* 21, 2.

Chatthip, N. (1991) The community culture school of thought. In: Manas, C. and Turton, A. (eds) *Thai Construction of Knowledge*. School of Oriental and African Studies, University of London, London. pp. 118–141.

Eng, P. (1997) Thai democracy: the people speak. *The Washington Quarterly* 20(4), pp. 169–188.

Enters, T. (1995) The economics of land degradation and resource conservation in Northern Thailand: challenging the assumptions. In: Rigg, J. (ed.) *Counting the Costs: Economic Growth and Environmental Change in Thailand*. Institute of Southeast Asian Studies, Singapore, pp. 90–110.

Eudney, A.A. (1989) 14 April 1986: eviction orders to the Hmong of Huai Yew Yee Village, Huai Kha Khaeng Wildlife Sanctuary, Thailand. In: McKinnon, J. and Vienne, B. (eds) *Hill Tribes Today: Problems in Change*. White Lotus, Bangkok, pp. 249–259.

FAO (Food and Agriculture Organization of the United Nations) (1999) *State of the World's Forests*. FAO, Rome.

Fox, J. (2000) How blaming 'slash and burn' farmers is deforesting mainland Southeast Asia. *Asia Pacific Issues* 47. Available at: http://eastwestcenter.org/stored/pdfs/api047.pdf

Hirsch, P. (ed.) (1996) *Seeing Forests for Trees: Environment and Development in Thailand*. Silkworm Books, Chiang Mai, Thailand.

ICRAF (World Agroforestry Centre) (2001) *Landscape Agroforestry in Northern Thailand: Impacts of Changing Land Use in an Upper Tributary Watershed of Montane Mainland Southeast Asia*. ICRAF, Chiang Mai, Thailand.

Kammerer, C.A. (1989) Territorial imperatives: Akha ethnic identity and Thailand's national integration. In: McKinnon, J. and Vienne, B. (eds) *Hill Tribes Today: Problems in Change*. White Lotus, Bangkok, pp. 259–302.

Kanok, R. and Benjavan, R. (1994) *Shifting Cultivation in Thailand: Its Current Situation and Dynamics in the Context of Highland Development*. IIED Forestry and Land Use Series, no. 4. International Institute for Environment and Development, London.

Kunstadter, P. and Tippawan, P. (2001) Changes in land use and environment in the northern Thai highlands. Paper presented at EUROSEAS Conference, School of Oriental and African Studies, London, 6 September 2001.

Kunstadter, P., Chapman, E.C. and Sabhasri, S. (1978) *Farmers in the Forest: Economic Development and Marginal Agriculture in Northern Thailand*. East-West Center, Honolulu.

Kunstadter, P., Tippawan, P, Bang-on, S, Auchara, S., Antika, T. and Chirasak, K. (2001) Pesticide exposure among Hmong farmers in Thailand. Paper prepared for *International Journal of Occupational Health,* June 2001.

Kwanchewan, B., Panadda, B. and Prasit, L. (2002) *A Study of the Socio-economic Vulnerability of Urban-based Tribal Peoples in Chiang Mai and Chiang Rai, Thailand*. Center for Ethnic Studies and Development and Social Research Institute, Chiang Mai, Thailand.

Lohmann, L. (1993) Thailand: land, power and forest colonization. In: Colchester, M. and Lohman, L. (eds) *The Struggle for Land and the Fate of the Forest*. Zed Books, London, pp. 198–227.

McKinnon, J. (1997) The forests of Thailand: strike up the ban? In: McCaskill, D. and Kampe, K. (eds) *Development or Domestication? Indigenous Peoples of Southeast Asia*. Silkworm Books, Chiang Mai, Thailand, pp. 117–131.

Michaud, J. (1997) A portrait of cultural resistance: the confinement of tourism in a Hmong village in Thailand. In: Picard, M. and Wood, R.E. (eds) *Tourism, Ethnicity, and the State in Asian and Pacific Societies*. University of Hawaii Press, Honolulu, pp. 128–154.

Mingsarn, K. (2000) *Ecosystem Management in Northern Thailand*. Occasional Paper, no. 3. Center for Sustainable Development Studies, Faculty of Economics, Chiang Mai University, Chiang Mai, Thailand.

Mingsarn, K., and Rutherford, T.J. (2002) Cross-sector linkages in mountain development: the case of northern Thailand. Report prepared for the FAO Forestry Department Policy and Planning Division, Policy and Institutions Branch, Rome, Italy.

Mingsarn, K., Acharee, S., Kobkul, R., Sombat, S., Pitsom, M., Pornpen, W., Jiraporn, P., *et al.* (2001) *Water Resources Management for Thailand: A Report Prepared for the Thailand Research Fund*. Thailand Research Fund, Bangkok.

Mingsarn, K., Nukul, K. and Akarapong, U. (2004) *Draft Main Report on Retained Value of the Tourism Industry of Thailand, Prepared for the Thailand Research Fund, August 2004* (in Thai). Thailand Research Fund, Bangkok.

Neef, A. (2000) Paper presented at the International Symposium II on Montane Mainland Southeast Asia (MMSEA) Governance in the Natural and Cultural Landscape, 1–5 July 2000, Chiang Mai, Thailand [CD-ROM].

Pasuk, P. and Baker, C. (1995) *Thailand Economy and Politics*. Oxford University Press, Kuala Lumpur.

Pratuang, N.N.A. (1996) Community forestry and watershed networks in Northern Thailand. In: Hirsch, P. (ed.) *Seeing Forests for Trees: Environment and Environmentalism in Thailand*. Silkworm Books, Chiang Mai, Thailand, pp. 116–146.

Prudhisan, J. and Maneerat, M. (1997) Non-governmental development organizations: empowerment and environment. In: Hewison, K. (ed.) *Political Change in Thailand: Democracy and Participation*. Routledge, London, pp. 195–216.

Renard, R.D. (2001) *Opium Reduction in Thailand, 1970–2000: A Thirty-Year Journey*. United Nations Drug Control Programme, Bangkok.

Sanitsuda, E. (1994) *Seeds of Hope: Local Initiatives in Thailand*. Post Publishing, Bangkok.

Thomas, D. (1996) *Forests for a Dynamic Kingdom: Support for the Emergence of Social Forestry in Thailand*. A Summary Report on the Ford Foundation's Thailand Social Forestry Program, Chiang Mai, Thailand.

Tongroj, O. (2002) Access questions to natural resources and their relevance to poverty and vulnerability. Paper presented at the International Symposium on Sustaining Food Security and Managing Natural Resources in Southeast Asia – Challenges for the 21st Century, Chiang Mai, Thailand, 8–11 January 2002.

Uraiwan, T.-K.-Y. (1995) *Muang Fai Communities Are for People: Institutional Strength and Potentials*. Social Research Institute, Chiang Mai University, Bangkok.

Vandergreest, P. (1993) Constructing Thailand: regulation, everyday resistance and citizenship. *Comparative Studies in Society and History* 35(1), 133–158.

Wassana, N. (2002) Citizenship for 200,000. *Bangkok Post,* January 9.

Watershed, 1998. Forum: Conflict or Resolution? People and Forests in Northern Thailand. *Watershed* 4(1), 10–28.

21 Public Policy Impacts on European Forest Sector Development

Franz Schmithüsen, Volker Sasse and Carsten Thoroe

Introduction

Under the auspices of the United Nations Economic Commission for Europe Timber Committee and the European Forestry Commission of the Food and Agriculture Organization of the United Nations (FAO), a series of studies has been undertaken over the last 50 years on the trends and prospects for the forest and forest industries sector. The most recent contribution to this series is the new European forest sector outlook study (UNECE/FAO, 2005). During its preparation, it was decided to carry out a special policy study providing information on relevant public policy domains that have an influence on the future development of forests and of the forestry sector (Thoroe *et al.*, 2004).

The policy study was initiated by a team of specialists, representing officially nominated national correspondents, major stakeholder groups in the forest sector and outlook specialists. Its objectives were as follows:

1. To identify public policy scenario areas with major impacts on the European forest sector;
2. To identify driving forces, the stakeholders involved and the changes expected to result from those forces;
3. To provide guidance for alternative (to the baseline policy scenario) projections of roundwood supply and demand for forest products.

The work started in 2001 with an intensive survey of publications and an analysis of policy documents by the secretariat, to provide a base for further analysis. Possible forest sector developments and relevant policy domains, as derived from the literature review, were initially structured into 19 scenario areas (Table 21.1), describing policy issues with relevance to the forest sector in Europe. The main criteria for defining policy scenario areas are significant in terms of policy domains, regional applicability and specifics of the role of stakeholders.

During an initial inquiry, experts familiar with forest sector development in Europe were asked to indicate their expectations about the importance of the identified policy scenario areas. On the basis of ranking provided by the experts, 13 scenario areas were marked with a high priority during the inquiry and retained for further analysis. A second round of expert assessment was organized in the form of a 'Delphi inquiry'. This was addressed to a broader group of forest sector stakeholders from various countries and international organizations. They were asked: (i) to evaluate the probability of occurrence of each of the policy scenarios; (ii) to estimate the expected impacts, in terms of variation from a business-as-usual baseline policy scenario, on various forest sector parameters (forest area, production, trade and consumption of forest products); and (iii) to identify the specific driving forces for each of the expected policy scenarios. The analytical outcomes were grouped

Table 21.1. Impact on forests and the forest industry sector by policy scenario area. (Thoroe *et al.*, 2004: p. 13.)

Policy scenario area	Impact 1 (highest priority) to 5 (lowest priority)	Rank	Standard deviation	Level			
				Local	National	Regional	Global
Nature conservation	2.09	1	0.98	x	x	x	
Nature-oriented forest management	2.09	1	1.02		x	x	x
Promotion of social/environmental benefits from forestry	2.14	2	0.76		x	x	
Competitiveness	2.17	3	0.92		x	x	x
Stabilization in Eastern Europe	2.46	4	1.20		x	x	
Changes in land use	2.48	5	1.09	x	x	x	
Energy policy	2.50	6	1.20		x	x	
Global trade liberalization	2.54	7	1.15		x	x	x
EU enlargement	2.64	8	0.74		x	x	
Waste management	2.67	9	0.98		x	x	
Innovations in wood-processing technologies and in wood products	2.79	10	1.08		x	x	
Climate change	2.79	11	1.26			x	x
Consequences of international negotiation processes	2.82	12	1.40		x	x	x
Pollution	2.91	13	1.16		x	x	
Social and demographic developments	2.96	14	1.25	x	x		
Changes in institutions and administration of the forest sector	3.02	15	1.39		x		
Innovations in harvesting and transport	3.15	16	1.09		x		
Trends in renovation and modernization of housing facilities	3.20	17	1.09		x		
Innovations in information technologies	3.26	18	1.30		x		

into five policy scenario packages. The main reason for this grouping was to combine similar scenario areas in terms of issues, concerned stakeholders and policy experts dealing with these issues.

Quantitative steering parameters were elaborated as an input to the econometric modelling of forest product markets and the simulation of developments of forest resources (e.g. gross domestic product growth rates, income, price elasticity, changes in forest cover). The respondents' answers were systematically differentiated among three major sub-regions: European Union (EU)/European Free Trade Association countries, Central and Eastern European countries and countries belonging to the Community of Independent States.

The following sections of the chapter focus on the findings of the research described above. It presents the profile of the public policy scenario packages and areas, the assessment of the probability of their occurrence, their likely impacts on specific forestry parameters and the driving policy and market forces that are likely to influence the future development of the European forest sector.

Five policy scenario packages, each of them with a number of policy scenario areas, were identified and retained for further examination:

- Biodiversity and nature conservation;
- Globalization, innovation and market structures;
- Countries with economies in transition;
- Regional development;
- Energy and environment.

Biodiversity and nature conservation

Conservation of biodiversity has been identified as an important objective in promoting sustainable forest management and an increase in forest area. In effect, the need to conserve biodiversity is one of the principal elements of several international agreements. At the global level, the most important of these are the Convention on Biological Diversity and the Forest Principles, or the non-legally binding authoritative statement for a global consensus on the management, conservation and sustainable development of all types of forest adopted in 1992 at the United Nations Conference on Environment and Development (UNCED). At the pan-European level, the 1993 Helsinki Ministerial Conference on the Protection of Forests in Europe adopted resolution H2 'General Guidelines for the Conservation of the Biodiversity of European Forests'. Subsequently, the Lisbon Ministerial Conference in 1998 endorsed the work programme 'Conservation and Enhancement of Biological and Landscape Diversity in Forest Ecosystems 1997–2000'. With a communication from the Commission to the Council and the European Parliament in 1998, the European Community Biodiversity Strategy was adopted, establishing a general framework for appropriate EU policies and instruments to meet the obligations of the corresponding UNCED Convention.

Due to the interplay between biodiversity preservation and sustainable wood supply, the commercial interests of wood producers and the demands of other forest land users are affected. A number of relevant stakeholder groups with differing interests are involved. Forest owners want to maximize the benefits, financial or otherwise, of their ownership. While private owners manage their forest for profitability, public owners could cover additional expenditure for biodiversity measures from public budgets. Government authorities responsible for policy-making are involved, both at the national and local levels, mainly because of their responsibilities for providing society with non-wood products and services. Non-governmental organizations (NGOs) – both those representing specific forest sector interests, such as forest owners and the wood-processing industries, as well as nature protection NGOs – play an important role in policy formation. Forest laws are being adapted to take account of countries' international commitments, and measures are being introduced to encourage their application in privately, as well as publicly owned forests. Three policy scenario areas, and for each of these a set of the most relevant driving forces, have been identified to assess the impacts on the forest sector over the long term (Box 21.1).

Globalization, innovation and market structures

In an increasingly competitive climate, it has become unavoidable for companies operating at

Box 21.1. Biodiversity and nature conservation.

1.1 More emphasis on nature conservation and promotion of biological diversity of forest ecosystems
Driving forces:
1.1.1 Increasing the area protected for nature conservation, reduction of harvesting in such areas
1.1.2 Building of ecological networks including core areas, corridors, buffer and restoration areas
1.1.3 Diversification of species composition and structure of ecological communities in forests
1.1.4 Intensified fire protection

1.2 More emphasis on nature-oriented forest management
Driving forces:
1.2.1 Eliminating/reducing clear-cutting, extending selection system of harvesting
1.2.2 Planting endemic/indigenous species, combinations of coniferous and broad-leaved species
1.2.3 Increasing rotation lengths
1.2.4 Abandonment/reduction of drainage systems
1.2.5 Reduction of use of chemicals (e.g. biocides) in forests

1.3 Demand for certification of forest management and wood products
Driving forces:
1.3.1 Certification of forest management
1.3.2 Certification of forest products
1.3.3 Certification in wood-processing industry

the international level to take advantage of the possibilities of locating their operations in an optimum way in relation to their markets and sources of inputs in order to reduce costs. In practice, this has often meant transferring production, or part of it, to places with a favourable mix of factors (labour, capital, know-how, energy, raw materials, etc.). The downside has included closure of industries in some traditional producing areas with loss of jobs, and social and environmental abuses, such as the use of child labour and uncontrolled emissions of pollutants. The capacity of a country's population and industrial sector to benefit from globalization is associated with the ability to develop new technology or to adopt it by importing it from elsewhere. This is linked to a country's capacity to attract and utilize effectively the appropriate factors of production, notably capital.

Most of the industrially advanced countries in Western Europe fall into the group of technological innovators, whereas some of the Central and Eastern European countries belong today to the group of technological adopters. The ability or inability to benefit from technological developments appears to be a major factor leading to the widening gap between rich and poor countries. The reversal of this trend requires the application of development strategies to ensure that the latter also have better access to technological innovation. Major stakeholders in the processes of globalization and innovation are the forestry enterprises and those in the wood-processing industries and trading enterprises, as well as their employees and their representatives, the trade unions. Governments – and in particular those departments dealing with company law, industrial development and negotiations on international trade – are also concerned. Scientific and educational institutions have a significant role to play in the development of new technology and its transfer and application. The EU, because of its programmes on innovation, and international organizations and financial institutions involved in development assistance and capital transfer, has a direct interest as well. Two scenario areas and related driving forces were retained in the globalization policy scenario package (Box 21.2).

Countries with economies in transition

The 12 countries of the Community of Independent States and the 15 countries in Central and Eastern Europe are in the process of transition from planned economies to various forms of market economy, following the breakdown of their former political regimes at the beginning of the 1990s. The countries with economies in transition account for 85% and the Russian Federation alone for 77% of the total forest and other wooded land area in Europe (1150 million ha). These countries' share of Europe's wood processing is more modest: in the case of sawnwood production, their share was about 35% in 2000 and less for other product groups.

The pace of transition, including the restitution and privatization of some forests and industries, has varied considerably from country

Box 21.2. Globalization, innovation and market structures.

2.1 Impact of globalization on the competitiveness of the European forest and forest industry sector
  Driving forces:
  2.1.1 Increasing international flows of capital
  2.1.2 International relocation of capacities
  2.1.3 International merging of companies

2.2 Intensified innovations and changes in competitiveness of wood products
  Driving forces:
  2.2.1 Innovations in harvesting techniques and facilities
  2.2.2 Innovations in wood-processing technologies
  2.2.3 Development of new products, e.g. engineered wood
  2.2.4 Progress in transport and logistics
  2.2.5 Innovations in information technologies
  2.2.6 Introduction of new non-wood commodities
  2.2.7 Development of new fields of application

to country, depending on the policies adopted, the potential for change and the need to seek access to external markets, especially in Western Europe. In the Russian Federation, for example, the forests have remained under state ownership, while much of the wood-processing sector has been transferred to private or joint stock companies. In Slovenia 70% of forests are now privately owned. The proportion in Poland, at 17%, is much the same as before the transition began, while in Slovakia it is 44% and will rise further. Privatization and restitution have resulted in a considerable increase in the number of small-sized forest units, many of which are owned by persons without sufficient forestry experience. Some have been tempted to exploit their forests for short-term gain without consideration of the environmental consequences or sustainability.

The principal stakeholders in the process of integration of the former planned economies into the global market are the populations of the countries involved and their governments, institutions, industries and trading companies. Also involved are countries and international organizations providing assistance to the process, in particular the EU and its member countries, who will also decide on its enlargement, the European Bank for Reconstruction and Development and the European Investment Bank. Other stakeholders are companies in partner countries, including those interested in direct foreign investment in the countries with economies in transition, as well as certain NGOs, particularly those concerned with environmental protection. A large part of the forest-product trade in the countries with economies in transition is conducted with companies located in the EU, which is consequently affected by business cycles in these countries. Two policy scenario areas have been defined for this scenario package (Box 21.3).

Regional and rural development

Many governments, as well as the EU, have spent large sums in policies to preserve social structures in rural areas, mainly through the subsidization of agriculture. Partly as a result of this, and partly through market forces, productivity has increased greatly, to the point at which employment in agriculture has fallen to account for only a small percentage of the total national labour forces and, especially in the EU, food production exceeds demand. The need to react to this situation has been recognized for some time, and moves have been initiated to reform the EU's Common Agricultural Policy. The practical difficulties of implementing such a major change in policy have proved immense, however, not least in finding ways to avoid damaging the social fabric in rural areas and to maintain employment and living standards. It has been envisaged that increased forestry activity, including afforestation, could be one of the solutions in certain areas. The fact that the social and environmental functions of the forest are mostly not income-generating (hunting, the commercial harvesting of non-wood forest products and some forms of recreation may be exceptions), coupled with the fact that in most western European countries the major part of the forest area is privately owned, raises the issue of how the provision of those functions is to be financed.

The main stakeholders in this policy scenario package are the owners of land, including forest owners, whether state, municipal, private or other; national and local authorities; forest services in their policy-making role and their counterparts dealing with other sectors of rural and urban development; farmers and other land users; those employed in the countryside, including farm and forestry workers, their labour unions and NGOs concerned with nature protection and landscape management. Members of the

Box 21.3. Countries with economies in transition.

3.1 Strengthening policies to develop the market framework in countries with economies in transition
Driving forces:
3.1.1 Recovery of forest and forest industry sector in countries with economies in transition
3.1.2 Changing ownership of forest land (e.g. privatization and restitution)

3.2 Progress in European Union enlargement
Driving forces:
3.2.1 Accession of the Central and Eastern European countries to the European Union
3.2.2 Accession of other European countries

general public, both town and countryside dwellers, are also stakeholders in their role of 'consumers' of the benefits that forests and landscapes provide. Three policy scenario areas have been considered within this policy scenario package (Box 21.4).

Energy and environment

UNCED in 1992 adopted the Framework Convention on Climate Change, in which signatory countries supported measures to bring the causes of climate change under control, notably by slowing down or arresting the net emissions of greenhouse gases. This could be achieved by greater efficiency in the use and conservation of energy, by increasing the use of alternative fuels (alternative to fossil fuels) and by taking measures to reabsorb the carbon dioxide (CO_2) from the atmosphere. Forestry and the forest industry sector could play a role in all of these. For example, the production of sawnwood and wood-based panels requires considerably less energy than that of competing construction materials, such as steel, concrete, glass and plastics, and the former have good thermal insulation properties and come from a renewable resource. The burning of wood is generally considered to be neutral so far as its impact on the environment is concerned, because the main emission is CO_2, which is taken back into biomass by the process of photosynthesis. Moreover, if established on a sufficiently large scale, plantations act as net carbon sinks. The use of wood for housing and for energy generation has a long tradition in rural areas of Eastern European countries, which could influence the forest sector in these countries.

The principal stakeholders in this policy scenario package are governments, which have the responsibility of setting the environmental standards for the production and use of energy, the producers and users of all types of energy, consumers and consumer groups, the wood-processing industries and users of wood products, such as the construction sector. Environmental NGOs play active roles in raising public awareness of the issues involved and in putting pressure on the policy-makers. The forestry sector is a major stakeholder as a provider of wood fuel and, potentially in the future, in the sequestration of carbon. In this policy scenario package, three policy scenario areas have been identified (Box 21.5).

Impacts on policy and market developments

Respondents were invited to indicate which driving forces were most probably having an impact on policy and market development in the five identified policy scenario packages and to ascertain to what extent regulatory instruments (laws and regulations), economic instruments and market forces would be of importance. The most significant driving forces identified for the scenario package 1, 'biodiversity and nature conservation', are the reduction in harvesting volumes caused by an increase of nature reserves and other protected areas (regulation and/or incentives);

Box 21.4. Regional and rural development.

4.1 Incentives for social/environmental benefits from forestry and wood products use
Driving forces:
4.1.1 Economic incentives for protective and recreational services of forests
4.1.2 Economic incentives for nature-oriented management of forests
4.1.3 Economic incentives for conversion of forests
4.2 Changes in agricultural, rural and regional development policies
Driving forces:
4.2.1 Changes in rates of subsidization in agricultural production and exports
4.2.2 Extension of Common Agricultural Policy elements, i.e. afforestation of agricultural land
4.2.3 Implementation of forestry measures in agriculture, e.g. biomass production
4.24 Promotion of forest and forest industry sector as an integral part of rural development
4.3 Social and demographic developments
Driving forces:
4.3.1 Migration of rural population
4.3.2 International emigration/immigration
4.3.3 Ageing of population

Box 21.5. Energy and environment.

5.1 Promotion of renewable energy sources
Driving forces:
5.1.1 Emphasizing use of wood biomass as a source of energy
5.1.2 Taxing fossil energy sources and utilization
5.1.3 Abandonment of nuclear power stations
5.1.4 Promotion of energy-saving technologies
5.2 Improvement of waste management and emission control
Driving forces:
5.2.1 Increasing recycling of waste paper and waste wood
5.2.2 Implementation of best practices in wood-processing industry
5.2.3 Implementing/extending integrated pollution control
5.2.4 Rationalizing use of wood products
5.2.5 Reduction of harvesting and transport losses of roundwood
5.3 Climate change
Driving forces:
5.3.1 Impacts of climate change on forest growth
5.3.2 Acceptance of forests as natural sinks for the compliance of emission reduction
5.5.3 Acceptance of wood products as carbon sinks

the elimination/reduction of clear-cutting areas and the limitation/prohibition of chemicals in forests (laws and regulations); the promotion of selective harvesting systems and silviculture practices close to nature (incentives and/or market forces); and the certification of sustainable forest management, as well as environmentally friendly wood-processing technologies (market forces).

The policy scenario areas in package 2, 'globalization, innovation and market structures', have been qualified as largely market-driven. Trade policies opening national markets and liberating international trade, high-quality research and educational facilities to foster innovation and competitive market access have to set the frame conditions for private sector activities. The inquiry findings referring to policy scenario package 3, 'countries with economies in transition', confirm that the main driving forces result from an overall recovery of the economy, from a restructuring of the public sector, from privatization of forestry operations and industrial wood processing and from the restitution of former private forest holdings. Building up an appropriate market framework in the countries with economies in transition needs to be induced by a balanced combination of new laws and regulations, economic instruments and private sector market forces. Driving forces for changes in policy scenario package 4, 'regional development', are expected to come from a reduction of subsidies in agriculture and a move to market forces. Incentives for protective and recreational services of forests have been considered as an important economic instrument to regional development policies. The replies concerning package 5, 'energy and environment', indicate that all three scenario areas are presumably very important as driving forces. This refers in particular to measures implementing pollution control, promoting higher levels of wood utilization by private and public consumers, fostering sustainable forest management and more efficient wood processing, as well as responding to the political demand to use production forests and afforestation/reforestation as CO_2 sinks in order to comply with internationally agreed emission reductions. More efficient utilization of forest resources, higher levels of wood processing and consumption and recycling of waste are expected to be mainly market-driven. The determination of emission standards and effective pollution reduction, on the other hand, first require consistent public policy frameworks and a concerted use of regulatory and economic instruments.

Probability occurrence of policy scenario areas

Table 21.2 shows the estimates of probability (in percentages), structured by policy scenario

Table 21.2. Probability occurrence estimates[a] of policy scenario areas (%). (From Thoroe *et al.*, 2004: p. 16.)

Package policy scenario area	EU-EFTA	CEE	CIS
1. Biodiversity, including nature conservation			
1.1 More emphasis on nature conservation and promotion of biological diversity of forest ecosystems	>90	50–70	50–70
1.2 More emphasis on nature-oriented forest management	North > 90; South 50–70	50–70	50–70
1.3 Increasing demand for certification of forest management and wood products	~50	20–30	20–30
2. Globalization, innovation and market structures			
2.1 Impact of globalization on the competitiveness of the European forest and forest industry sector	50	60	70
2.2 Intensified innovations and changes in competitiveness of wood products	60	70	75
3. Countries with economies in transition			
3.1 Strengthening policies to develop market framework in countries with economies in transition	75[b]	90	80
3.2 Progress in European Union enlargement	80[b]	100	50
4. Regional development			
4.1 Incentives for social/environmental benefits from forestry and wood products use	65	60	50[b]
4.2 Changes in agricultural, rural and regional development policies	80	80	40[b]
4.3 Social and demographic developments	50	70	90[b]
5. Energy and environment			
5.1 Promotion of renewable energy resources	100	100	100
5.2 Improvement of waste management and emission controls	100	100	100
5.3 Climate change	100	100	100

[a]Identified by the respondent of inquiry II and revised by the working groups at the December 2001 meeting of the Team of Experts

[b]Secretariat estimates

EU-EFTA = European Union-European Free Trade Association; CEE = Central and Eastern Europe; CIS = Community of Independent States.

areas, packages and sub-regions. It summarizes the results of the research process outlined in the introduction, as well as the review provided by the meeting of expert working groups. An assessment of 100% probability means that, in the opinion of the inquiry addressees and the working group experts, a given policy scenario is most likely to occur. An estimate of 20–30%, on the other hand, means that the probability of occurrence of a given policy scenario is low. It is striking that the only three scenario areas indicated in Table 21.2 with the highest probability rating throughout the European region are those in policy scenario package 5, 'energy and environment'. At the other extreme, demand for certification of forest management and wood products obtained a relatively low rating in all sub-regions. On a weighted basis, the policy scenario areas with the highest probability ratings were 5.1 (renewable energy resources), 5.2 (waste management and emission control) and 5.3 (climate change). These were followed by policy scenario areas 1.1 (nature conservation and biological diversity), 2.2 (innovation and competitiveness), 4.2 (agriculture, rural and regional development policies), 4.3 (social and demographic developments), 1.2 (nature-oriented forest management), 2.1 (globalization and European competitiveness) and 4.1 (incentives for social and environmental benefits). One may also note the policy scenario areas 3.1 (policies to develop the market framework) and 3.2 (EU enlargement policies), with their high scores in the Central and Eastern European countries and the Community of Independent States sub-regions. Without overlooking the other policy scenario areas entirely, this list can provide guidance on which policy scenario areas should receive particular attention in considering impacts on forestry and wood processing. Policy scenario areas by sub-regions with an estimated probability of 70% and higher are expected to occur in most of the countries in the sub-region and will have significant impacts on the future development of the European forest sector.

Policy impacts on forest sector parameters

Estimating the extent of future policy impacts on specific forest sector parameters was probably the hardest part for the respondents and the expert working groups. Estimates were made of the policy impacts and market developments in each scenario area for the following four sector parameters:

- Area of forest available for wood supply;
- Roundwood removals and production of wood products;
- Consumption of forest products;
- Trade (exports and imports).

With regard to the parameter 'area of forest available for wood supply', it can be stated that on the whole, no drastic changes in comparison with the baseline assumptions are to be expected. A lower growth rate has been estimated due to policy impacts from policy scenario area 1.1 (nature conservation and promotion of biological diversity) in the EU/European Free Trade Association and Central and Eastern European countries sub-regions, and from policy scenario area 1.3 (certification of forest management and wood products) in all sub-regions. On the other hand, the policy effects of policy scenario area 4.2 (agriculture, rural and regional development policies) are assessed as having a positive impact on forestry land use, causing higher growth and increased production in the EU/European Free Trade Association and Central and Eastern European countries sub-regions. Positive effects on wood production in the Central and Eastern European countries and those with economies in transition are also expected from developments in policy scenario area 3.1 (strengthening policies to develop the market framework in countries with economies in transition).

Effects on the parameter 'roundwood removals and production of processed products' are anticipated in a rather similar manner. The estimated impacts for most of the policy scenario areas are fairly consistent with the projected baseline developments. Only in a few cases are additional growth (higher or much higher) or decrease (lower) in comparison with the baseline assumptions considered as likely. In the longer-term perspective, increasing nature conservation measures, mainly in the EU/European Free Trade Association sub-region, could slow down roundwood removals, a trend that should be seen in relation with the possible reduction of the forest area available for wood supply. Positive impacts on an increase in the production of processed wood products are estimated for policy scenario package 2 (globalization, innovation and market structures) for all sub-regions. Similar expectations have been articulated in relation to policy scenario package 3 (countries with economies in transition), indicating a higher growth potential as consistent with the recent dynamic developments for the sub-regions of the Central and Eastern Europe countries and countries with economies in transition. With regard to further strengthening of policies to develop appropriate market frameworks and further progress in EU enlargement policies, the positive effect of roundwood removals has been assessed as high and the effects on the production of processed forest products as much higher than at the baseline. For policy scenario areas 5.1 (promotion of renewable energy resources) and 5.3 (climate change), important positive policy impacts were articulated by the respondents.

The inquiry results for the parameter 'consumption of forest products' indicate that all policy scenario areas in package 2 (globalization, innovation and market structures) will have some impact, although a limited one, on the consumption of wood products in the EU/European Free Trade Association countries and a more pronounced impact in the countries with economies in transition. A rather low impact is estimated for the policy scenario areas in packages 1 (biodiversity, including nature conservation) and 4 (regional development). On the other hand, the respondents are of the opinion that the impacts of policy scenario area 3.1 (strengthening policies to develop the market framework in countries with economies in transition) are likely to increase the consumption in the EU/European Free Trade

Association sub-region and even more in the Central and Eastern European countries.

No significant impacts on the parameter 'trade (exports and imports)' are expected from the scenario areas in package 1 (biodiversity including nature conservation) and in package 4 (regional development). 'Much higher' trade is estimated for policy scenario area 3.1 (strengthening policies to develop the market framework in countries with economies in transition), probably affecting all sub-regions. The same observation refers to policy scenario 3.2 (EU enlargement), again for all sub-regions. Policy scenario area 5.1 (promotion of renewable energy resources) could have a significant impact on forest-product trade in Europe as a whole.

Conclusions

The study provides evidence for the growing and complex framework of public policies dealing with multiple economic, social and environmental benefits from forest land use and wood production in modern societies. Scenario areas, such as biodiversity conservation and multifunctional nature-oriented forest management receive a fairly high estimation with regard to their probability of occurrence, but a rather low assessment of the anticipated impact. On the other hand, the empirical inquiry results provide indications that the main public policy impacts for future developments in the European forest sector are likely to come from cross-sectoral policy scenarios focusing on changes in agriculture and regional policies, globalization of trade and liberalization of consumer markets and from environmental and energy policies. Changes in such policy areas will increasingly influence the sector's market framework in the future. Based on the assumptions about the anticipated impacts of the five policy scenario packages that have been identified, the study depicts two broad alternative trends that are significant for policy formation and implementation.

The *conservation trend* assumes further forest sector-specific progress towards biodiversity and nature conservation in forests, nature-oriented forestry management and certification of forestry practices. There will be increasing public pressure for further reduction of negative effects on forest stands from air pollution, as well as for more efficient environmentally friendly techniques in wood-processing and waste management. Agriculture and regional policies will shift gradually towards incentives for forest land use in order to maintain and increase social and environmental benefits. The sustainable energy trend anticipates a significant increase of the use of renewable energy sources based on policy, technological innovations and changing attitudes among consumers. The forest area for wood supply is expected to increase, particularly due to additional short-term plantations. Regulatory, as well as incentive measures for reducing the pollution contributing to climate change will become increasingly important.

The *globalization trend* focuses on macroeconomic assumptions of accelerated international trade and further market liberalization. This assumes additional economic growth, caused by a significant move towards technological progress and strengthening human capital, an increase in the resources devoted to research and strong efforts to provide a high level of education and training. The macroeconomic assumptions for the EU/European Free Trade Association countries are that policies aimed at accelerating technological progress and enhancing human capital will have to deal in particular with public environmental concerns. Given the high stock of capital per employee, combined with a comparatively low marginal productivity of capital, the increase in gross domestic product has been assessed as rather slow in the immediate future. For the two sub-regions of Central and Eastern Europe and the Community of Independent States, accelerated progress towards the market economy in conjunction with economic, social and political stabilization is to be expected. Policies enhancing savings and investment, improving human resources and facilitating technological catching-up will lead to accelerated convergence. The policy scenario area 'strengthening policies to develop the market framework in countries with economies in transition' did in fact receive the highest values in terms of expected probability of occurrence and impacts.

Taken together, the findings of the policy study allow the following conclusions:

1. Natural and economic potentials in Eastern Europe and in Russia in particular, in combination with further progress towards a market economy are expected to result in further growth in the forest sector in Europe. This trend will have a notable impact on forest-product trade, as well as on forest land use.

2. Forest sector stakeholders should realize that important forest-related decisions are also being taken in other policy domains. They have to engage in dialogue with stakeholders in other policy areas, such as energy, trade, environment, renewable resources and regional development. They have to highlight to politicians and citizens the benefits from sustained forestry land uses and timber utilization and the way in which these contribute to economic and social advancement.

3. It is essential to combine the innovative energy of the European forest industries and to create concerted actions among all forest-sector branches. The goal is to make the necessary initial investments available for new infrastructure and innovations in forestry, wood processing and recycling.

4. We live in a period of globalization of companies, NGOs and international economic, social and environmental agreements and processes. A more proactive approach to promoting European experience in sustainable forestry management is warranted in a worldwide context.

References

Thoroe, C., Peck, T., Guarin Corredor, H. and Schmithuesen, F. (2004) *The Policy Context of the European Forest Sector*. Geneva Timber and Forest Discussion Paper 34, United Nations, Geneva, 109 pp.

UNECE/FAO (2005) *European Forest Sector Outlook Study 1960–2000–2020 – Main Report*. Geneva Timber and Forest Study Paper 20, United Nations, Geneva, 231 pp.

22 Comparative Analysis of Framing the 'Forest Sector': Case Studies from Austria and the Netherlands

Evelien Verbij, Esther Turnhout and Heiner Schanz

Introduction

The worldwide call for cross-sectoral coordination in today's forest governance seems more than self-evident. During the last decade, forest policy-makers have put cross-sectoral coordination at the heart of a political agenda aimed at achieving sustainable forest management (FAO, 1994, 1996, 1999; UN-CSD-IPF, 1996; MCPFE, 2003). This call is based on a conviction that key problems in forests, such as deforestation and forest die-back, cannot be adequately tackled within sectoral boundaries. Instead, the causes underlying these key problems have to be addressed in sectors other than the forest sector (Ellefson, 1991; De Montalembert, 1992, 1995; De Montalembert and Schmithüsen, 1993). In the European context, the growing attention to cross-sectoral coordination can be attributed to the changing role of forests in Europe in the light of processes of urbanization and rural development – the urbanizing lifestyles of forest owners and the declining importance of primary production (Hoogstra *et al.*, 2004). Cross-sectoral coordination would thus mean counteracting the failure of the traditional forest sector to recognize these social changes and adopt appropriate new strategies (Hellström and Reunals, 1995). The aim of this chapter is to contribute to the discussion by illustrating the ways in which changes in the relevant framework of the social actors involved in the forest sector challenge the notion of cross-sectoral coordination.

Approach

Existing studies on cross-sectoral coordination in forestry have mainly interpreted forest sectors as social and institutional systems (Schmithüsen *et al.*, 2001; Hogl, 2002; Schmithüsen, 2003; Krott and Hasanagas, 2006). In accordance with the theory of social systems, the impediments to cross-sectoral coordination are seen to be rooted in differences in organizational history, culture, interests and belief systems between different social entities and economic sectors (Shannon and Schmidt, 2002). Consequently, studies to date have focused on describing and analysing linkages, interfaces and coordination mechanisms in the forest sector in relation to other sectors (FAO, 2003).

As Verbij and Schanz (2002) have argued in their review on cross-sectoral coordination, the problem with interpreting sectors as social systems is that the boundaries of the sectors are drawn according to the observer's analytical focus and not by the actors themselves. Strictly speaking, it is the observer who determines whether interactions have to be seen as either sectoral or intersectoral. However, actors in practice are themselves continuously engaged in setting, accepting, maintaining and adapting the boundaries between groups of other actors, as these boundaries help them to define, structure and distinguish the complexity of their political and social interactions (Schanz, 1999). It therefore seems reasonable to assume that the way actors themselves frame the forest sector, by drawing boundaries around their social system, has at least as important an impact on coordination as the differences in the characteristics of the resulting social entities. This is in line with Andrian *et al.* (2002), who argued that the definition of sectors, and more precisely of sector boundaries, should be the starting-point for analysing cross-sectoral coordination issues.

This chapter illustrates the ways in which changes in the relevant social actors' perception of the 'forest sector' challenge the notion of cross-sectoral coordination. For this purpose, qualitative case studies in Austria and in the Netherlands were considered to be most appropriate (Yin, 1994). Austria and the Netherlands were selected on the basis of their similarities with regard to political culture, as both countries originally had a corporatist tradition in the field of forest policy. At the same time, they differ widely with regard to topography, natural conditions and thus also forest tradition. Although Austria can generally be regarded as a typical forest country, the Netherlands is one of the most densely populated and urbanized areas in Europe, with an absolutely and relatively small forest area, consisting mainly of young forest stands (Table 22.1).

Conceptually, the study draws on framing theory as a special form of discourse analysis. Schön and Rein (1994) assert that frames determine the way in which parties see issues, policies and policy situations, and that these frames embody different prescriptions for action. Consequently, the study assumes that the 'forest sector' is an expression of competing or convergent frames, which social actors use to (re)construct a social subsystem favouring and justifying their own policy positions. The idea of 'sectorization' is introduced here in order to make it clear that actors are involved in continuous processes of framing a policy sector. The assumption is thus that there is not a single sector definition, but rather that the meaning of the 'forest sector' changes over time, is subject to specific conditions of a certain locality and differs between social actors. As framing theory shows, these abstractions are not formed arbitrarily by the actors (Fisher, 1997). Instead, they serve as a starting point to emphasize specific policy matters and offer a particular interpretation of situations

Table 22.1. Characteristics of Austria and the Netherlands.

Country characteristics	Austria (2002)[a]			The Netherlands (2002/2004)[b]		
Forest cover	3,960,000 ha (47%)			360,000 ha (10%)[c]		
Population	8.2 million			16.3 million		
Private ownership	*Size (ha)*	*Owners (n)*	*ha*	*Size (ha)*	*Owners (n)*	*ha*
	<200	*c.* 170,000	2,130,000	5–250	1301	32,098
	>200	*c.* 1400	1,111,000	>250	39	20,298
	Total private ownership share 82%			Total private ownership share 20%		
National Forest Service	591,000 ha (15%)			85,555 ha (31%)		
Public land (municipalities, provinces, public services)	129,767 ha (3%)			66,575 (24%)		
Private nature conservation organizations	–			61,751 (23%)		
Community forests	348,886 (not included in above overview)			–		
Other owner categories	–			5,753 (2%)		
Wood stock (m^3)	1095 billion m^3			58 million m^3		
Annual growth	31.28 million m^3			2.2 million m^3		
Annual felling	18.8 million m^3, 60% of annual growth			0.9 million m^3, 41% of annual growth		

[a]From BFW (2005).
[b]From Probos (2005) and Bosschap (2005).
[c]Included are approximately 85,000 ha, owned by about 15,000 owners who own less than 5 ha. As these are rough estimates, we have calculated ownership shares in accordance with the number of 272,030 registered hectares of forest.

and events (Triandafyllidou and Fotiou, 1998). With regard to cross-sectoral coordination, divergent frames result in different challenges for policy coordination, as different social actors will act according to their specific policy demands.

These frame constructs are often crystallized in 'generative metaphors' – i.e. in ideas or symbols that are empirically reflected in the motifs used by the actors to describe a situation. Motifs are defined by Broekhans (2003) as powerful symbols or concise terms which, through suggestion and possible multiple interpretation, can bond or bring together actors. Through motifs, actors legitimize their existence and create their identity. In the following, motifs are interpreted as a manifestation of different sector identities and are used to draw conclusions regarding changing patterns of actors and sector boundaries. The findings of the study presented are based on 59 semi-structured interviews that were conducted in 2004 with actors involved in forestry and forest policy in the two case-study countries. In addition, secondary sources and policy documents were used to complement the analysis. An extended version of the country case studies that have been undertaken will be published in a doctoral thesis (Verbij, in preparation).

Case Study Results from Austria and the Netherlands

Austria

Forest cover in Austria is relatively high, at over 47% of the land area, partly explained by the fact that the Austrian landscape is dominated by the Alps, most of the slopes of which are covered with forests, which are very important for avalanche and torrent control. Over 85% of the Austrian forests are privately owned by some 1400 large owners (> 200 ha) and 170,000 smaller owners, mostly of farm forests. The remaining 15% is publicly owned and managed principally by the Austrian Federal Forest Service. The origin of the dominance of private ownership dates from the 19th century, when the state sold more than half of its productive forests to private entrepreneurs in order to pay off high debts caused by several wars. As a result, the majority of the remaining state forests are today located in the less productive higher Alpine region. Commercial use of forests is deeply rooted in the Austrian society.

Until the revolution of 1848, many forest areas in Austria were subject to strict regulations to serve the mining and salt-production industries. In 1853, these binding regulations were abolished, and the state was no longer able to control the management of the predominantly privately owned Austrian forests. However, part of state supervision continued by virtue of the 1852 Imperial Forest Act, justified by the goal of 'safeguarding the protective function of forests'. On the other hand, the main focus of the 1852 law was to stimulate the commercial use of forests. This is most clearly reflected in the obligation for private owners with holdings of more than 500 ha to appoint professionally trained foresters who had to pass an obligatory state examination in forestry (Weiss, 2001).

As a result, the group of forest professionals, often referred to as the 'forestry family', expanded. The term 'family' in this context indicates a high level of trust and informality in mutual relationships amongst a rather small group. In particular, the role of the Austrian Forest Association was crucial, since forest owners and foresters became closely associated through the meetings it organized. After the Second World War, a corporatist system of social partnership facilitated the formalization of these relationships. The Chambers of Agriculture established compulsory membership for farmers and foresters and granted exclusive powers, such as the right to participate in law-making to these social partners. As forestry played a fairly subordinate role in the Chambers of Agriculture, the Austrian Federation of Forest Owners' Associations was established, mainly representing the interests of large forest owners. Close cooperation between these three actors resulted in dominance of the 'forestry' motif – leading to a heavy emphasis on the commercial utilization of forests. The motif of 'forestry' thus identifies a relatively straightforward perception of the forest sector as an economic resource. The economic function of forestry served as a solid basis for legitimizing the forest sector, its importance, its objectives and its boundaries. This characterization of the 'forestry' motif is in accordance with the studies by Pregernig (1999), Hogl (2000) and Voitleithner (2002), who describe the Austrian forest sector in terms of 'sector corporatism' and as a policy community characterized by a limited number of participants.

During the 1980s, the 'forestry' motif came to be challenged. First, development towards a post-industrial society in Austria led to new demands from the public to have an influence on policy processes, to attempts to increase the role of the Austrian Parliament and to efforts to reduce the role of civil servants and the corporatistic chambers (Tálos and Kittel, 1996). Secondly, the society started to value forests for their amenity, recreation and nature-conservation values. The 'forest family', which up to this point had been a rather closed group, was forced to open up and enter into dialogue and joint projects with other social organizations, such as nature-conservation groups and hiking and Alpine organizations. New, frequently informal, relationships were established with actors who did not actively support the economic use of forests with the implied 'forestry' motif. Thirdly, not only did the number of forest professionals decrease, but a new group of forest owners also appeared, who were no longer exclusively interested in the commercial use of their forests (Hogl et al., 2003; Kvarda, 2004). As a result of these societal developments, two new motifs emerged in Austria, which can be referred to as the 'forest-based industry' motif and the 'forest' motif, reflecting different perceptions of the forest sector and its definition. It is interesting to see that these new motifs did not supersede the old 'forestry family'. The relationship between the two new motifs is also of interest, as some actors can be found using both motifs simultaneously.

The 'forest-based industry' motif is based on a shared interest in increasing domestic wood production. In addition to the wood-processing industry and large forest owners, some small farm-foresters also make use of it. Declining incomes from agriculture led to farmers shifting their focus to more intensive utilization of their forest-land. With increasing international competition, the wood-processing industry hopes to enhance its competitiveness by stimulating Austrian forest owners to increase harvesting activities. This common move towards an increase in domestic wood production led to the establishment of farm-forest cooperatives, which by 2005 extended over a quarter of the Austrian forests. The public forest authorities support and make use of the 'forest-based industry' motif, since the industry contributes strongly to the Austrian economy. It currently holds second place in the national trade balance and represents about 110,000 jobs, largely in rural areas (FPP, 2005). Adherents of the 'forest-based industry' motif seek social legitimization for their activities and emphasize the social implications of the motif, since exclusively economic legitimizations are no longer accepted by society.

The 'forest' motif reflects a multifunctional perspective on forests that contrasts with a one-dimensional focus on wood production. This type of perspective implies involvement of a broader range of actors in forest-related decision-making processes. As an example, the forest authorities in April 2003 initiated what is known as the Austrian Forest Dialogue, with the aim of improving communication among an increasing number of stakeholders. More than 50 concerned parties are currently involved in this forum, together with the established forest organizations. One of the results was the publication of an Austrian Forest Programme at the end of 2005 (BMLFUW, 2006). The public forest authorities, as well as the traditional forestry actors, became involved in developing this type of programme, as it is capable of serving as a building-block for a possible European forest policy that would be adhered to by several 'traditional' forest countries within the European Union (EU). The 'forest' motif is thus used not only to facilitate a broad dialogue on forests in Austrian society, but also to legitimize a forestry perspective at the level of the EU. At the same time, however, the 'forest' motif is not necessarily changing the prevailing pattern of the actors involved. So far, most non-governmental organizations (NGOs) appear to be rather sceptical with regard to their ability to influence the outcome of this dialogue. They are concerned about possible domination by groups acting in accordance with the traditional 'forestry family' and 'forest-based industry' motifs that are resistant to new initiatives that might restrict forest utilization and timber harvesting by private forest owners.

The Netherlands

In contrast to Austria, the Netherlands can hardly be described as a typical forest country, with forest cover representing only 10% of the country. A total of 60% of the Dutch forests are publicly owned, while the remaining 40% belong to private owners and nature-conservation organizations. The existence of private nature conservation organizations dates back to the beginning of the 20th century; several of these organizations have in the meantime become large landholders, purchasing properties from private owners. Up to the Second World War, the Dutch government was hardly involved in forestry and limited its activities mainly to managing state forestland through the National Forest Service and to stimulating afforestation on public land. When the Forest Law came into force in 1922, its main purpose was to support an increase in the country's forested area without interfering substantially with the interests of individual forest owners, as it was feared that this would be counterproductive to reforestation. Due to enormous wood shortages during the First and Second World Wars, the government extended its influence – e.g. by subsidizing reforestation on private land. It was only at this time that both private forest owners and the government started to develop a 'forestry' motif, reflecting a shared focus on wood production. Close informal relationships between civil servants from the Forest Service and influential private forest owners (Buis *et al.*, 1999) used the 'forestry' motif to establish a whole range of institutions aimed at increased wood production – for instance, the Industrial Board for Forestry and Silviculture. This is a corporatist-like statutory trade organization established by law, with compulsory membership for all owners with more than 5ha of forestland. However, in contrast to Austria, the 'forestry' motif – based solely on the economic function of forests – was predominant in the Netherlands only for a very short period of time.

By the mid-1960s, desperate financial circumstances forced private forest owners and the government to reconsider the one-dimensional focus on wood production and explore a possible multifunctional focus on forests. Both the wood-processing sector and in particular the paper industry, as well as a society oriented towards conservation and recreation, became increasingly interested in model management of Dutch forests. At the same time, under pressure from ecologists, ideas amongst forestry professionals started to shift towards a more ecosystem-oriented form of forest management (van der Windt, 1995; Zevenbergen, 2003). In effect, the 'forestry' motif started to shift towards a 'forest' motif based on a multifunctional view of forests. The meaning of 'forest' allowed different actors to make use of the motif. On the one hand, the Dutch paper industry, private forest owners and the National Forest Service were able to emphasize the importance of the wood-production function. They argued in favour of expanding the forest area and increasing the country's self-sufficiency in wood. On the other hand, the 'forest' motif allowed for active involvement by all forest owners in providing society with recreation and nature-conservation services. In forests managed by the forest service, the government was able to move easily towards a more recreation-friendly forest management policy. Outside the state forests, government needed to provide financial resources to private forest owners and expected in return that the owners would open their forests up for recreational use. Due to their financial difficulties, the private owners took considerable advantage of these opportunities.

Despite public subsidies, many private owners were forced to sell their properties, which were subsequently purchased by the National Forest Service or by private nature-conservation organizations. However, the growing sizes of properties, in combination with low wood prices, made it difficult for the new owners to finance the increasing costs of managing the properties. The large private nature-conservation organizations in particular proved capable of capitalizing on society's growing concern over and interest in the environment. During the last 10 years, there have been no fundamental disagreements between conservation NGOs and the national policy objectives for forest resources. The conservation NGOs have established clientele relationships with the ministry, preferring direct contacts and lobbying to public action or campaigning at the national level. In this way, they have been able to foster broad support in society by maintaining their image as protectors of nature and of the Dutch landscape. Wood harvesting, widely perceived as destructive by the public, does not suit this image particularly well. In addition, the Ministry of Agriculture, Nature and Fisheries published a Nature Policy Plan (Ministerie van LNV, 1990) that prioritized financial support for nature conservation organizations and paid little attention to the wood-production issue.

During the 1990s, the Ministry of Agriculture, Nature and Fisheries was forced to further strengthen its focus on nature conservation in order to implement international agreements on biodiversity, such as the international Convention on Biodiversity and the Habitats and Birds Directives of the EU. At the same time, a considerable proportion of forest expertise in the Ministry was diluted due to decentralization trends resulting from the forest law and privatization of the National Forest Service. This disrupted the formerly close relationships between forest owners, forest professionals and their advocates at the ministry. In addition, the wood-processing industry has been losing interest in Dutch forests, due to the low proportion of domestic raw material supply to the Dutch wood-processing industry. It appears that there is little interest on the part of private forest industry in becoming involved in discussions on the management of forest resources in the Netherlands. In accordance with the same logic, the image of Dutch forestry is of less relevance to the private forest industry than the image of forestry in the principal supplier countries.

These developments make it increasingly difficult for the actors involved to continue using and making sense of the 'forest' motif to mobilize the forest sector. Only a small, shrinking group of individuals appear to be continuing to rely on this motif and are having great difficulty in finding a legitimization for promoting wood harvesting. Instead, most organizations involved in forest management no longer identify themselves with forests or forest management activities, but rather with the role that forests play in the overall landscape. What seems to be developing amongst these actors is a shared use of the 'landscape' motif. In 2000, the previously separate policy fields of forest and nature conservation were combined with the publication of an integrated policy document (Ministerie van LNV, 2000). As a result, non-forest nature areas and their managers came to the fore. The 'landscape' motif holds together all these different actors and, comparable to the 'forest' motif, allows different management functions, including wood production, to be combined.

In sharp contrast to the 'forestry' motif, the 'landscape' motif no longer suggests a stable common identity. Instead, cooperation happens in smaller ad hoc and structural groups, differing in identity and membership. Furthermore, most of the actors appear to value and promote their identity as separate actors. This means that the 'landscape' motif does not frame a 'forest sector' in the traditional sense. Instead, it seems to be a (low-profile) motif that provides some structure for solving common problems and at the same time allows for actions of individual actors to reach individual goals. More structural, long-lasting common problems are typically dealt with in shared institutionalized forums, while ad hoc issues tend to be dealt with in coalitions specially established for the particular occasion. An example of an institutionalized forum of this type is the Industrial Board for Forestry and Silviculture, which is now starting to include non-forestland properties in order to become a forum for landscape management. Although the area represented by forestland is still increasing in the Netherlands, the framing of a 'forest sector' is currently limited to a small group of individuals. With blurring boundaries, the requirement for cross-sectoral coordination is no longer self-evident. It is becoming obvious that the process of cross-sectoral coordination presupposes different entities that need to be coordinated and does not necessarily depend on the policy problems at stake.

Discussion

Both the Austrian and Dutch case studies make it clear that the motifs used by social actors to frame the 'forest sector' differ over time, between different places, and also between different groups. However, despite the differences in the natural, economic and social characteristics involved, the processes of re-framing the 'forest sector' by the social actors remain largely the same. In both countries, changes in 'forest sector' frames reflect wider modern developments. First, the general move away from a one-dimensional focus on wood production is in line with changes in society. Increasing international market competition and decreasing prices have resulted in challenges to the previously dominant motifs. In both countries, it is emphasizing the social relevance of forests of being considered increasingly important, but different motifs are being used to promote social legitimization. Secondly, the changing role of the government observed in both cases again reflects wider social processes. Both countries have a corporatist tradition in the field of forest policy, but are moving – although at a different pace – towards more inclusive and interactive forms of policy-making. Policy-making is now relying on the formation of broader social groups in which responsibilities can be shared, and it is also relying on government having more of a facilitating role. This is in line with the current governance debate, in which the central role of the state is being questioned and in which responsibilities are increasingly being distributed between state, market and civil-society actors (Jordan, 2001; Hajer, 2003).

Nevertheless, there are significant differences between the two countries with regard to the development of forest sector frames. In Austria, there is still a relatively clearly defined and recognized group of actors today that is perceived as representing the 'forest sector'. In the Netherlands, the 'forest sector' has been shrinking to only a few individuals trying to obtain recognition and create a common identity for what they feel is a legitimate and viable, but barely acknowledged, sector. Differences in physical characteristics, such as forest cover percentages, the role of forests for society – e.g. with regard to protective functions – as well as different historical paths obviously play an important role in explaining these differences. On the other hand, it could also be argued that the momentum resulting in re-framing processes has differed in the two countries. In accordance with the findings of Triandafyllidou and Fotiou (1998) in their research on EU policy-making, actors' shifts in frame in both countries have been driven by the resources available to them. Setting and maintaining 'sector boundaries' obviously depends on opportunities to mobilize resources, with social legitimacy being the most valuable one. Dwindling social legitimacy forced Dutch forest actors to recognize the 'blindness induced by their own ways of framing the policy situation' (Schön and Rein, 1994: 187) and consequently to adopt much broader sector frames, resulting in the disappearance of a 'forest sector'. In Austria, in contrast, it could be argued that social legitimacy has been changing but has still remained strong enough for the existing actors' frames not to be questioned.

Conclusions

The results of this study indicate that the absolute quality of the widespread call for cross-sectoral coordination in forest policy needs to be questioned. As has been shown in the case of the Netherlands, cross-sectoral coordination presupposes different entities that are in need of coordination, and does not automatically depend on the policy problems at stake. Social or political entities that require coordination depend on the way the actors concerned frame these entities and thus draw the boundaries around their social systems.

'Sectors' are not objectively given structures. Instead of searching for a general model of cross-sectoral coordination in the rationalist and pluralist tradition, the findings of this study suggest that one should focus on coordination processes across different frames. Instead of focusing on the multiple meanings of the forest sector, it appears to be more fruitful to focus on shared frames associated with a specific problem or issue at hand and the social boundaries they create in the specific context. This automatically implies a focus on more context-specific issues. Understanding of the issues involved will be enhanced by applying empirical scrutiny to what needs to be integrated or coordinated.

This implies that cross-sectoral coordination is no more special than any other coordination process. It is neither essentially different nor more or less difficult and important. The challenge simply involves managing the frame alignment processes that occur between different actors such as frame bridging, frame amplification, frame extension and frame transformation (Snow et al., 1986). Given the continuous trend towards integration, and in view of the blurring of functional, structural and territorial boundaries – in short, with the rise of governance processes – an instrumental approach to frame alignment processes will become increasingly important.

References

Andrian, G., Musumeci, L., Pettenella, D. and Secco, L. (eds) (2002) *Cross-sectoral Linkages in Forestry. Country Case Study Summary Review*. FAO Forestry Department, Rome.

BFW (2005) *Österreichische Waldinventur 2000/2002*. Austrian Forest Research Institute. Available at: http://web.bfw.ac.at/

BMLFUW (2006) *Österreichisches Waldprogramm*. Bundesministerium für Land- und Forstwirtschaft, Umwelt und Wasserwirtschaft (BMLFUW), Vienna.

Bosschap (2005) *Members of Bosschap*. Available at: http://www.bosschap.nl

Broekhans, B. (2003) Hoe Milieukunde Geschiedenis Werd. Demarcatie van een Maatschappelijk Relevante Wetenschap 1970–2000. Dissertation, Catholic University of Nijmegen.

Buis, J., Verkaik, J.P. and Dijs, F. (1999) *Staatsbosbeheer 100 Jaar. Werken aan een Groen Nederland*. Matrijs, Utrecht.

De Montalembert, M.R. (1992) Intersectoral policy linkages affecting the forestry sector. In: Gregersen, H.M., Oram, P. and Spears, J. (eds) *Priorities for Forestry and Agro-Forestry Policy Research. Report of an International Workshop*. International Food Policy Research Institute, Washington DC, pp. 79–92.

De Montalembert, M.R. (1995) Cross-sectoral linkages and the influence of external policies on forest development. *Unasylva* 46(3), 25–37.

De Montalembert, M.R. and Schmithüsen, F. (1993) Policy and legal aspects of sustainable forest management. *Unasylva* 44(4), 3–9.

Ellefson, P.V. (1991) Integration of forest sector plans with national plans for development: a strategic necessity. *Journal of World Forest Resource Management* 5, 73–83.

FAO (Food and Agriculture Organization of the United Nations) (1994) *Readings in Sustainable Forest Management*. Forestry paper 122. FAO, Rome.

FAO (1996) *Formulation, Execution and Revision of National Forest Programmes. Basic Principles and Operational Guidelines*. FAO, Rome.

FAO (1999) *Status and Progress in the Implementation of National Forest Programmes – Outcome of an FAO World-Wide Survey*. FAO, Rome.

FAO (2003) Dubé, Y.C. and Schmithüsen, F. (eds) *Cross-Sectoral Policy Impacts between Forestry and Other Sectors*. FAO Forestry Paper, 142. Food and Agriculture Organization of the United Nations, Rome. Available at: http://www.fao.org/docrep/006/y4653e/y4653e00.htm

Fisher, K. (1997) Locating frames in the discursive universe. *Sociological Research Online* 2(3). Available at: http://www.socresonline.org.uk/2/3/4.html

FPP (2005) *Arbeitsplätze/Betriebe rund ums Holz in Österreich*. Forst-Platte-Papier. Available at: www.fpp.at

Hajer, M.A. (2003). Policy without polity? Policy analysis and the institutional void. *Policy Sciences* 36, 175–195.

Hellström, E. and Reunals, A. (1995) *Forest Conflicts from the 1950s to 1983: A Review of a Comparative Study on USA, Germany, France, Sweden, Finland, and Norway*. EFI Research Report, no. 3. European Forest Institute, Joensuu.

Hogl, K. (2000) The Austrian domestic forest policy community in change? Impacts of the globalisation and Europeanisation of forest politics. *Forest Policy and Economics* 1(1), 3–13.

Hogl, K. (2002) Reflections on inter-sectoral coordination in National Forest Programmes. In: Tikkanen, I., Glück, P. and Pajuoja, H. (eds) *Cross-Sectoral Policy Impacts on Forests*. EFI Proceedings, no. 46. European Forest Institute, Joensuu, pp. 75–90.

Hogl, K., Pregernig, M. and Weiss, G. (2003) *Wer sind Österreichs WaldeigentümerInnen? Einstellungen und Verhalten traditioneller und 'neuer' Waldeigentümergruppen im Vergleich*. Discussion Paper, Institute of Forest Sector Policy and Economics, University of Natural Resources and Applied Life Sciences (BOKU), Vienna.

Hoogstra, M.A., Schanz, H. and Wiersum, K.F. (2004) The future of European forestry – between urbanization and rural development. *Forest Policy and Economics* 6, 441–445.

Jordan, A. (2001) The European Union: an evolving system of multi-level governance ... or government? *Policy and Politics* 29(2), 193–208.

Krott, M. and Hasanagas, N.D. (2006) Measuring bridges between sectors: causative evaluation of cross-sectorality. *Forest Policy and Economics* 5, 555–563.

Kvarda, E. (2004) Non-agricultural forest owners' in Austria – a new type of forest ownership. *Forest Policy and Economics* 6, 459–467.

MCPFE (2003) *Vienna Resolution 1: Strengthen Synergies for Sustainable Forest Management in Europe Through Cross-Sectoral Co-operation and National Forest Programmes*. Ministerial Conference on the Protection of Forests in Europe, Warsaw.

Ministerie van LMV (1990) Natuurbeleidsplan. Ministerie van Landbouw, Natuur, en Visserij (LNV), Den Haag.

Ministerie van LNV (2000) *Natuur voor Mensen, Mensen voor Natuur. Nota Natuur, Bos en Landschap in de 21e Eeuw*. Ministerie van Landbouw, Natuurbeheer en Visserij (LNV), Den Haag.

Pregernig, M. (1999) Austria. In: Gluck, P., Oesten, G., Schanz, H. and Volz, K.-R. (eds) *Formulation and Implementation of National Forest Programmes, Vol. II: State of the Art in Europe*. EFI Proceedings, no. 30. European Forest Institute, Joensuu, pp. 13–44.

Probos (2005) Available at: http://www.bosschap.nl

Schanz, H. 1999. *What is a Forest? What is Forestry? Science on Boundaries*. University of Wageningen, Wageningen.

Schmithüsen, F. (2003) Understanding cross-sectoral policy impacts – policy and legal aspects. In: Dubé, Y.C. and Schmithüsen, F. (eds) *Cross-Sectoral Policy Impacts between Forestry and Other Sectors*. FAO Forestry Paper, 142. Food and Agriculture Organization of the United Nations, Rome, pp. 5–44.

Schmithüsen, F., Bisang, F. and Zimmerman, W. (2001) *Cross-Sectoral Linkages in Forestry: Review on Available Information and Consideration on Future Research*. FAO, Rome.

Schön, D.A. and Rein, M. (1994) *Frame Reflection: Towards the Resolution of Intractable Policy Controversies*. Basic Books, New York.

Shannon, M.A. and Schmidt, C.H. (2002) Theoretical approaches to understanding intersectoral policy integration. In: Tikkanen, I., Glück, P., Pajuoja, H. (eds), *Cross-Sectoral Policy Impacts on Forests*. EFI Proceedings, no. 46. European Forest Institute, Joensuu, pp. 15–26.

Snow, D.A., Rochford, E.B., Worden, S.K. and Benford, R.D. (1986) Frame alignment processes, micromobilization, and movement participation. *American Sociological Review* 51(4), 464–481.

Tálos, E. and Kittel, B. (1996) Roots of Austro-corporatism: institutional preconditions and cooperation before and after 1945. In: Bisschof, G. and Pelinka, A. (eds) *Austro-Corporatism: Past – Present – Future*. Contemporary Austrian Studies, 4. Transaction Publishers, New Brunswick, pp. 21–52.

Triandafyllidou, A. and Fotiou, A. (1998) Sustainability and modernity in the European Union: a frame theory approach to policy-making. *Sociological Research Online*, 3(1). Available at: http://www.socresonline.org.uk/3/1/2.html

UN-CSD-IPF (1996) *Implementation of Forest-Related Decisions of the United Nations Conference on Environment and Development at the National and International Levels, Including an Examination of Sectoral and Cross-Sectoral Linkages. Report of the Secretary-General*. United Nations Commission on Sustainable Development/Ad Hoc Intergovernmental Panel on Forests, New York.

Verbij, E.E.M. and Schanz, H. (2002) Intersectoral co-ordination: state of the art and beyond. In: Tikkanen, I., Glück, P. and Pajuoja, H. (eds) *Cross-Sectoral Policy Impacts on Forests*. EFI Proceedings, no. 46. European Forest Institute, Joensuu, pp. 91–102.

Verbij, E.E.M. (forthcoming) *Understanding Inter-Sectoral Coordination in Forest Governance*. Dissertation, University of Wageningen.

Voitleithner, J. (2002) The National Forest Programme in the light of Austria's law and political culture. *Forest Policy and Economics* 4, 313–322.

Weiss, G. (2001) Mountain forest policy in Austria: a historical policy analysis on regulating a natural resource. *Environment and History* 7(3), 335–355.

Windt, H.v.d. (1995) *En Dan: Wat is Natuur Nog in Dit Land? Natuurbescherming in Nederland 1880–1990*. Boom, Amsterdam/Meppel.

Yin, R.K. (1994) *Case Study Research: Design and Methods*. SAGE, London.

Zevenbergen, M.P. (2003) *De Teloorgang van de Bosbouw? Een Overzicht van de Discussie over de Zelfvoorzieningsgraad van Hout in Nederland*. Online discussion paper 'Nature Forest in Society', no. 1. Forest and Nature Conservation Policy Group, University of Wageningen, Wageningen.

23 Cross-sectoral Linkages between Forestry and other Sectors in Romania

Ioan V. Abrudan

Introduction

In order to promote coherent national policies and development planning, and to eliminate deficiencies in intersectoral coordination in Romania, various institutional mechanisms, such as interministerial committees, cross-sectoral coordination groups and the Council for Economic and Financial Coordination were developed at the governmental level during the last decade. The structure of the present government includes ministries and public authorities, which play an important role in the coordination and integrated development of all economic sectors (Table 23.1). The government's General Secretariat has been made responsible for coordinating the activities of the line ministries and ensuring that the correct legal procedures are observed in the elaboration and endorsement of draft legislation. In terms of intersectoral coordination mechanisms, the option currently used by Romanian public authorities is defined by Governmental Decision 555/2001 (Monitorul Oficial al Romaniei, 2001). This decision regulates the procedures for the development and submission of legislation/regulations to the Cabinet for discussion and approval, and specifies the administrative procedures for cooperation during legislation/regulations drafting if cross-sector issues are envisaged.

Since the end of 1999, the European Union (EU) accession process has had a major positive impact on policy formulation, development planning, intersectoral coordination, participation and transparency. In December 2000, a Secretary of State responsible for European integration was appointed in each line ministry, and these Secretaries of State have met regularly in an interministerial committee and with the representatives of the Ministry for European Integration. At the civil servant level, interministerial working groups have been established to deal with the preparation of each negotiating chapter. All these mechanisms have improved formal and informal intersectoral coordination, as well as the quality of Romanian preparations for EU accession, leading to the closing of the accession negotiations in December 2004.

The complex network of public policies and legislation affecting forest sector development is structured in three categories: (i) policies establishing the institutional framework: economic growth, privatization, public finance, employment, rural development and landscape planning; (ii) policies related to specific economic sectors: agriculture, game management, wood harvesting, transport and processing, mining, energy, infrastructure and tourism; and (iii) policies promoting development: environment protection, water management, nature conservation, education and research (Schmithüsen, 2003). This chapter presents an analysis of current linkages between the Romanian forest sector and other economic sectors. The most important public policies, and the direction and scale of their impacts are discussed for each of these categories.

Table 23.1. Ministerial structure of the Romanian Government. (From Abrudan, 2002.)

Intersectoral coordination	Cross-sectoral responsibilities	Multisectoral responsibilities	Single-sector responsibilities
• Development and prognosis	• Public finance	• Work and social solidarity	• Foreign affairs
• Public administration	• Small and medium-sized enterprises	• Industry and resources	• Justice
• European integration	• Public information	• Agriculture, food and forests	• National defence
• General Secretariat of the Government[a]		• Waters and environmental protection	• Interior
• Relations with parliament[a]		• Public works, transport and housing	• Tourism
		• Education and research	
		• Culture and religion	
		• Health and family	
		• Youth and sports	
		• Communication and information technology	

[a]Activity coordinated by a minister (member of the Cabinet), although there is no distinct ministry.

Overview of the Romanian Forest Sector

Romania's forests cover 27% of the country and include some of the largest tracts of natural and old-growth forests still remaining in Europe. Forests have traditionally played an important role in Romania's social and economic development, providing a major source of rural employment and income from logging, wood processing and non-timber forest product industries. Despite existing inefficiencies, the forestry sector is still a significant contributor to the Romanian economy. The annual export value of forest products in the last 5 years, including processed timber, represented approximately 11% of all national exports and the sector's contribution to the gross domestic product (GDP) was approximately 4.5%. The non-pecuniary values of Romanian forests are, however, considerably larger than the financial values (World Bank, 1999).

Political initiatives to restore property to its previous owners and re-establish the free market economy became a reality at the beginning of the 1990s. As an initial measure, under Law 18/1991, approximately 350,000 ha of forest land were returned to around 400,000 pre-1948 individual owners. With the absence of sufficient institutional capacity and an adequate legal and regulatory framework to ensure sustainable management of forests on private lands, there were considerable negative environmental impacts in many regions (Abrudan *et al.*, 2002).

In 2000, another land restitution law (Law 1/2000) was passed by Parliament and implementation was initiated. In accordance with this law, more than a third of the Romanian forests were to be restored to their previous owners. As a consequence of the much larger scale of restoring forest lands that has been implemented, the state's role needs to change rapidly from that of an owner–manager of forests to that of a 'safeguarder' of the public interest in the new free market and land ownership system. Responsibilities would include ensuring sustainable forest management practices on private, as well as state forest lands, conservation of environmental services and support to the new private sector actors in forestry and related industries.

Institutionally, the regulatory, supervisory and support functions of the state are now separated from the ownership and management functions of the state-owned forest property. The Department of Forests in the Ministry of Agriculture, Forests and Rural Development is the public authority responsible for forests in Romania and has two directorates. The regulatory function is held by the Directorate for Forest Strategy, Policy and Legislation, while the supervisory and support functions are concentrated in the Directorate for Forest Regime. The organizational development of the Department of Forests is showing typical teething difficulties, in terms of reporting arrangements, standard operating procedures, codes of practice, dual reporting and some overlapping of functions.

The National Forest Administration manages the state forests and was established as a legal state-owned entity with an essentially commercial mandate. It operates as a financially autonomous organization performing management and silvicultural operations, as well as engaging in non-timber forest products and services. It undertakes a wide range of public purpose activities and is responsible for the management of protected forest areas and national parks. Forest restitution has reduced the National Forest Administration area by up to 40%, with consequent reductions in revenues and greater impacts on fixed costs.

The Association of Private Forest Owners is a national umbrella organization established in 1998 and represents all categories of private forest owners in Romania. It is a registered legal entity, and at present its membership includes local and county associations, communes, town halls and individual members. The association is facing difficulties, as it neither has a business development plan adapted to private forest sector realities, nor has it addressed the precise range of services it should provide to its members.

The last decade has been characterized by an almost continuous dispute between the ministry responsible for forest management and the public authority responsible for wood harvesting and processing. Although the former has taken measures to achieve better use of forest resources, free competition for wood resources and harvesting methods fulfilling ecological requirements, the latter has been fighting and lobbying for cheap resources and advantageous contractual terms for wood harvesting. In early 2001, the Cabinet approved the supervision of the reserve price for standing wood that the National Forest Administration sells by auction. This has been perceived as a state intervention to control the standing wood price in areas where the competition for wood resources is low.

In recent years, many investors have been lobbying for an improvement of the auctioning system, to allow long-term harvesting contracts. Wood harvesting is seen as the most critical activity in the Romanian forest, with a potentially high environmental impact. Due to obsolete equipment, improper harvesting techniques and poor law enforcement, the remaining trees, existing regeneration, soil and waters are, in many cases, negatively affected by harvesting, especially in mountain areas. Long skidding distances associated with the low road density also contribute to the negative environmental impact (World Bank, 2001).

Impacts of Public Policies on Forest Sector Development

The network of public policies and legislation, which directly and indirectly affect the development of the forest sector in Romania, includes: (i) policies establishing the institutional framework: macroeconomic, privatization, land tenure, etc.; (ii) policies related to specific economic sectors: agriculture, timber processing, transport, tourism, etc.; and (iii) policies promoting development: environmental, nature protection, education, research, etc. Apart from these public policies, the development of the forest sector is also influenced by the international commitments of Romania and the EU accession process (Table 23.2).

During the last decade, the country's economic situation significantly affected the forestry sector, including forest management. The years of economic decline have had a negative impact on the activities of the logging and processing companies; have diminished the volume and quality of forest operations; and have considerably reduced the income of the National Forest Administration and its investment capacity. Also, the budget allocation for forestry has been limited and many of the governmental programmes related to forestry have only been partially achieved.

Restitution of forest land and privatization of wood harvesting, transport and the processing sector have probably had the highest impact on the development of forestry and forest management in Romania. The size of the restituted forest areas according to Law 18/1991, which in many cases represented only part of the prenationalization individual ownership, created frustration among forest owners. In addition, the poor capacity to enforce forest legislation and to raise forest owners' awareness of sustainable forest management resulted in significant environmental damage in private forests.

Table 23.2. Public policy impacts on forestry sector development. (From Abrudan, 2002.)

Public policy domain	Impact	
	Main direction[a]	Impact scale[b]
Public policies establishing the institutional framework		
• Economic growth	Forest policy impacted	***
• Privatization	Forest policy impacted	****
• Public finance	Forest policy impacted	***
• Employment	Forest policy impacted	**
• Rural development and land use planning	Reciprocal links	**
Public policies related to specific economic sectors		
• Agriculture and game management	Forest policy impacted	***
• Wood harvesting, transport and processing	Reciprocal links	****
• Mining and energy	Forest policy impacted	**
• Infrastructure	Forest policy impacted	**
• Tourism	Reciprocal links	***
Public policies promoting development		
• Environmental protection and water management	Reciprocal links	****
• Nature conservation	Reciprocal links	***
• Education and research	Forest policy impacted	**

[a]Possible direction: forest policy is interfered with; reciprocal link; forest policy interferes.
[b]On a scale from * = little impact to **** = very significant impact.

Although private forest management structures have increased significantly since 2002, the general opinion is that in the short term, the forests restituted according to the second restitution law promulgated in 2000 are poorly managed. The reasons for this include: a lack of capacity and knowledge, a vested interest in gaining immediate economic benefits and inadequate law enforcement capacity (Bouriaud and Abrudan, 2004).

The almost completed privatization of wood harvesting, transport and processing has had mainly positive effects on forest management. Privatization resulted in a higher competition for wood resources and increased prices for standing wood, with direct financial benefits for the National Forest Administration. The public financing has a direct impact on the development of the forestry sector as the regulatory, control and extension functions of the state depend on the annual budget allocation. The budgetary allocation for the forest sector has been relatively small during the last decade, thus particularly affecting the control and extension functions, as well as the staff quality and commitment in public authorities.

Agricultural policies and legislation have had some important influences on forest sector development. An important aspect is, for instance, that any agricultural policy and regulation must not lead to the reduction of the public forest area. Indeed, afforestation of degraded agricultural land is a priority in the present government policy to increase the forest cover. This priority also is consistent with the EU agricultural and rural development policies, given that Romania's forest cover per capita is at present lower than the average among member countries. Some agricultural policies, as well as agricultural activities have negative effects on forests and forest management. Despite being forbidden by law, grazing represents by far the main problem. Legislation on game management and hunting is also affecting forest management. In accordance with the existing legislation, the central public authority for game management assigns game management rights to the legally established hunting organizations. This provision has created some conflicts between hunting organizations and private agricultural and forest land owners.

During the period 1990–2000, forests, environmental protection and water management were under the same public authority (a ministry), and as a result of this situation there were no apparent major conflicts between these sectors.

After 2000, responsibility for the forest sector was transferred to the public authority responsible for agriculture. Since then, some conflicts between the forest sector and environmental protection have been recorded, especially in the field of nature protection. The Medium-Term Environmental Protection Strategy includes strategic objectives with direct positive influences on forest development: extension of the forest area, establishment of forest belts in areas exposed to desertification, afforestation of degraded agricultural land and improvement of the legislation on forest protection. Other positive influences on forest development have resulted from the provisions of the Environmental Protection Law. Close cooperation has developed in recent decades between forest and water management bodies. However, the effect of water management on forests appears to be much lower than the effect of forest management on water quality and management. Special forest management practices are recommended in order to protect waters and watersheds. New wood harvesting and transport guidelines tend to include water protection requirements.

There are many linkages between forestry and nature conservation. Although intensive logging had negative impacts on biodiversity preservation during the first half of the 20th century, the close-to-nature approach in silviculture that has been practised since the 1950s in Romanian forestry has reduced such impacts. Unique or rare forest ecosystems and wildlife species have been preserved due to the efforts of the foresters and forest organizations. During the last decade, they have been involved in most of the processes, programmes and activities related to nature and landscape protection. This includes, for instance, the establishment and management of most of the protected areas that exist today in Romania (11 national parks and 5 natural parks). Despite this fact, there have also been situations in which forestry operations have had negative cumulative effects on protecting water flow and water quality, and preserving flora and fauna. However, in recent times foresters have increasingly become more open to the dialogue with conservation organizations and the public on nature conservation issues.

Strong linkages exist between tourism and forestry, especially in the Carpathian region, and cooperation among the public authorities responsible for tourism and forestry has improved significantly. Thus, the national authority for tourism has participated actively in the development of the current National Forest Policy and Strategy. Ecotourism has become a priority for both the forestry sector and tourism agencies. Although forestry now appears to be having quite positive effects in providing a number of external services in tourism, the latter produces a relatively negative impact on forests: clear-felling to allow development and the construction of hotels, restaurants, skiing facilities, etc.; garbage left in the forest by tourists; illegal camping and picnicking; and above all, forest fires caused by negligence on the part of tourists are relevant examples.

Development and modernization of the road infrastructure (public and forest roads) have had both negative and positive impacts on forest sector development. The negative impacts result from forest clear-felling to make room for new public roads or motorways. In many cases, the government has approved clear-felling and exempted such development from the land use change tax. On the other hand, the development of the transport infrastructure may have positive impacts on forest management, as it provides better access to forest resources, particularly for wood harvesting, tending of forest stands and maintenance operations.

During the last decade, impacts of the effects of mining on the forestry sector have been reduced significantly. There are now plans to afforest many closed quarries. Oil extraction has influenced the forest sector both by reducing the forest area and by increasing the risk of accidental pollution. However, until now the effects of this have been relatively low and the affected areas small. On the whole, the energy sector has had both positive and negative effects on forestry development. The expansion of the gas network in rural areas is contributing to a reduction in the demand for the firewood traditionally used for heating and cooking. This reduces pressure on forest resources in forest-poor areas (plains and hilly regions). In forest-rich areas (mountains), the expansion of the gas network has led to a decrease in the price of firewood and to a lower income for the National Forest Administration.

Although there is no separate public authority responsible for rural development, this sector is highly important, especially within the framework of the EU accession process. Romania has developed a National Plan for Agriculture and Rural Development, and significant financial support is available from the EU for the implementation of this during the period 2002–2006. Some of the measures eligible for funding are directly related to forestry and will have a positive influence on forest sector development. Typical measures funded under this programme are afforestation of land inappropriate for agriculture; establishment of forest nurseries; construction of forest roads, wood harvesting and wood processing; and the establishment of local associations of private forest owners (CEC, 2004).

Education and research play an important role in the development of the forestry and wood-processing sectors. Specialized forestry high schools and higher education institutions have provided the necessary technical and professional staff employed by the sector and have been engaged in research and demonstration projects. The recent rapid increase in the number of graduates from both middle and higher educational institutions has both positive and negative consequences for the forestry and wood-processing sectors. The quality of education was negatively affected by the increased number of students, while at the same time stronger job competition has led to the employment of better-qualified staff. In recent years, efforts have been made to adapt the work of the National Forest Research Institute to the expanding needs of the sector to manage forests in a sustainable and multifunctional manner in the conditions of the liberalized market economy. However, the institute, like many other organizations, with a mandate for forest research, is finding it increasingly difficult to secure funding from public and private sources. Although it has the professional and technical expertise to implement management planning and ecological research, it does not yet have sufficient experience and capacity to identify international funding sources and participate in collaborative research networks.

Conclusions

After the fall of Communism, the conditions for policy-making changed fundamentally in Romania. The change has been determined not only by national processes in the context of the democratization of society and the transition to a market economy, but also by international processes and trends towards globalization. A significant improvement in cross-sectoral policy formulation and development planning has occurred, which has stimulated decision-making processes that can foster stakeholder participation and greater transparency. Nevertheless, there are still many improvements to be achieved in order to develop the dialogue between sectors and interest groups, as well as within policies that determine the role, structure and functioning of public administrative bodies that are directly or indirectly concerned with forest protection and forestry development. Romania's international commitments, as well as the EU accession process, are important factors affecting the forest sector.

The institutional changes needed to maximize positive public policy impacts and minimize negative ones depend on internal factors specific to the forest sector, as well as on external factors related to more general mechanisms at the political decision-making level. The strengthening of the Department of Forests within the Ministry of Agriculture, Forest and Rural Development should be a priority, in order to increase its capacity to actively influence decisions at the political level and to foster synergies with macroeconomic policies that have a direct or indirect impact on forest sector development. The establishment of interministerial committees and working groups as mechanisms for addressing cross-sectoral issues appears to be having beneficial effects, at least in relation to the following sectors: agriculture, timber processing, environmental protection, water management, nature conservation, tourism, research and education. In the legislation development process, the Department of Forests should enhance its collaboration and dialogue with the main forest stakeholders and interest groups (private forest owners, private wood-processing industries, research and education institutions and non-governmental organizations (NGOs) engaged in environmental, nature and landscape protection) in order to be able to adequately reflect and represent their opinions and interests in political decision-making processes. The department will play an important role in the establishment and development of the private forest management structures. It needs to increase its capacity to coordinate and monitor the development and functioning of the private forest districts and should be able to provide extension and technical services through its territorial units to private land owners. There is an urgent need for the establishment of a strong unit to coordinate the activities related to private forests, in view of the recent forest restitution process.

While in the near future a significant proportion of Romanian forests will be in private hands, the role and mandate of the National Forest Administration should be adapted to its new position in the Romanian forestry sector. The managers of state forests should be able to act fully in accordance with their commercial mandate, as they have to face the private sector competition. The National Forest Administration needs to improve its organizational, operational and commercial efficiency and should be in a position to optimize its contribution to the economy through the sustainable management of state forest resources. There is a strong need for public awareness campaigns targeted at the general public, as well as at key stakeholders, with particular emphasis on communities living in forested areas, private forest owners (individuals and cooperatives), public decision-makers and influential groups, such as the church and NGOs. The National Association of Private Forest Owners increasingly needs public support in fulfilling its role in the sustainable development of restituted forests and providing services to new forest owners. Introducing appropriate regulations, including financial mechanisms to support sustainable forest management, as well as the development of alternative income-generating activities in rural areas, can provide methods of maintaining the ecological functions of the restituted forests. This requires a coordinated effort on the part of the Department of Forests, the Forest Inspectorates and the Association of Private Forest Owners.

References

Abrudan, I. (2002) *Cross-sectoral linkages in Romanian Forestry. Country Case Study Report prepared for FAO Forestry Policy and Institutions Service*. FAO Rome.

Abrudan, I., Tamas, S. and Ionascu, G. (2002) Privatization in Romanian forestry. In: *Proceedings of the International Seminar: Privatization in Forestry*. Belgrade, 2002, pp. 59–72.

Bouriaud, L. and Abrudan, I.V. (2004) Recent forest policy developments in Romania and the main challenges ahead. In: *Forests in Transition II: Challenges in Strengthening of Capacities for Forest Policy Development in Countries with Economies in Transition*. United Nations University Press, Belgrade, pp. 159–174.

Commission of the European Communities (CEC) (2004) *Regular Report on Romania's Progress Towards Accession, SEC (2004) 1753*. European Commission, Brussels.

Monitorul Oficial al Romaniei (2001) *Hotararea Guvernului Nr. 555/2001 pentru Aprobarea Regulamentului Privind Procedurile pentru Supunerea Proiectelor de Acte Normative spre Adoptare Guvernului*. Monitorul Oficial al Romaniei, Bucharest.

Schmithüsen, F. (2003) Understanding cross-sectoral policy impacts—policy and legal aspects. In: Dubé, Y.C. and Schmithüsen, F. (eds) *Cross-Sectoral Policy Impacts between Forestry and Other Sectors*. FAO Forestry Paper, 142. Food and Agriculture Organization of the United Nations, Rome, pp. 5–44.

World Bank (1999) *Economic Evaluation and Reform of the Romanian Forestry Sector. Final Report*. World Bank, Washington, DC.

World Bank (2001) *Project Appraisal Document for Romania Forest Development Project. Draft Document*. World Bank, Washington, DC.

24 Importance of the Turkish Forestry Sector in the National Economy: an Input–Output Analysis

Mustafa F. Türker

Introduction

Forests in Turkey occupy 21.2 million ha or 27.2% of the total land area. They are classified either as productive forests (48%) or as degraded forests, rangelands and eroded areas (Muthoo, 1997; Konukçu, 1998). The state is the sole owner of the total forest area, and forest management activities are carried out by 217 State Forest Enterprises under the General Directorate of Forests of the Ministry of Environment and Forestry (Anonymous, 2006). Forest management plans have been prepared according to the principle of maximum sustained yield, and multiple uses of forests have generally been ignored. The forestry sector faces many challenges, such as inadequate recording and monitoring, management planning and transformation of its raw material.

The main purpose of this chapter is to determine the importance of the forestry sector in the national economy in terms of production, income, added value, trade and its impacts on the other sectors of the national economy. Another objective is to improve knowledge and awareness among national and forestry sector planners and decision-makers about the important role the forestry sector plays in the Turkish economy. It is interesting to note that the results of the analysis confirmed the importance of the forestry sector without taking into account the non-market economic aspects, the social and environmental goods and services provided by forests. In other words, the results of the analysis, although positive, grossly undervalued the overall importance of forests in the national economy. The study argues that awareness among politicians and decision-makers about the total economic value of forests needs to be improved. It also proposes the development of methods of improving the recording of this value in the national accounts.

Methodology

National economic development plans have been prepared using input–output tables since 1963 in Turkey (Erdoğmuş, 1988). Input–output tables are tables showing the flows of goods and services between industries (Leontief, 1966; Çakir, 1987). They are one of the most suitable models for analysing the structural relationships between a country's different sectors. These models describe the general economic structure of a country and the relations between all sectors of its economy. In addition, input–output models play an important role in dealing with economic problems, such as production, income and employment (Türker, 1999).

In this study, input–output tables have been used to determine the importance of the forestry sector in the national economy. The data used were taken from the tables of inter-industry treatment tables, the input–output coefficient matrix and the technological coefficient inverse matrix provided by the State Institute of Statistics. The most recent input–output tables, prepared in 1996 and published in 2001, were used in this study (SIS, 2001).

The coefficient matrix is obtained by dividing each entry in the input–output table by the corresponding column total. Each input coefficient measures the proportion of input value that an industry has to purchase from other industries in order to produce one unit of its output. With regard to the technological coefficient inverse matrix, the highest values in the sums of the row and column elements indicate that the demand for the outputs of the corresponding sectors is increasing. The inverse matrix is also useful for calculating other useful indicators, such as forward and backward linkage rates, or in estimating production, income and employment multipliers. Forward and backward linkage rates are obtained by dividing the sum of intermediate products sold or bought from other sectors by one sector by its total production. They measure the importance of a sector in the economy. Similarly, the multiplier analysis shows the impact of one unit of increase in the final demand for each sector on production, household income and employment in the national economy (Morrison and Smith, 1977; Çakir, 1987).

Table 24.1 shows the conceptual framework of the input–output table. Each industry is included twice, once as input user and once as output producer. Along the row for any sector, the distribution of products into various industries and final demands is shown. Column elements represent the inputs from other industries to make its products. This table consists of four main fields: field I represents inter-industry flows, field II shows final demands, field III presents added values and field IV depicts direct factor inputs for final demand (Bocutoğlu, 1982).

Results and Discussion

Table 24.2 shows the share of the forestry sector's production and other sectors in the national economy. It also shows that the forestry sector is an important supplier of intermediate goods to other sectors of the national economy. Timber, mining pole, wire pole, pulpwood, fibre and wood chips, industrial wood, fuelwood, resin, styrax (liquidambar oil), etc., were taken into account in estimating the share of the forestry sector's total production in the total production of the national economy. However, if private-sector wood production and illegal fuelwood production are included in the calculation, the share of the forestry sector in the national economy would rise to 1.76% (Çakir, 1984). In addition, the forestry sector also provides important environmental services, such as erosion control, water production and regulation of water to extend the economic life of dams or meet the recreational needs of urban citizens. These cannot be valued in monetary terms. If these services were also to be included in the calculation, the share of forestry in the national economy would be much higher. In effect, according to the results for Turkey of the country report by the Mediterranean forest evaluation project, only the monetary values of wood forest products (42%),

Table 24.1. Input–output table.

		Intermediate demand							Final demand						Total output		
		1	2	...	*j*	...	N	K	C	I	G	E	S	Y	Z	–M	X
	1	X_{11}	X_{12}	–	X_{1j}	–	X_{1n}	K_1	C_1	I_1	G_1	E_1	S_1	Y_1	Z_1	$-M_1$	X_1
	2	X_{21}	X_{22}	–	X_{2j}	–	X_{2n}	K_2	C_2	I_2	G_2	E_2	S_2	Y_2	Z_2	$-M_2$	X_2
			Field I								Field II						
Sectors	i	X_{i1}	X_{i2}	–	X_{ij}	–	X_{in}	K_i	C_i	I_i	G_i	E_i	S_i	Y_i	Z_i	$-M_i$	X_i
	n	X_{n1}	X_{n2}	–	X_{nj}	–	X_{nn}	K_n	C_n	I_n	G_n	E_n	S_n	Y_n	Z_n	$-M_n$	X_n
	U	U_1	U_2		U_j		U_n	U = K									
			Field III								Field IV						
	V	V_1	V_2	–	V_j	–	V_n	V_K	V_C	V_I	V_G	V_E	V_S	V = Y			
	X	X_1	X_2	–	X_j	–	X_n	X	C	I	G	E	S		Z	–M	X

Table 24.2. Input–output analysis results.

Indicator	National economy (97 sectors)	Forestry sector
Value of production (million TL)	31,715,306,845	106,189,963
Share of forestry and other sectors (%)	–	0.003348
Export (million TL)	3,653,236,279	350,862
Import (million TL)	4,129,894,553	11,819,075
Export/import rate (%)	88	3
Intermediate demand (supply) (million TL)	11,752,352,805	85,686,740
Total demand (supply) (million TL)	31,715,306,845	106,189,963
Share of intermediate demand in total supply (%)	37	81
Gross added value (million TL)	15,833,059,488	81,413,926
Total production (million TL)	31,715,306,845	106,189,963
Share of added value in total production (%)	50	77
The sum of column J elements	–	1.222
Order of forestry in the 97 sectors	–	93
The sum of row 'i' elements	–	1.868
Order of forestry in the 97 sectors	–	24
Forward linkage rate	–	0.81
Backward linkage rate	0.37	0.12
Production multiplier	0.37	1.222
Place of forestry in the 97 sectors	–	93
Income multiplier	–	1.377
Place of forestry in the 97 sectors	–	89
Employment multiplier	–	0.291
Place of forestry in the 97 sectors	–	17

TL = Turkish lira.

non-wood forest products (8%), hunting (3.4%) and recreation (0.2%) are reflected in forestry sector and national balance sheets. In other words, the values of the other components of the total economic value of the forestry sector that were also estimated in that project, such as grazing (21%), carbon storage (14.8%), pharmaceuticals (10.5%) and biodiversity conservation (0.1%), are not included in the sectoral and national balance sheets (Türker *et al.*, 2005).

Non-wood forest product exports represent about 98% of the total exports of the sector and rank Turkey as the third major exporter of these products worldwide after China and India (Türker *et al.*, 2005). However, the shares of forestry sector's exports and imports in the national economy's total exports and imports are not significant. The majority of the sector's production is consumed domestically, and imports of wood products are necessary to satisfy the overall demand for forest products. Consequently, it can be said that forestry is not an export sector in Turkey.

The forestry sector is also characterized by few backward linkages but important forward linkages, providing inputs to other sectors, which make it a strategic sector (Geray, 1993). The fact that the share of added value in the forestry sector's total outputs is high in comparison with other sectors is explained by the high forward linkages and the use of labour-intensive technology. With regard to the production multiplier, the increase in the final demand of the forestry sector has the lowest influence on the total production increase in the national economy. With regard to the income multiplier, the income effect of the forestry sector is not significant. With regard to the employment multiplier, the values were obtained from research carried out in the Trabzon subregion located in the northern part of Turkey (Özyurt, 1982). According to this study, the forestry sector ranked 17th among 64 sectors with regard to the employment effect in the Trabzon subregion's economy. This result is not too surprising when one considers that the forestry sector is a labour-intensive industry.

In summary, the findings of this study are:

- The forestry sector does not have an important effect on production in the regional and national economy. However, the environmental and ecological services provided by the forestry sector contribute to increase its overall importance.
- The forestry sector mainly supplies forest products to domestic markets. Thus, its exports are negligible as a proportion of the national economy.
- The forestry sector produces important added value. It can therefore be used as a tool for reducing unemployment and income inequities.
- The forestry sector's sectoral linkages show that forestry favourably affects other sectors by providing them with strategic raw material.

Conclusions

This study has found that it is difficult for the forestry sector to demonstrate its real effect on the national economy, due to a lack of understanding, information and methods of accounting for its real contribution in the national accounts. In effect, only 53.46% of forestry-sector output is included in the national balance sheets, and the balance of positive effects is not shown. These figures show that the share of the forestry sector in the national economy is lower than its real value. If the real effects of the forestry sector could be calculated and reflected in the national accounts, its profitability would increase and therefore greater investment would be likely to flow into the sector. In addition, sound policies and strategies regarding the effectiveness of the use of forest resources could be established from a sustainable development point of view.

The Turkish forestry sector plays an important role in the national economy. In addition, it provides social and environmental services that are vital for the sustainable development of the national economy. There is therefore no reason why it should be perceived by politicians and decision-makers as a marginal sector in the national economy. In addition, the awareness of policy-makers should be increased regarding the importance of the forestry sector, particularly in terms of:

- Being a relatively independent sector and important supplier of raw material to many other national industries. The fact that it makes a strategic contribution to most other sectors should be taken into consideration when allocating public funds.
- Providing essential wages to mountain dwellers and forest villagers, and thereby contributing to the alleviation of poverty.
- Being a low-profit sector, with a need to pay for (invest in) the environmental services provided by forests in order to compensate the producers of these services to society.
- Contributing to eliminating interregional income disparities.
- Making the production and income effects clear to decision-makers, and incorporating unrecorded flows of goods and services into the national accounts and budgets.

References

Anonymous (2006) General Directorate of Forests web page. Available at: http://www.ogm.gov.tr

Bocutoğlu, E. (1982) Inter-industries approach in analyzing Turkish imports: input–output and linear programming models and their application (in Turkish). Postdoctoral thesis, University of Trabzon, Turkey.

Çakir, M. (1984) *The Importance of Forestry Sector in the National Income* (in Turkish). Forest Engineers Association Publication, no. 9. Forest Engineers Association, Ankara.

Çakir, M. (1987) *Regional Planning and the Importance of Forestry Sector (Bolu Regional Directorate Example)* (in Turkish). Technical Bulletin Series, no. 189. Forestry Research Institute, Ankara.

Erdoğmuş, Ş. (1988) Input–output model in planning the agricultural sector and Turkish example (in Turkish). MSc Thesis, Ankara University, Eskiçehir.

Geray, U. (1993) Problems concerning the management of forest resources in Turkey (in Turkish). *Journal of Forest Faculty* 43, 15–30.

Konukçu, M. (1998) *Statistical Profile of Turkish Forestry*, State Planning Organization, Ankara.

Leontief, W. (1966) *Input–Output Economics*. Oxford University Press, New York.

Morrison, W.I. and Smith, P. (1977) *Input–Output Methods in Urban and Regional Planning: a Practical Guide*. Pergamon, Oxford, UK.

Muthoo, M.K. (1997) *Forests and Forestry in Turkey*. Ministry of Forestry, Ankara.

Özyurt, H. (1982) Trabzon subregion input–output model and its application to sub-region planning (in Turkish). Postdoctoral thesis, University of Trabzon, Turkey.

SIS (State Institute of Statistics) (2001) *The Input–Output Structure of the Turkish Economy, 1996*. Publication, no. 2488. State Institute of Statistics, Ankara.

Türker, M.F. (1999) The determination of the importance of forestry sector in the national economy. *Turkish Journal of Agriculture and Forestry*, 23, Additional Issue 1, 229–237.

Türker, M.F., Pak, M. and Öztürk, A. (2005) Turkey. In: Merlo, M. and Croitoru, L. (eds) *Valuing Mediterranean Forests: Towards Total Economic Value*. CAB International, Wallingford, UK, 448 pp.

25 The Experience of the US Forest Sector in Cross-sectoral Impact Analysis

Denise C. Ingram, Ralph J. Alig and Karen L. Abt

Introduction

In the USA, forest resources, and the land on which they occur, are in themselves multiple-sector production units. They can provide, *inter alia*: recreation and leisure; potable water for rural and urban populations; wood and fibre for construction, education, health and hygiene; transportation routes; places to live and work, and valuable cultural and religious significance to the nation's citizens. The very definition of forest resources belies an intrinsic mandate for planning and management across the social, economic and political boundaries of natural resource-based sectors. Understanding these relationships is an important goal of the stewards of forests and of the public and private planners of sustainable development.

Historically, the methodologies and analytical tools applied to the US forest sector focused primarily on input–output studies, partial and general equilibrium models, comparative analysis, environmental quality indicators and natural resource assessments. A look at a number of complementary approaches reveals the challenges of understanding the level and extent of cross-sectoral impacts in the forest sector. Land is the primary mechanism by which policies from one sector influence the forestry sector, and thus land use models are a major tool for cross-sectoral analysis. The outputs of various sectors are linked through employment and income, providing a second tool, input–output analysis. Finally, several types of assessments use these tools. This discussion also highlights some caveats in the development of cross-sectoral analysis for future strategies on the conservation and management of forest resources.

Four methods used to analyse cross-sector impacts in the USA are discussed below. The first is a bottom-up look at the impacts of changes in the quantity of forest land, particularly as land shifts to and from forest land and agriculture. A second bottom-up evaluation method is the use of input–output analysis, which has been consistently used for cross-sector analysis in many countries, including the USA. The third and fourth methods are top-down methods, which often include bottom-up land use change methods and input–output methods as inputs. These methods include national and bioregional assessments of forest conditions, and environmental and resource accounting.

Land Use Modelling

Forest and agriculture sectors analysis is used in order to determine land market linkages that have important influences on the forest sector, affecting natural resource sustainability options, costs and benefits. Intersectoral land use changes among agricultural, forest and developed (e.g. urban) uses in the USA have annually totalled more than 1 million ha in recent years (Alig *et al.*, 2003). Cross-sectoral impacts warrant coordinated forest and non-forest policy deliberations to improve efficiency and equity aspects of changing land resource conditions. Intersectoral coordination would include analysis of human and natural dimensions of land base changes and conditions, given the associated broad range of goods and services that include forest-based environmental services, agricultural food and fibre, many places to live and work, as well as valuable cultural and religious features for US citizens. Below, we evaluate the usefulness and application of two land use modelling methods for addressing and understanding cross-sector impacts. Both model types address changes across sectors that use land, primarily agriculture, forestry and the 'urban' aggregate. However, in terms of sector, most of our economy is in the 'urban' aggregate (manufacturing, service, retail and wholesale, transportation, etc.), which limits the ability to address impacts from outside the forestry and agriculture sectors. Urban-related uses typically sit at the top of the economic hierarchy of land use and, at certain scales, are often satisfied first, before land use competition between forest and agriculture occurs. For land use modelling, it is important to recognize the fixed nature of the land base, often involving the use of constraints to ensure zero sum accounting.

Theoretical land use models are derived from assumptions governing human behaviour and natural processes and explore land use patterns associated with these assumptions. Empirical land use models explain observed patterns implied by the theoretical models using data to quantify the hypothesized behavioural relationships (Alig, 1983; Alig *et al.*, 1983). Once estimated, empirical models can be used to project how land use will change in response to changes in economic conditions and policies. Empirical land use models use econometrics, mathematical programming or simulation, each varying in their relative strengths and spatial and temporal scales. A fourth approach – incorporating land use into national and global trade models (e.g. Ianchovichina *et al.*, 2001) – is highly aggregated and generally provides insufficient spatial detail with which to examine specific forest resource concerns. We limit our discussion to econometric and mathematical programming models, because they have been used in recent large-scale assessments that support analyses of cross-sectoral impacts.

Econometric land use models use statistical methods to quantify relationships between land uses and their hypothesized determinants, on the basis of the observed or revealed landowner behaviour. Landowners are assumed to allocate land parcels to the use that generates the highest profit or utility. Models are estimated with data describing land use decisions and economic benefits derived from alternative land uses (Ahn *et al.*, 2000). Additional variables may be included to control land use policies and other factors that may influence land use decisions. Because they measure the way in which landowners actually respond to observed changes in economic and other decision variables, econometric models capture the combined influence of several factors motivating land use decisions that may otherwise be difficult to discern. Projections with such models assume that the relationships between such factors and land use changes will continue (Alig *et al.*, 2003, 2004).

Although estimation on the basis of historical data is the strength of econometric models, it is also a weakness. Econometric models become less accurate when used for predictions outside observed historical ranges, although the extent of the error is measurable. Thus, econometric models can primarily be used to understand the extent and direction of relationships between various forest and non-forest sectors. These models can also be used for short-term projections of primarily marginal changes in policies or other assumptions. Econometric models are not typically used to evaluate the consequences of major market, technology or policy changes. Additional examples of econometric land use models include Alig (1986), Hardie *et al.* (2000) and a series of econometric studies that offer insights about determinants of afforestation (Plantinga *et al.*, 1999) and reforestation activities (Alig *et al.*, 1990; Lee *et al.*, 1992; Kline *et al.*, 2002).

Mathematical programming describes land use allocations using numerical optimizing techniques to find the multi-market price and quantity vectors that either maximize or minimize the value of an objective function, resulting in an optimal land use allocation (Adams *et al.*, 1996; Parks *et al.*, 2000; Alig *et al.*, 2004). Maximization or minimization of the objective function can be subject to sets of constraints that characterize the resource commodity production over time, initial land and resource conditions, availability of fixed resources (e.g. land) and policy constraints. Land is modelled as an input resource that moves among different sectors depending on relative land rents.

Advantages of the mathematical programming approach include its theoretical basis of market equilibrium. Mathematical programming models also allow for intertemporal optimization, given the dynamic economic behaviour pertaining to land use. Such models are better able to handle economic conditions outside of historical ranges, which facilitates evaluating new phenomena, such as global climate change and new policies. Programming models can also incorporate a wealth of physical structural detail, which is important when physical structure strongly influences the behavioural response, such as for existing forest age class structure and timber harvest behaviour. Disadvantages of mathematical programming land use models include a high level of spatial aggregation, behavioural relationships on the basis of assumed or revealed behaviour and difficulty in calibrating to recent historical trends where appropriate.

The Forest and Agricultural Sectors Optimization Model (*FASOM*) is a linked model of the forest and agricultural sectors (Adams *et al.*, 1996). This model considers whether the net present value of the future returns to land in one sector outweighs those earned if land remains in the other sector, plus any adjustment costs involved with land transfers. If this is the case, then land will transfer, and this rate of transfer will continue until the land markets equilibrate such that the value of the marginal hectare in either sector differs by no more than the transfer costs. In forestry, this calculation considers the returns to re-establishing a stand, plus the value of all future forest stands that would succeed the stand now being considered and the terminal value of any unharvested trees at the explicit model end. On the agricultural side, this considers the returns to land in cropping and/or livestock uses in the current condition, plus all future periods including the final, perpetual one.

An extended version of FASOM was developed to aid the US Environmental Protection Agency and US Department of Agriculture (USDA) Forest Service in examining greenhouse gas policy options. The policy analyses include the way in which various policy instruments may differentially affect the various sectors of the economy, especially the forest and agricultural sectors. The extended version of FASOM has a composite objective function that spans greenhouse gas, agricultural sector, forestry sector and intersectoral transfer submodels. When policies that provide incentives or require adjustments are imposed, the objective function causes the model to consider the possibility of adjusting activity, such as increased use of no-tillage agriculture to increase carbon sequestration in soil, anywhere within the forestry and agricultural sectors (Alig *et al.*, 2002). In addition, the amount of land transfers may be affected (e.g. afforestation or land use change from agriculture to forest use), as well as commodity movements across sectors. Such activities are also adjusted temporally and geographically. In turn, the model re-equilibrates the land, commodity and factor markets across all sectors, time periods and geographic locations.

This model also allocates forest and agricultural land to developed uses (e.g. urban use) in view of a growing and, on average, wealthier US population (Alig *et al.*, 2004) that will demand more developed area for living space, transportation infrastructure and other developed uses. Developed uses sit at the top of the economic hierarchy of land uses, such that prices for developed land often dominate prices for forest uses in areas suitable for development (Alig and Plantinga, 2004). Policy deliberations recognize that the growth in developed area is not likely to be arrested, but rather may be diverted or relocated by policies offering preferential tax assessment or other traditional programmes (Alig *et al.*, 2002, 2004).

Input–Output Analysis

A second method of analysing cross-sector impacts through individual sectors uses input–output modelling. This is the most common method of determining jobs and income effects that derive from forests and forest uses, and thus provides a link from forests to the larger measured monetary economy of a region or country. The method is data-intensive, subject to severe limitations due to assumptions on technology and is at times referred to as ‘looking through the rear-view mirror' to see how economic sectors are interrelated.

The use of demographics, the economic behaviour of producers and consumers, employment, income and other economic multipliers in input–output models results in the added benefit of defining the structure of the economy in question. However, with a reliance on intensive data interpretation and compilation, input–output models present numerous challenges to users. The accuracy and correct application of the multipliers require an intimate understanding of the economic system, the economic sectors and their interrelationships in order to avoid common mistakes, such as multiple accounting of effects or misspecification of a multiplier (Deller *et al.*, 1993). Input–output models can also suffer from the lack of adequate disaggregation of data. Although reliable secondary data and proxy values for input multipliers are sometimes used, the data intensity of input–output models requires as little deviation as possible between the model parameters and reality. Input–output models are also restrictive, in the sense that all resources are considered fully employed (Alward and Palmer, 1983), such that the output from the model does not allow a redistribution of resources within the economic system. Although this feature is a positive element with regard to tracking imports and exports for an economy, it probably does not reflect many economic decisions that would result in some redistribution of resources. Input–output models also assume fixed employment-to-output ratios that do not allow responses to economies of scale and substitution effects (Deller *et al.*, 1993).

For the US forest sector, detailed data and a sophisticated modelling framework are available for evaluating existing sector interrelationships, for evaluating marginal changes in sectors that affect forests and forestry and how marginal changes in forestry will affect other sectors. This framework is known as the impact analysis for planning modelling system (IMPLAN), developed by the USDA Forest Service (Alward and Palmer, 1983). The model has since been modified into a regional input–output tool (Alward *et al.*, 1989), with several applications for evaluating forest contributions to regional, state and local economies. IMPLAN and its results are often referenced in forest resource assessments, which are described in the following section. Applications include Deller *et al.* (1993), who modelled a county-level economy in Wisconsin to assess the impacts of changes in the demand for paper products on employment in the services sector. Munn (1998) and Aruna and Cubbage (2000) analysed the contribution of forestry to state level economies. Hughes and Hinson (2000) consider the contributions of the green industry (floriculture, turfgrass production and environmental horticulture) to the economy of the state of Louisiana, including impacts on the combined sectors of agriculture, forestry and fishery. The range of scale that is supported in its application reflects the flexibility of the model to analyse real world dynamics and complexities.

National and Regional Forest Resource Assessments

The focus of national-level forest assessments prior to 1974 was timber supply and demand (Haynes, 2003). After passage of the Forest and Rangeland Renewable Resources Planning Act (RPA) of 1974, other resources were included, and the most recent RPA assessment, in 2000, specifically addressed water, range, recreation, wildlife and minerals, in addition to timber. In complement to the resource documents, specific questions of interest were included in the assessment that reflect common interchanges between forest resources and other resource and economic sectors, as well as among forest resources. The RPA assessment has not, in previous years, been used specifically to evaluate cross-sectoral impacts, although some analysis occurs as a by-product of understanding forests and projecting future trends. The evaluation of land use trends in the assessment includes descriptions and understanding of the linkages between agriculture, forestry and urbanization. In addition, the contribution of forests to jobs and income are included in the RPA assessment. Alternative futures are also represented in the assessment that include linkages to other sectors, such as recycling rates (waste management) or demand for housing (construction). Because the structure of the assessment is forest-centric, including more evaluation of cross-sectoral policy impacts would require including other sector-specific questions in the planning of future RPA assessments.

A second type of national assessment addressed the Montreal process criteria and indicators for sustainable forest management (USDA Forest Service, 2003). Criterion 7 addresses the legal, institutional and economic framework, and comes closest to an evaluation of cross-sectoral impacts. These impacts are alluded to in the assessment, but the timing of the national report precluded sufficient analysis of policies in other sectors that could affect forests and forestry. Specific indicators that allow for the inclusion of cross-sectoral questions include indicator 49, which evaluates the extent to which policies recognize the full range of forest values including coordination with relevant sectors, and indicator 54, which evaluates the capacity to conduct assessments, including cross-sectoral planning coordination. Although neither indicator specifically addresses policies in other sectors that affect forestry or forestry policies that influence other sectors, instead focusing on forestry policies, conditions and assessments, the development of the assessment was accomplished with input and coordination from other federal agencies and through public workshops, which provided multiple-resource viewpoints. One challenge for further development of the criteria and indicators and sustainability assessments includes the need for cross-sectoral impact analysis. One suggestion is to add the addition of forest land values to the set of US indicators.

A third type of assessment conducted in the USA is the bioregional assessment. A bioregional assessment is one whose extent is defined by an ecoregion or some ecological need, rather than a jurisdictional assessment whose extent is defined by government boundaries, such as the RPA or the national sustainability assessments. Bioregional assessments tend to be resource-centric, in much the same way as the RPA and sustainability assessments. The focus is on the condition of the forest, specifically addressing the wildlife, fish or timber aspects that instigated the need for a regional assessment.

Perhaps the most famous of these is the Forest Ecosystem Management Assessment (FEMAT, 1993), which ultimately resulted in the North-west Forest Plan and made striking changes to the management of public forest lands in parts of the Pacific Northwest. This plan provided analysis of regional declines in old-growth forests with regard to the northern spotted owl and other late-seral forest-dependent species. Its efforts compelled the management of public and private lands to change to protect habitat and sustainability.

The Sierra Nevada Ecosystem Project (University of California–Davis, 1996) was a response to a congressional request to implement scientific review of late-successional forests, key watersheds and significant natural areas (e.g. areas with sensitive species) on federal lands, 'including their social, economic and ecological components' in the Sierra Nevada ecoregion, including California and western Nevada. The results contributed to a gap analysis of the state of California. In another example, the Interior Columbia River Basin Assessment (Quigley and Bigler-Cole, 1997) linked landscape, aquatic, terrestrial, social and economic characterizations to describe biophysical and social systems for eastern Washington and Oregon, and Idaho, and contributed to planning on national forests in the region.

These three assessments were established specifically to address the future of public lands, mostly national forest lands. The intention was to address either an issue, such as an endangered species that crossed jurisdictional boundaries, or to address issues common to redeveloping forest plans for national forests in an ecoregion or bioregion. The efforts were led by USDA Forest Service employees, with assistance from other federal scientists.

The most recently completed regional assessment is the Southern Forest Resource Assessment (Wear and Greis, 2002), a collaboration of the USDA Forest Service, US Environmental Protection Agency, US Fish and Wildlife Service and the Tennessee Valley Authority. It assesses the trends and conditions of southern forests, as well as the factors that drive changes. This work underscores the role and importance of individual forest landowners, ecosystem services and the significance of ongoing restoration and protection efforts. This is not strictly a bioregional assessment, in that the focus is on southern forest types that cover parts or all of 12 southern states (Virginia, North Carolina, South Carolina, Georgia, Florida, Kentucky, Tennessee, Alabama, Arkansas, Mississippi, Louisiana and Texas); however, some parts of the assessment follow ecoregion or watershed boundaries, rather than state and county lines. All USDA Forest Service sponsored assessments required extensive public input in the development of the key questions and in framing the extent of the analysis and evaluation. Although neither the sole purpose nor the original intent of these assessments was to evaluate cross-sectoral impacts, they provided both a stakeholder forum for broaching locally significant impacts and a planner forum for evaluating potential cross-sectoral linkages and challenges.

Most of the policies that have influence beyond the forestry sector are national in scope, thus limiting the ability of regional assessments to evaluate the impacts or potential policy solutions to current issues. Local stakeholders may not even recognize the significance of other sectors on forestry, and most policy analysts ignore these impacts. In part, this is because the analysis is complex, but there may also be some fatalism associated with the lack of interest in these questions. National policies, such as mortgage interest deductions and agricultural price supports are influential in affecting changes in forest land area – more than any other national-level policy. It is unlikely, however, that the results from forest resource assessments could lead to timely changes in these types of national policies.

National and bioregional assessments combine qualitative and quantitative analysis of relationships among biophysical, social and economic systems. Their usefulness is also found in the 'whole-systems' view of these relationships. Rather than point to historical trends in static characteristics and shocks to systems, these assessments push policy-makers to understand the complexity and interdependence of systems, which in turn help to refine the quantitative definitions used in the modelling approaches described earlier.

Environmental and Natural-Resource Asset Accounting

Increased scarcity of natural resources, whether perceived or real, influences the demand for information on land values, forest values and the trade-offs among choices for forest managers and users. Traditional commodity values that drive land use, and forest land use decisions, remain important to overall economic and social well-being of human populations. However, the untapped valuation of other critical forest resources, that all populations have been dependent upon for as long as life has existed, currently surfaces in the use of accounting methodologies of natural resource assets.

The linkages among natural resources, ecosystem services and environmental values from forests are reflected in environmental accounting approaches that derive comparative market values from the flux of resources and services provided by forests. The values themselves reflect, and are driven by, changes in other sectors through economic, institutional, social and political linkages. Environmental accounting methodologies generally include accounts of natural resource assets, material flows, pollutants, expenditures on environmental protection and conservation, and macroeconomic aggregate data (Lange, 2003). The combination of these four primary components facilitates the likelihood of sustainability within an economic and environmental system. Thus, environmental accounting facilitates the understanding of physical, social and economic interactions representing sectors of an economy at different scales.

Similar to resource assessments, natural-resource asset accounts are not specifically designed to analyse cross-sectoral impacts, and like the modelling approaches, accounts are data-intensive. Often, the data and information used to construct environmental accounting and natural resource asset accounts are derived for use in other methodologies (e.g. input–output models or econometric models). However, the critical information used to define relationships among sectors in models leads to processes of understanding impacts and can guide policies to address future changes (Rodenburg, 2000). Understanding the distribution and trends of a nation's total wealth (produced, natural and human) helps to identify linkages to specific sector changes that contribute to natural capital stocks, such as forest assets.

Although the maintenance of natural-resource asset accounting data alone does not elucidate specific sector-to-sector linkages or impacts, the underlying objective of tracking sustainable levels of resource use can form the basis of significant policy developments. For example, Hecht (2003) points out that non-declining national wealth does not guarantee sustainability; however, declines in national wealth should prompt direct attention. Indeed, concerns for the decline in habitat (a natural resource asset – old growth forests) for the spotted owl in the Pacific

Northwest region of the USA led to significant policy changes and management decisions on close to 10 million ha of federal lands.

Relationships between environmental indicators that measure resource trends and conditions, indicators of health and human conditions, and management activities, are not well understood; therefore, it is difficult to establish effective, efficient institutional and infrastructure strategies. However, by linking environmental indicators to information in economic indicators, environmental accounting is a potential tool for examining cross-sectoral impacts *on* the forest sector and those generated *by* the forest sector (USGAO, 2004).

Conclusions

Cross-sector impacts from and on the forest sector will be driven by a number of factors, including forest land values and the choices of societies and consumers among other land use options. Markets for carbon credits, water values, conservation easements, recreation access, critical habitats and cultural and aesthetic assets are examples of growing competitive uses of forest land that require a better understanding of the driving factors for the existence and consumption of forest resources.

The common land base is competed for by different sectors of the economy, thereby complicating efforts to view sustainability as solely a forest issue. A broad view could foster resolution of important forest and rangeland health issues through design and leadership of integrated decision systems that involve ecological, political, economic and institutional factors. Increased attention to forest land conditions and land values can aid efforts to improve analyses of cross-sectoral impacts. Focal areas include ecological risk assessment, socio-economic modelling and policy/decision analysis – particularly as they relate to forest and rangeland productivity, fish and wildlife habitat and human communities. Crucial to this endeavour is the ability to bridge multiple disciplines, recognize multiple values and analytically evaluate risks and trade-offs associated with various scenarios for improving the health of forests, rangelands and related natural resources. The ability to articulate prospective outcomes for both short-term and long-term timeframes to policy-makers and resource managers is also pivotal. This includes planning for urban developments and open space (Kline *et al.*, 2004) and the valuation of environmental services from forests, with modelling ties to domestic and global impacts.

The analysis of the forest sector affects further advances, with periodic assessments under consistent frameworks versus infrequent exercises that are not usually replicated and in which public investments lose the added value of cumulative knowledge about forest resources and forest sectors. National and bioregional assessments can contribute additional insights if the measured indicators are used to help refine the relationships that are essential to the strictly quantitative approaches, such as input–output, mathematical programming and econometric models. This feedback relationship benefits decision-makers, as well as analysts, when considering impacts among sectors.

Developments (from decision-making) in forestry and other sectors, outside US geopolitical boundaries, also impact decision-making by public and private groups for US forest resources. Global trade in forest products and the growth in environmental services, such as carbon credits, are forces that impact the overall costs of production on a given area of land, and in turn influence the global balance of supply and demand for forest goods and services. The globalization of the forest sector requires a better understanding of policy and management decisions affected by such trends, such as international mergers among forest companies, transboundary impacts from forest policy and management practices, the spread of invasive species, and more bilateral and multilateral trade agreements.

The use of analytical approaches to evaluate forest sector impacts is continuing to evolve. For example, new planning regulations for the USDA Forest Service, implemented in 2004, require changes in monitoring and reporting systems of forest resource decision-making on federal lands. Private and public planners and analysts differ with regard to information, consultation and problem-solving needs. Only recently have we focused on measures of environmental quality and non-commodity values. Nevertheless, from resource-specific estimations of commodity and land values to comprehensive qualitative and quantitative assessments of resource systems, we are continuing to identify critical intersector and intrasector relationships for further study.

References

Adams, D.M., Alig, R.J., Callaway, J.M., McCarl, B.A. and Winnett, S.M. (1996) *The Forest and Agricultural Sector Optimization Model (FASOM): Model Structure and Policy Applications*. Research Paper PNW-RP-495. USDA Forest Service, Portland, Oregon.

Ahn, S., Plantinga, A.J. and Alig, R.J. (2000) Predicting future forest land area: a comparison of econometric approaches. *Forest Science* 46, 363–376.

Alig, R. (1986) Econometric analysis of forest acreage trends in the Southeast. *Forest Science* 32, 119–134.

Alig, R. and Plantinga, A. (2004) Future forestland area: impacts from population growth and other factors that affect land values. *Journal of Forestry* 102, 19–24.

Alig, R., Plantinga, A., Ahn, S. and Kline, J. (2003) *Land Use Changes Involving Forestry for the United States: Recent Trends and Projections*. (General Technical Report 587.) USDA Forest Service, Portland, Oregon.

Alig, R., Kline, J. and Lichtenstein, M. (2004) Urbanization on the U.S. landscape: looking ahead in the 21st century. *Landscape and Urban Planning* 69, 219–214.

Alig, R.J. (1983) Forest acreage model for the US southeast. In: Bell, J.F. and Atterbury, T. (eds) *Proceedings of the Conference on the Renewable Resource Inventories for Monitoring Changes and Trends*. Oregon State University, Corvallis, Oregon, pp. 242–246.

Alig, R.J., Adams, D.M. and Haynes, R.W. (1983) Long-range projection of forest area change: new approaches are needed. *Journal of Forestry* 81, 723–727.

Alig, R.J., Lee, K.J. and Moulton, R.J. (1990) *Likelihood of Timber Management on Nonindustrial Private Forests: Evidence from Research Studies*. General Technical Report SE-60. USDA Forest Service, Asheville, North Carolina.

Alig, R.J., Adams, D.M. and McCarl, B. (2002) Projecting impacts of global change on the U.S. forest and agricultural sectors and carbon budgets. *Forest Ecology and Management* 169, 3–14.

Alward, G.S. and Palmer, C.J. (1983) An input–output analysis system for forest service planning. In: Seppala, R., Row, C. and Morgan, A. (eds) *Forest Sector Models: Proceedings of the First North American Conference*. AB Academic Publishers, Oxford, pp. 131–140.

Alward, G., Siverts, E., Olson, D., Wagner, J., Senf, D. and Lindall, S. (1989) *Micro-IMPLAN*. USDA Forest, Fort Collins, Colorado.

Aruna, P.B. and Cubbage, F.W. (2000) Regional economic analyses of the forest products and tourism sectors in North Carolina. *Southern Center for Sustainable Forests Working Paper* 9, 1–14.

Deller, S.C., Sumathi, N.R. and Marcouiller, D. (1993) *Regional Economic Models for the State of Wisconsin: An Application of the Micro-IMPLAN Modeling System*. Center for Community Economic Development Staff Paper 93.6. University of Wisconsin–Madison/Extension, Madison, Wisconsin.

FEMAT (1993) *Forest Ecosystem Management: An Ecological, Economic, and Social Assessment Report of the Forest Ecosystem Management Assessment Team*. US Government Printing Office 793-071, IX-24. Washington, DC.

Hardie, I., Parks, P., Gottleib, P. and Wear, D. (2000) Responsiveness of rural and urban land uses to land rent determinants in the U.S. South. *Land Economics* 76, 659–673.

Haynes, R.W. (technical coordinator) (2003) *An Analysis of the Timber Situation in the United States: 1952–2050*. General Technical Report PNW-GTR-560. USDA Forest Service, Portland, Oregon.

Hecht, J.E. (2003) *Lessons Learned from Environmental Accounting: Findings from Nine Case Studies*. IUCN: The World Conservation Union, Washington, DC.

Hughes, D.W. and Hinson, R. (2000) Estimating the value of the green industry to Louisiana's economy. *Journal of Agribusiness* 18, 207–220.

Ianchovichina, E., Darwin, R. and Shoemaker, R. (2001) Resource use and technological progress in agriculture: a dynamic general equilibrium analysis. *Ecological Economics* 38, 275–291.

Kline, J.D., Butler, B.J. and Alig, R.J. (2002) Tree planting in the South: what does the future hold? *Southern Journal of Applied Forestry* 26, 99–107.

Kline, J., Alig, R. and Garber-Yonts, B. (2004) Forestland social values and open space preservation. *Journal of Forestry* 102, 39–45.

Lange, G. (2003) *Policy Applications of Environmental Accounting*. Environmental Economic Series Paper, no. 88. The World Bank Environment Department, Washington, DC.

Lee, K.J., Kaiser, H.F. and Alig, R.J. (1992) Substitution of public for private funding in planting southern pine. *Southern Journal of Applied Forestry* 16, 204–208.

Munn, I.A. (1998) *Forestry in Mississippi. The Impact of the Forest Products Industry on the Mississippi Economy: an Input–Output Analysis*. Forestry and Wildlife Research Center Bulletin, no. 087. Mississippi State University, Starkville, Mississippi.

Parks, P.J., Hardie, I.W., Tedder, C.A. and Wear, D.N. (2000) Using resource economics to anticipate forest land use change in the U.S. mid-Atlantic region. *Environmental Monitoring and Assessment* 63, 175–185.

Plantinga, A.J., Mauldin, T. and Miller, D.J. (1999) An econometric analysis of the cost of sequestering carbon in forests. *American Journal of Agricultural Economics* 81, 812–824.

Quigley, T.M. and Bigler-Cole, H. (1997) *Highlighted Scientific Findings of the Interior Columbia Basin Ecosystem Management Project*. General Technical Report PNW-GTR-404. USDA Forest Service and USDOI Bureau of Land Management, Portland, Oregon.

Rodenburg, E. (2000) Material flow accounting – methodology and frameworks. In: OECD Working Group on Environmental Information and Outlooks (WGEIO) (eds) *Special Session on Material Flow Accounting, Paris, 24 October 2000*. OECD, Paris, pp. 79–84. Available at: http://www.oecd.org/dataoecd/2/20/4425421.pdf

University of California–Davis (1996) *Summary of the Sierra Nevada Ecosystem Project Report*. Report no. 39. Centers for Water and Wildland Resources, University of California, Davis, California.

USDA Forest Service (2003) *National Report on Sustainable Forests – 2003*. FS-766. United States Department of Agriculture, Washington, DC.

USGAO (2004) Environmental indicators: better coordination is needed to develop environmental indicator sets that inform decisions. *GAO Highlights,* GAO-05-52.

Wear, D.N. and Greis, J.G. (eds) (2002) *Southern Forest Resource Assessment*. General Technical Report SRS-53. USDA Forest Service, Asheville, North Carolina.

26 Cross-sector Challenges for Public Land Management in the Western Canadian Sedimentary Basin

Marian L. Weber

Introduction

Canada's boreal region is one of the largest remaining intact forest ecosystems in the world. It is an important habitat for migratory birds, and some of the world's largest populations of woodland caribou, wolves and bear. The boreal also contains vast timber and mineral resources. The Western Canadian Sedimentary Basin (WCSB) is one of the largest hydrocarbon resources in the world. Until recently, the remoteness of the region isolated it from developmental pressure. However, over the last 20 years, improvements in technology and depletion of resources elsewhere have led to unprecedented rates of forest disturbance. During the 1990s, large areas of productive forest were allocated to the forest industry. At the same time, increases in world energy prices and improvements in recovery technology led to rapid expansion of the energy sector, particularly in Alberta, where the surface impacts of energy activities will soon surpass those from forestry. For example, between 1995 and 2002, the average annual area cleared by the energy sector was approximately 47,000 ha.[1] By way of comparison, in 1999 the total harvested area for industrial roundwood was 42,000 ha (CFS, 2001). Furthermore, the cumulative impacts of energy development can be far-reaching due to the prolonged use of linear features, such as roads, and the absence of reforestation requirements.

The lack of a framework for integrated land management has led to increased conflict over access to resources and protection of ecosystem services. Industrial development has already compromised the ecological integrity of Alberta's forest in some areas (AEP, 1998). In addition, local fibre shortages have been predicted in areas in which the annual allowable cut was allocated without due consideration of the joint effects of disturbance from fire and energy (Schneider *et al.*, 2003; Cumming and Armstrong, 2004). Conflicts with the energy sector have also increased land management costs for forest companies since time-consuming efforts to plan for caribou and other ecological objectives are jeopardized by subsurface activities over which they have limited control. In a recent British Columbia survey, 70% of the respondents indicated that their business was adversely affected by conflicts over

[1]Between 1980 and 2002, an average of 3700 wells were drilled annually and a total of 134,000 km of pipeline was laid in Alberta's boreal region (AEUB, 2005a). The footprint estimate assumes a seismic ratio of 3 km/well and seismic width of 1 m (Schneider *et al.*, 2003); access road ratio of 2.5 km/well and road width of 8 m (J.B. Stelfox, Forem Technologies, Edmonton, 2006, personal communication); and pipeline width of 15 m (Holroyd and Retzer, 2005). Although production of conventional oil in the WCSB peaked in 1997, this will be offset over the next 25 years by the development of gas and oil sand reserves (AEUB, 2005b).

tenured land use (MSRM, 2004). Similarly, overlapping and shared tenure was identified by the Forest Stewardship Council as a key challenge in the development of forest certification standards. Moreover, it is increasingly difficult for the energy sector to negotiate permission for access with other tenures on public land. The review process for energy projects can be lengthy, and companies are often required to participate in public consultation and other planning processes, which deal with issues outside the scope of their line of business.

In this chapter, we examine the evolution of integrated land management as a policy approach to managing cross-sectoral linkages between forestry, energy and the environment. The dynamics of land policy are driven by the emergence of new ideas and actors, as well as changes in the nature of the resource constraints under which actors operate (Howlett, 2001). We conduct a comparative analysis of integrated land management in British Columbia and Alberta in order to understand the influence of different actors, ideas and resource constraints on land policy. Although such a concept has been around in various forms over the last 20 years, past approaches have been flawed in that they have focused primarily on managing public expectations for environmental protection within existing forest licenses and agreements. The emergence of the energy sector as an economically significant actor in forest management has been a catalyst for approaches to integrated land management that are focused on institutional change. However, different regulatory and market structures in the energy and forest sectors make it difficult to combine their activities in a way that achieves an optimal balance between economic and ecological objectives. In particular, energy policy objectives have traditionally been viewed as best delivered by a competitive market, whereas forest policy objectives have traditionally been delivered through public planning processes, with market forces playing a secondary role.

Cross-sectoral Challenges for Integrated Land Management

The majority of forest land in Canada is publicly owned and managed by the provinces. The institutional structure for managing environmental, timber and energy resources is fragmented between individual provincial government departments, which in the past have not coordinated in issuing dispositions or in operational planning. Forest policy in Canada has been described as public forest management for commercial fibre production. Most fibre is allocated through long-term bilateral area-based agreements negotiated between the provinces and large forest companies to grow and harvest timber on a sustained-yield basis. Awarded initially to encourage private investment in silviculture, over time the de facto management responsibilities for area-based tenure holders have increased to include management of public values, such as biodiversity and scenic landscapes. These responsibilities are implemented through numerous planning and regulatory requirements, such as those promulgated by the Forest Practices Code in British Columbia. Forest companies are responsible for creating strategic and tactical forest management plans, which are vetted through public consultation processes. The planning process drives the activity of the forest industry and its impact on public land and environmental objectives.

The planning approach for managing forestry and environmental values can be compared to the market approach for managing the exploitation of energy resources. The objective of energy policy is to maximize competitiveness in order to ensure that all economically feasible reserves are discovered and mined, and generate a continuous royalty stream for the provinces. The energy development cycle begins with geophysical exploration in order to obtain information about the value of energy deposits. Areas of interest for further exploration are then submitted to the government, which posts the parcels and conducts an auction for the petroleum and natural gas rights. These rights are sold by vertical strata, so for a given surface area there can be multiple overlapping mineral rights. The competitiveness of markets for petroleum and natural gas rights has two implications for the ecological footprint of the energy sector. First, the competitive acquisition of information may lead to redundant exploration effort from a global perspective. In addition, where vertical layers are leased by multiple firms, there may be benefits to coordination between firms through sharing roads and other infrastructure. An important question for land policy is whether there exist incentives that would maintain the competitiveness of the energy sector while coordinating development in such a way as to minimize the externalities from surface impacts on forestry and the environment.

Originally, forest policy focused only on industry and government actors, but this paradigm has been under attack since the 1960s from groups, such as independent loggers, environmentalists and First Nations, who formed a coalition questioning the allocation of public lands and the effectiveness of forestry regulation (Howlett and Rayner, 2001). During the 1970s and 1980s, provincial governments embarked on a number of land use planning exercises in response to public demands. Although land use planning opened the door to new actors, the processes were hampered by the limited scope through which outside actors could address concerns. Instead, integrated land management was conducted as a means of assuaging public concerns within existing resource requirements, and in any case the resulting land use plans lacked legal status (Howlett and Rayner, 2001; Kennett, 2002). In the 1990s, a number of factors converged to push it back to the forefront of the policy agenda. First, high energy prices and the depletion of conventional energy reserves elsewhere brought the energy sector into increasingly remote forest areas, such as the Muskwa-Kechika Management Area in British Columbia, and into direct conflict with forest tenures and the public concerns over conservation. Opportunities to offset emerging land constraints were limited by the lack of flexibility in forest tenures, and the energy sector was left to defend its activities in the forest on a project by project basis. As a result of this situation, the Canadian Association of Petroleum Producers called on the government to establish clear and competitive policies for access to land and address conflicting land tenure issues, as well as implement an integrated policy framework and one-window regulatory approach (CAPP, 2004).

Within the government, the emerging idea of 'smart regulation' provided momentum for adopting a broader vision in land policy.[2] Smart regulation is a set of principles emphasizing results-based management, policy coordination, transparency, inclusiveness and evidence-based decision making. These principles are reflected in the current focus of integrated land management on institutional reform, in particular coordination of the disposition system in order to optimize resource use for a broad range of economic, social and environmental objectives (NFSC, 1998; CILMC, 2005). The primary instrument is land use planning, which in addition to identifying opportunities for cooperation between sectors is seen as the key mechanism for developing land use objectives that can then be used to inform the allocation and regulatory process. Land use planning consists of hierarchical objective setting (provincial to regional to local plans) and the development of regional targets for environmental, social and economic indicators. Objectives are implemented through zoning and best practices. The limitations of this approach are illustrated by examining the way in which it has played out in British Columbia and Alberta. In both cases, there has been only partial integration of the energy sector in forest management, and Alberta has failed to identify land use planning objectives that support ecological outcomes. Furthermore, neither province has utilized economic instruments to evaluate trade-offs and implement land use objectives. There is therefore still no mechanism for ensuring that land is allocated between activities in a way that is cost-effective and maximizes public benefits.

British Columbia Experience

In British Columbia, all dispositions on Crown land, except for oil and gas, are allocated through the Ministry of Sustainable Resource Management under the authority of the Land Act. The Oil and Gas Commission, which operates under the Ministry of Energy and Mines, is a one-stop shop for issuing oil and gas dispositions, including rights for surface access. The Commission has a pre-tenure review process for all posting requests to identify access constraints and potential land use conflicts. Participants involved in the review include other government ministries, local governments, First Nations and environmental organizations. Restrictions and issues identified by reviewers may be included as caveats in the posting, or, in some cases, lead to a deferral to allow for resolution of land use issues. Concerns may also be included as conditions in the approvals (MEM, 1998). Postings for leases occur once a month and

[2]See EACSR (2004) for a discussion of the government's perspective on smart regulation.

there is a 7-week period between the closing date for postings and the tender notice (OGC, 2005). Petroleum and natural gas rights include rights to surface access over Crown land, although permission for surface access must still be obtained from companies with timber rights. Land use conflicts are resolved through negotiation, and there are no mechanisms for compensation of forest tenures on public land (MSRM, 2004). However, companies pay stumpage to the Ministry of Forests for trees that are removed (OGC, 2004).

In the early 1990s, the British Columbia government initiated the Commission on Resources and the Environment, a multi-stakeholder process that resulted in land and resource management plans for 80% of the province. Although the plans do not have legal status, they provide objectives for land management and are influential guides for regulatory decisions (CIRL, 2004). Most oil and gas development in British Columbia occurs in the north-east, so the discussion of the framework for integrated land management in the province reflects the context of this region, particularly the Muskwa-Kechika Management Area. This is one of the largest wilderness areas in North America, unique for its lack of roads, as well as its cultural, ecological and geographical significance. In many respects, it represents an ideal example of integrated land management. As a result of the planning processes for Fort Nelson and Fort St. John, the Muskwa-Kechika Management Area was legislated as a special management area in 1998 under the Muskwa-Kechika Management Act. Under the Act, its management plan has legal status for guiding land use decisions. In addition, a key innovation was the adoption of tiered quantitative thresholds for a number of indicators (RMC/SC, 2004). Pre-tenure planning is required for all regulatory approvals for oil and gas operations except geophysical exploration, to ensure that they are consistent with the plan. Geophysical exploration, though exempt from pre-tenure planning, still requires the use of best management practices and engagement in public consultation for approval.

Alberta Experience

Land use planning in Alberta is less advanced than in British Columbia, and authority for energy dispositions is fragmented between more departments. In Alberta, public land is managed and disposed of by the Department of Sustainable Resource Development, while subsurface mineral rights are disposed of by the Department of Energy. Energy activities are regulated by the Alberta Energy Utilities Board, which is an independent, quasi-judicial tribunal. Dispositions for subsurface rights are issued every 2 weeks by public tender. Once companies obtain subsurface rights, they apply for surface rights through the Department of Sustainable Resource Development. At this point, reviews are undertaken to assess potential impacts and take an average of 13 days (MacKendrick *et al.*, 2001). In order to acquire surface rights, companies must obtain permission for access from forest companies with area-based tenures. Under the Surface Rights Act, forest companies are given the right to negotiate payments in return for granting access. If parties fail to come to an agreement, the Surface Rights Board can grant a right-of-entry order and award compensation to the forest company. No tenure agreement is necessary for geophysical exploration, which means that seismic activity is subject only to permission for access from surface owners. Once dispositions are allocated, companies apply for licenses and approvals from the Alberta Energy Utilities Board, which reviews all projects in order to ensure that they are in 'public interest' (Kennett and Wenig, 2005). Approval of a project is considered to be routine unless there are objections from the public. However, since the Energy and Utilities Board review takes place after the disposition has been allocated, it is difficult to preclude opportunities to exercise rights, particularly where millions of dollars are involved in a transaction (Kennett and Wenig, 2005). In addition, without policy and planning guidance on landscape level objectives and trade-offs, the Energy and Utilities Board is unable to evaluate whether the trade-offs from any individual application are acceptable.

To date, there is no legally defined integrated land management planning process in Alberta, and resource management objectives that have been approved in some areas of the province have no legal standing (CIRL, 2004). In any case, recent oil sand developments in north-east Alberta have rendered Alberta's framework for land management obsolete. This policy vacuum was explicitly recognized in the Draft of the Mineable Oil Sands Strategy, released in November 2005 (Government of Alberta, 2005). The Strategy proposes managing the mineable oil sands area as a coordinated development zone in which oil sands mining will have the highest priority, and is a clear admission that not every objective can be achieved on every landscape. In particular, the policy states that 'specific conditions related to the protection of wildlife habitat within the [mineable oil sands development zone] will not be implemented prior to or during oil sands mining' and that to 'ensure current and future generations continue to enjoy wildlife resources, habitat and population objectives will be achieved within a regional and provincial context … [placing] a greater emphasis on other habitat locations at a regional and provincial scale' (Government of Alberta, 2005). The strategy appears to be predicated on the existence of regional and provincial land use plans that do not yet exist; and although Alberta is currently developing a provincial land use framework, the process for identifying desired land use objectives and the linkages to dispositions and the regional integrated planning processes are not established. The government's current integrated land-management programme is intended to be a vehicle for regional implementation of the land use framework. However, its focus is on post-disposition management, particularly managing and reducing the industrial footprint, rather than on broader institutional change (ASRD, 2005).

Comparative Analysis

Opportunities for integrated land management vary according to the relative influence of the energy and forest sectors in the provincial economy. In British Columbia, the forest industry has been the dominant economic driver for the province throughout the 20th century. Energy development is fairly recent and confined largely to the north-east area of the province. The situation is reversed in Alberta, where the energy sector has dominated since the 1950s. The forest industry that emerged in the boreal during the 1990s was largely driven by a downturn in the energy sector and a desire for diversification of Alberta's economy. The highly competitive energy sector is not amenable to the structured planning regime that characterizes forest policy, and Alberta's weaker framework for land use planning reflects the greater historical influence of the energy sector on resource policy in the province. In Alberta, oil and gas dispositions are processed more frequently and with shorter review and approval periods than in British Columbia. Furthermore, in contrast to British Columbia, the review of energy projects takes place once dispositions have been allocated, and without guidance from land use plans (Kennett and Wenig, 2005). In terms of integrating energy sector activities within a land use planning framework, Alberta clearly lags behind British Columbia. However, Alberta has stronger incentives for energy companies to directly engage forest companies in integrating activities and employing best practices. Where land use conflicts arise between the two sectors in Alberta, the damage awards to timber companies clearly give protection to forest tenures and provide a financial incentive for tenure holders to integrate their activities. It is important to note that direct environmental interests are left out of this equation, although regulatory requirements may drive the motivations of firms to cooperate. This can be compared to British Columbia, where firms are encouraged to resolve their conflicts through negotiation without the additional financial incentives provided by compensation.[3]

The review of integrated land management illustrates the lack of attention that has been paid to mechanisms, beyond increasing regulatory requirements, for implementing land use objectives. An outstanding policy dilemma for both provinces is that even if projects are evaluated against quantifiable thresholds, the approval for projects still takes place on a first-past-the-post or first-come, first-served basis. There are no guarantees that land will be allocated to the highest-benefit use under this system. Once a decision on a project has been made, there is no opportunity to reallocate land afterwards, as resource tenures are not transferable or flexible. This means, for example, that higher-value land uses cannot be accommodated by purchasing past allocations. In addition, conservation organizations are unable to express values for ecological objectives on

[3]Although energy companies pay stumpage to the Ministry of Forests in British Columbia, the incentive for them to integrate their activities with forestry activities is not as strong.

behalf of the public by purchasing or negotiating limitations on industrial tenures. In order for integrated land management to succeed, further attention to flexible mechanisms for reallocating resource rights is required. For example, tradable rights to forest disturbance and offset credits have been proposed as ways of allocating surface rights more efficiently (Schneider, 2002; Weber and Adamowicz, 2002). Under a trading system, rights to disturb land up to a threshold are allocated to firms. Trading between existing tenures reallocates surface activities to ensure that conservation objectives are met at a minimal cost. The rights can also be purchased by new entrants or by conservation interests as long as they find a party willing to sell. Cost-effective mechanisms reduce the trade-offs associated with achieving ecological objectives and can lead to higher levels of both conservation and economic benefits than hierarchical approaches (Weber, 2004).

Quantitative land use objectives are required in order to utilize market instruments for integrated land management. At the same time, setting objectives involves complex social, economic and ecological trade-offs. Although participatory processes have increased public involvement in land management, the public has lacked tools to quantify trade-offs systematically. As a result, most land use plans fail to identify thresholds for indicators, and where thresholds exist they tend to be qualitative rather than quantitative. This is certainly the case in Alberta, where the Energy and Utilities Board has stated that the absence of thresholds against which to measure ecological effects makes it difficult to evaluate the acceptability of individual projects (Kennett and Wenig, 2005). In order to establish thresholds, we need information on how people value trade-offs when making choices between alternative land uses. There are numerous hypothetical approaches for eliciting values associated with land-management decisions that could be used to identify objectives (Ananda and Herath, 2003). At the same time, a market-based system, which allowed the public to directly express trade-offs through contracts restricting industrial rights would reveal information about trade-offs between ecological and economic outcomes, and create opportunities for the public to improve the environment beyond established thresholds.

Conclusions

Most forested land in Canada is public land. Forest companies have land-based tenures that include responsibilities for managing ecological objectives. However, these tenures do not include rights to restrict access to subsurface resources, which has resulted in conflict between energy, forestry and environmental interests. The policy dynamic on public land has created momentum for a new approach to integrated land management that encompasses changes to the allocation process in order to achieve multiple land use objectives. None the less, there are several cross-sectoral challenges to integrating competing objectives on public lands – particularly the different policy paradigms for managing environmental, energy and forestry interests. As a result, integrated land management in practice tinkers at the margins of land policy and resembles more a set of best-management practices for coordinated planning rather than the broad institutional change required to address the allocation issues that underlie land use conflicts.

A comparative analysis between Alberta and British Columbia illustrates how the distribution of economic and political influence, as determined by the history of resource development, has influenced the evolution of land policy in each province. In both cases, it suffers because of its focus on the planning process per se, rather than on mechanisms for evaluating trade-offs and implementing land use objectives. Both provinces attempt to achieve ecological objectives by layering new requirements on tenures within the existing disposition system. This approach is costly for industry and does not necessarily satisfy public concerns over the allocation of land. The analysis shows that we need a better understanding of the way in which flexible market instruments and economic incentives can be harnessed to reallocate public lands between uses and to integrate various activities better in order to achieve multiple-use objectives. Changes in policy that promote side-agreements between the different sectors provide opportunities for different actors to express their preferences directly through market transactions, rather than through a politicized planning process. This is not to suggest that public lands should be privatized, but rather that the allocations and dispositions made on public lands could retain attributes of private-market instruments in order to reveal information about public land values, and also to increase access to public lands by non-traditional sectors.

In Alberta, where the energy sector has had a much greater role in policy development, attempts at integrated land management have failed to generate regional land use objectives and plans. This is in part due to the difficulty of delivering effective energy policy within a planning framework. At the same time, there are greater economic incentives for coordination between the forest and energy sectors in Alberta due to the timber damage assessment, a charge that the energy sector is required to pay to forest companies for timber removed during exploration and development. This is in contrast to the situation in British Columbia, where the forest sector has much stronger political and economic linkages. There, energy projects undergo a pre-tenure review subject to land use planning objectives, but there are no direct financial incentives that would encourage side-agreements between forestry and energy companies when there are tenure conflicts. The legacy of resource development and comparative advantage has led to the divergence of provincial land policies in the past. However, increased competition for access to resources, along with the emergence of the energy sector as an important actor in British Columbia, is likely to lead to policy convergence in the future, due to the need for both provinces to incorporate greater responsiveness to market pressures from the energy sector in public land policy. Finally, flexible approaches to integrated land management cannot work in the absence of quantified regional thresholds. It is therefore necessary for participatory planning models to incorporate approaches for eliciting public preferences about trade-offs in order to quantify regional management objectives.

References

Alberta Energy and Utilities Board (AEUB) (2005a) *General Well Data File*. Electronic Data Services, Calgary, Alberta.

Alberta Energy and Utilities Board (AEUB) (2005b) *Alberta's Reserves and Supply/Demand Outlook 2005–2014*. ST 98-2005. Alberta Energy and Utilities Board, Calgary, Alberta.

Alberta Environmental Protection (AEP) (1998) *The Boreal Forest Natural Region of Alberta*. Government of Alberta, Edmonton, Alberta.

Alberta Sustainable Resource Development (ASRD) (2005) *Integrated Land Management*. Government of Alberta, Edmonton, Alberta. Available at: http://www3.gov.ab.ca/srd/land/u_ILM.html

Ananda, J. and Herath, G. (2003) Incorporating stakeholder values into regional forest planning: a value function approach. *Ecological Economics* 45, 75–90.

Canadian Association of Petroleum Producers (CAPP) (2004) *Policy Direction for Alberta's Oil and Gas Industry*. CAPP, Calgary, Alberta. Available at: http://www.capp.ca/raw.asp?x=1&dt=NTV&e=PDF&dn=80955

Canadian Forest Service (CFS) (2001) *The State of Canada's Forests: 2000/2001*. Canadian Forest Service, Ottawa, Ontario. Available at: http://nrcan.gc.ca/cfs/proj/ppiab/sof/common/latest.shtml

Canadian Institute for Resources Law (CIRL) (2004) *The Legal and Policy Framework for Managing Public Access to Oil and Gas Corridors on Public Lands in Alberta, Saskatchewan and British Columbia*. Publication 2004-0012. Canadian Association of Petroleum Producers, Calgary, Alberta.

Canadian Integrated Land Management Coalition (CILMC) (2005). *Integrated Landscape Management – Applying Sustainable Development to Land Use*. Discussion paper. Available at: http://ilm.law.uvic.ca/PDF/ILM_concept_e.pdf/

Cumming, S.G. and Armstrong, G.W. (2004) Divided landbase, overlapping tenures, and fire risk. *Forestry Chronicle* 80, 478–484.

External Advisory Committee on Smart Regulation (EACSR) (2004). *Smart Regulation: A Regulatory Strategy for Canada*. Government of Canada, Ottawa. Available at: http://www.pco-bcp.gc.ca/smartreg-regint/en/08/sum.html

Government of Alberta (2005) *Mineable Oil Sands Strategy – Discussion Document*. Alberta Energy, Edmonton, Alberta. Available at: http://www.energy.gov.ab.ca/docs/oilsands/pdfs/MOSS_Policy2005.pdf

Holroyd, P. and Retzer, H. (2005) *A Peek into the Future: Potential Landscape Impacts of Gas Development in Northern Canada*. Pembina Institute, Edmonton, Alberta.

Howlett, M. (2001) Introduction: policy regimes and policy change in the Canadian forest sector. In: Howlett, M. (ed.) *Canadian Forest Policy*. University of Toronto Press, Toronto, pp. 3–20.

Howlett, M. and Rayner, J. (2001) The business and government nexus: principal elements and dynamics of the Canadian Forest Policy Regime. In: Howlett, M. (ed.) *Canadian Forest Policy*. University of Toronto Press, Toronto, pp. 23–62.

Kennett, S.A. (2002). *Integrated Resource Management in Alberta: Past, Present and Benchmarks for the Future*. Occasional Paper, no. 11. Canadian Institute for Resources Law, Calgary, Alberta.

Kennett, S.A. and Wenig, M.W. (2005) Alberta's oil and gas boom fuels land-use conflicts – but should the AEUB be taking the heat? *Resources* 91, 1–8.

MacKendrick, N., Fluet, C., Davidson, D.J., Krogman, N. and Ross, M. (2001) *Integrated Resource Management in Alberta's Boreal Forest: Opportunities and Constraints*. Project Report 2001–2002. Sustainable Forest Management Network, Edmonton, Alberta.

Ministry of Energy and Mines (MEM) (1998) *Petroleum and Natural Gas Tenure Review Process*. Information Letter EMD98-05. Government of British Columbia, Victoria, British Columbia.

Ministry of Sustainable Resource Management (MSRM) (2004) *A Practical Guide to Effective Coordination of Resource Tenures*. Government of British Columbia, Fort St. John, British Columbia. Available at: http://srmwww.gov.bc.ca/rmd/crt/pdf/TENURE.pdf

National Forest Strategy Coalition (NFSC) (1998) *National Forest Strategy 1998–2003: Sustainable Forests – A Canadian Commitment*. National Forest Strategy, Canadian Council of Forest Ministers. Available at: http://nfsc.forest.ca/strategies/strategy4.html#pre_e

Oil and Gas Commission (OGC) (2004) *Area-Based Stumpage Rates for the Oil and Gas Industry*. (Backgrounder.) Government of British Columbia, Victoria, British Columbia.

Oil and Gas Commission (OGC) (2005) *2005/2006 Petroleum and Natural Gas Rights Disposition Dates*. Information Letter: Titles 05-05. Government of British Columbia, Victoria, British Columbia.

R. McManus Consulting Ltd. and Salmo Consulting Inc. (RMC/SC) (2004) *Conserving Canada's Natural Capital: the Boreal Forest Muskwa-Kechika Case Study*. National Round Table on the Environment and Economy. Available at: http://www.nrtee-trnee.ca/eng/programs/Current_Programs/Nature/Boreal-Forest/Documents/200407-Muskwa-Kechika-Case-Study/200407_Muskwa-Kechika-Case-Study_1-Introduction_E.htm

Schneider, R. (2002) *Alternative Futures: Alberta's Boreal Forest at the Crossroads*. Federation of Alberta Naturalists, Edmonton, Alberta.

Schneider, R.R., Stelfox, J.B., Boutin, S. and Wasel, S. (2003) Managing the cumulative impacts of land uses in the Western Canadian Sedimentary Basin: a modeling approach. *Conservation Ecology* 7(1): 8. Available at: http://www.consecol.org/vol7/iss1/art8/

Weber, M.L. (2004). *Assessing the Effectiveness of Tradable Landuse Rights for Biodiversity Conservation: An Application to Canada's Boreal Mixedwood Forest*. FEEM Working Paper no. 29. SSRN Electronic Publishing Inc. Available at: http://ssrn.com/abstract = 504524.

Weber, M. and Adamowicz, W. (2002) Tradable landuse rights for cumulative effects management. *Canadian Public Policy*, 28, 581–595.

27 Urbanization and Timberland-use Change in the Southern USA

Rao V. Nagubadi and Daowei Zhang

Introduction

The rates of population increase in some southern states are among the highest in the USA and have increased the demand for natural resources in the region. For example, between 1990 and 2000, Georgia, Florida, Texas and North Carolina experienced population growth between 21.4% and 26.4%, far exceeding the national average of 13.2% population growth in the USA (US Census Bureau, 2001). Higher population levels have stimulated urban development, urban sprawl and concomitant loss of forest cover (Wear, 2002). Changes in land use have implications for a wide variety of policy issues, such as maintenance of water quality, protection of species biodiversity, preservation of open space, mitigation of global climate change and other economic and environmental effects and ecosystem services. Changes in forest type could imply a significant impact on the condition of forests and consequently the various services and amenities provided by the forests. Better predictions on land use changes by forest types are needed to allow the necessary measures to be taken to enhance the services and amenities provided by these forests.

The southern USA is highly significant in terms of forest resources. In 2002, the 13 states in the southern USA had 202.7 million acres of timberland and contributed about 58% of all roundwood products harvested in the USA (Smith *et al.*, 2004). In the USA, 51% or 181.4 million acres of private timberland is located in these southern states. Put in a global perspective, the southern USA accounts for 18% of the world's industrial roundwood with just 2% of the world's forest land and roughly 2% of the world's forest inventory (Harris *et al.*, 2003). Further, 145.5 million acres or 71.8% of all timberland in the region is under the ownership of millions of non-industrial private forest (NIPF) landowners, who often respond to market and cross-sectoral forces, such as urbanization and policy instrument in their land use decisions.

In fact, tremendous changes in timberland, experienced in the second half of the last century in the region, were largely driven by cross-sectoral forces, such as population change, economic growth, changes in relative returns to forestry and agriculture and government subsidy. Although the first two factors led to decreases in both forest land and agricultural land due to urbanization, the third factor resulted in some agricultural land being converted to forest land use due to favourable returns to forestry relative to agriculture. Furthermore, subsidized tree planting under cost-share programmes, such as the Forestry Incentives Programme, Stewardship Incentives Programme, Soil Bank Programme, Agricultural Conservation Programme and Conservation Reserve Programme, has stemmed the decline of timberland by converting agricultural land use into forest land use (Nagubadi and Zhang, 2005).

The purpose of this study is to quantify the influence of urbanization on timberland use in the southern USA. Over the last two decades, a number of studies have attempted to model and develop projections of land use changes among aggregated groups, such as forestry, agriculture and urban. Long-term timberland and forest area projections have been a regular feature and an important component of the Forest and Rangeland Renewable Resource Planning Act (1974) assessments and other forest policy analyses (Alig *et al.*, 2003). However, little effort has been given to modelling land use with respect to more disaggregated use groups, such as considering the effects on timberland by forest type.[1] By pooling timberland across forest types, earlier studies may have assumed that economic, demographic and other factors have equal effects across forest types (Alig, 1986; Mauldin *et al.*, 1999; Hardie *et al.*, 2000; Ahn *et al.*, 2002, 2003; Lubowski *et al.*, 2003). Thus, earlier studies may have imposed an implicit restriction on changes in timberland use – all forest types respond in the same magnitude to various factors.[2] These implicit restrictions may have hampered more realistic predictions of land use changes needed by policy-makers engaged in enhancing the services provided by the forests to the society. There is little information on how various variables affect the timberland use by forest type.

In this study, we apply a modified multinomial logit technique to develop a predictive model of land use by forest type, using county-level data in 11 states of the southern USA. In addition to population and income, we have included some urbanization variables, such as the urban–rural continuum (CONTI) code and median housing values, and a knowledge variable – educational attainment – in order to capture the impact of urbanization and socio-economic variables on timberland use by forest type. We generate marginal effects and elasticities for the variables on specific timberland use by forest type. We also develop preliminary projections of timberland use to the year 2050 on the basis of our model estimates and certain assumptions on the likely behaviour of some variables included in our analytical model.

The next section presents the analytical framework used in the study, followed by a description of data. The remaining sections present the results by forest type, develop projections based on our model estimates and draw some relevant conclusions. Our results confirm that various factors influencing the timberland use differ in their effects significantly by forest types. Our projections indicate that there would be a larger loss of land under timberland use, particularly land use under the softwood forest type, under the pressures of urbanization in the southern USA.

Multinomial Logit Model

In theory, land is devoted to the use that provides the greatest value to its owners in terms of the present value of the stream of returns expected in future years. Land use choice often involves complex interactions between factors that include characteristics of land, landowner and the economic, social and policy context in which the choice is made.

Modern land use theory builds on the contributions of Ricardo, who developed the concept of land rent in rural land use, and von Thünen, who developed a location rent model for urban land use.[3] Miller and Plantinga (1999), and Hardie

[1]FIA defines timberland as forest land producing or capable of producing more than 20 ft^3/acre/year of industrial wood crops under natural conditions, which is not withdrawn from timber utilization and not associated with urban or rural development.

[2]Nagubadi and Zhang (2005) made an attempt to see whether the coefficients in the three different forest type equations were significantly different, with the help of dummy variables and dummy variable interactions with the variables from the softwood forest type equation, by stacking the observations from dependent and independent variables one below the other. The results showed that the dummy intercept and dummy variable interactions with PSTPR, OSTPR and land quality variable were significantly different from zero, all at the 1% significance level. For the mixed forest type equation, dummy variable interactions with PSTPR and OSTPR were found to be significant at the 1% level, and dummy interaction term with cost share acres variable was found to be significantly different at the 5% level.

[3]While Ricardo attributed rent to differences in fertility, von Thünen developed a complementary theory of rent based on differences in location with respect to a central market and postulated that land rent decreased as distance from the urban centre increased in concentric rings. The Ricardian model has been extensively applied in agricultural economics literature, while the von Thünen model has served as a standard-bearer in the analysis of urban economic growth.

et al. (2000) developed a model of land use change by combining Ricardian and von Thünen land rent models depicting the landowners' decision problem as one of allocating a fixed amount of land to alternative uses. The empirical model is formulated as a modified multinomial logit model (Amemiya and Nold, 1975; Parks, 1980; Hardie and Parks, 1997). The details of the model specification are given in Nagubadi and Zhang (2005). The multinomial logit specification is convenient because it constrains the predicted land use shares between zero and one and requires that they sum to one. The logarithmic transformation and the use of cross-sectional data induces the problem of heteroscedasticity from one or more explanatory variables. As a correction to this problem of heteroscedasticity, maximum likelihood estimates are obtained by using a multiplicative heteroscedastic regression method (Harvey, 1976; Greene, 1995, pp. 264–267).[4]

Since the dependent variable is the logarithm of the ratio of proportion of land uses, it is difficult to interpret the coefficients directly. Consequently, marginal effects and elasticities are estimated at mean levels of the continuous explanatory variables, and a value of one for dummy variables. Marginal effects are estimated as in Greene (1993, p. 666) and acreage elasticities as in Wu and Segerson (1995, p. 1037).

Marginal effects are interpreted as the change in the probability of land being in a particular use for a unit change in the independent variable, *ceteris paribus*. Acreage elasticity is interpreted as a percentage change in a particular land use due to a percentage change in the value of the independent variable. Marginal effects and elasticities for the multinomial logit function are not monotonic, but depend on the point of evaluation and may vary in sign from the sign of the estimated coefficients, because of complex equations involving several coefficients, values of independent variables and land use proportions. The standard errors for marginal effects and elasticities are computed using the delta method (Greene, 1993, p. 297).

[4]In the multiplicative heteroscedastic regression model, ordinary least squares estimates are obtained in the first step and in the second step maximum likelihood estimates are obtained using the set of variables or forms of variables, which cause most heteroscedasticity as weights.

Data and Variables

This study uses county-level data from 11 states (Alabama, Arkansas, Florida, Georgia, Louisiana, Mississippi, North Carolina, South Carolina, Tennessee, East Texas and Virginia) in the southern USA over the period from 1964 to 2003.[5] Within this period, land use in timberland declined the least, agriculture land use suffered the largest decline and land use in urban and other uses increased dramatically (Table 27.1). However, within the timberland use, land use under the hardwood forest type registered a slight increase.

For the purpose of this analysis, the total land area of a county is defined as the sum of acreage under timberland, crop, pasture, urban and other uses, and excludes water area, unproductive forests and productive reserved forests. Timberland area by forest type is obtained from Forest Inventory and Analysis (FIA) surveys conducted in different years.[6] Land in agricultural use includes cropland, pastureland and rangeland reported in the various agricultural censuses approximately at 5-year intervals by the National Agricultural Statistics Service (NASS) and is linearly interpolated to conform to the FIA years. Land in the urban and other category includes urban land and land devoted to roads,

[5]1964 and 2003 are the earliest and latest FIA years included in our analysis. Kentucky, Oklahoma and western counties of Texas are excluded from the analysis, since these areas belong to different ecozones. The number of FIA surveys and years for the states are:
Alabama (4) 1972, 1982, 1990, 2000.
Arkansas (5) 1969, 1978, 1988, 1995, 2001.
Florida (4) 1969, 1980, 1987, 1995.
Georgia (5) 1972, 1982, 1989, 1997, 2000.
Louisiana (5) 1964, 1974, 1984, 1991, 2003.
Mississippi (4) 1967, 1977, 1987, 1994.
North Carolina (4) 1974, 1984, 1990, 2002.
South Carolina (4) 1978, 1986, 1993, 2000.
Tennessee (4) 1971, 1980, 1989, 1999.
Texas (East) (5) 1965, 1975, 1986, 1992, 2002.
Virginia (4) 1977, 1984, 1992, 2001.

[6]The FIA programme periodically collects, analyses and reports information on the status and trends of the USA's forests – i.e., how much forest exists, where it exists, who owns it and how it is changing, as well as how the trees and other forest vegetation are growing and how much has died or has been removed in recent years.

Table 27.1. Historical land use in the southern USA.

Land use	1970[a]		2000[b]		Change	
	1000 acres	%	1000 acres	%	1000 acres	%
Softwood	66,264.2	21.2	64,747.4	20.7	−1,516.8	−2.3
Pine–oak mixed	29,009.1	9.3	27,671.4	8.9	−1,337.7	−4.6
Hardwood	90,958.4	29.1	91,550.7	29.3	592.3	0.7
Timberland[c]	186,231.7	59.5	183,969.5	58.9	−2,262.2	−1.2
Agriculture[d]	85,350.8	27.3	74,030.2	23.7	−11,320.6	−13.3
Urban/other[e]	41,522.2	13.3	54,349.3	17.4	12,827.1	30.9
Land area	313,104.7	100.0	312,349.0	100.0	−755.7	−0.2
Misc. area	4,724.1		5,217.2		493.1	
Water area	24,316.2		24,378.0		61.7	
Total area	342,145.0		341,944.2		−200.9	

[a]1970 figures are based on AL (1972), AR (1969), FL (1969), GA (1972), LA (1964), MS (1967), NC (1974), SC (1978), TN (1971), TX (1965) and VA (1977); Texas includes 43 counties of Eastern Texas only.
[b]2000 figures are based on AL (2000), AR (2001), FL (1995), GA (2000), LA (2003), MS (1994), NC (2002), SC (2000), TN (1999), TX (2002) and VA (2001); Texas includes 43 counties of Eastern Texas only.
[c]Timberland statistics are from Forest Inventory and Analysis (FIA) surveys.
[d]Agriculture statistics are from National Agricultural Statistics Service (NASS), interpolated to correspond to the FIA years.
[e]The land under urban and other uses is derived by subtracting timberland, agriculture land and miscellaneous land from the total land area.

rural transportation and other special uses, and is estimated as a residual by subtracting timberland and agricultural area from the total land area in each county.

We use Timber Mart-South (TMS) prices for pine sawtimber and oak sawtimber in dollars per thousand board feet to represent the returns to timberland use for the years from 1977 onwards (Norris Foundation). For the period before 1977, pine sawtimber and oak sawtimber prices are constructed by tracing backward from 1977, using the percentage changes in Louisiana sawtimber stumpage prices (Howard, 2003). As county level prices are not available, we use prices for two TMS regions for each state. Three area TMS prices before 1992 are converted to two area prices using conversion weights developed by Prestemon and Pye (2000). These prices are deflated using the producer price index (PPI) for all commodities 1982 = 100. As a proxy for agricultural returns, we use county level net agricultural returns obtained from the NASS. Net agricultural returns are computed as the total cash receipts from all crops and livestock minus total production expenses. Per acre net agricultural returns are obtained by dividing the county net agricultural returns with county acreage under crops and pasture. Economic returns are expected to help explain timberland, agricultural or urban land use.

We use population density as a demographic variable. Population density is estimated as the number of persons per thousand acres of land area using the US Census Bureau's mid-year population estimates from the Census Bureau's Regional Economic Information System (REIS) (Bureau of Economic Analysis, 2005). Population density is a very general measure of human influence, but helps to define where population effects may be most concentrated. As population density increases, we expect negative impact towards land use of all types of timberland and agriculture land use. County level per capita personal income is also obtained from the REIS. The per capita income data are deflated using the consumer price index (CPI) for urban areas 1982–1984 = 100. We hypothesize that per capita income would negatively affect timberland (including all forest types) use and agricultural land use relative to urban and other land use.

Data on land quality is obtained from the US Department of Agriculture (USDA). The ratings for a land parcel range from 1 to 8, in which 1 is the most productive and 8 is the least productive (Klingebiel and Montgomery, 1961). The average land quality index (AVLCC), calculated as a weighted average of acres in each land class in the county, is used in the analysis. By construction, this variable is expected to have a negative effect on the agricultural land use and a positive effect on the timberland use. The values of this land quality variable for each county are the same for all years.

We have two variables in our model to represent the urbanization process and real-estate markets in the southern USA – the CONTI code and median house values (MHVAL). The CONTI code (1–9; 1 is highly metropolitan and 9 is highly rural) is a classification system developed by the Economic Research Service, USDA, available for the years 1974, 1983, 1993 and 2003 (assigned to the nearest FIA years) and it distinguishes metropolitan counties by the population size of their metro area and non-metropolitan counties by the degree of urbanization and adjacency to a metro area or areas. Data on MHVAL for the years 1960–2000 are obtained from the Census Bureau and are interpolated for intercensual years and extrapolated for the years after 2000 using a compound growth rate formula. Real values of MHVAL are obtained by deflating with CPI for urban areas (1982–1984 = 100). We expect that the variable CONTI is positively related and that MHVAL is negatively related to all types of timberland use. However, since location theory states that agricultural land is located nearer or next to urban locations, these variables are hypothesized to influence the agricultural land use in an opposite way to the timberland use.

We use one variable to reflect the knowledge level of the counties – educational attainment as indicated by the percentage of persons over 25 years completing a bachelor's degree or higher (BEDU). Higher educational attainment signals a higher knowledge level and higher learning attitudes, which may influence agricultural land use negatively, since knowledge of improved practices would help in producing the same quantity of output from a lesser area of land. On the other hand, higher educational attainment may also indicate those who have higher educational degrees pursuing employment in urban areas, thus leaving agricultural land to be converted to timberland use because of declining real agricultural returns relative to timberland returns, and also inducing higher ownership of timberland as part of investment portfolio.

Table 27.2 summarizes information on the variables used in the analysis, including definitions, sources and descriptive statistics for the balanced sample of 2436 observations used in the estimation as against a total sample of 3909 observations.[7] The remaining counties could not be included because of zero values for some land uses in certain years, which create problems in the logarithmic transformation of the dependent variables in our analysis.

Estimation Results

The model is estimated using a maximum likelihood method with multiplicative heteroscedastic correction. State dummy variables are included to capture variation related to state-specific policies and regulations affecting land use. The estimated results suggest reasonably good fits, with conventional adjusted R^2 values ranging from 0.36 to 0.42, considering the large size of sample. Table 27.3a–d provides results related to land use distinguished by forest types – softwood, oak–pine mixed, hardwood and agriculture land uses. We discuss estimation results in terms of significant marginal effects and elasticities of variables towards the respective land use.

The different signs of significant marginal effects and elasticities for pine stumpage price (PSTPR), oak stumpage price (OSTPR) and net agricultural returns (NETAGRET) in the softwood type, hardwood type and agriculture land use equations, generally in line with expectations, indicate that returns to respective land use significantly affect the probability of land use change between forestry and agriculture. Exceptions are the positive and significant signs of marginal effects and elasticities for the NETAGRET in the mixed type equation, PSTPR in the hardwood forest type equation and OSTPR in the agricultural equation contrary to our expectations. This may have been caused by the stumpage prices and agricultural prices moving in tandem in most periods. Another problem was

[7]'Balanced sample' here means that the same counties should be present in all the survey years. Zero values for some land uses and the consequent loss of observations in the sample is one of the disadvantages in the estimation of the multinomial logit model.

Table 27.2. Descriptive statistics of variables in the analysis (n = 2436).

Variable	Description	Unit	Source	Mean	Std. dev.	Min.	Max.
AR	State dummy variable	1 or 0		0.06	0.24	0.00	1.00
FL	-do-	-do-		0.07	0.26	0.00	1.00
GA	-do-	-do-		0.24	0.43	0.00	1.00
LA	-do-	-do-		0.06	0.23	0.00	1.00
MS	-do-	-do-		0.10	0.30	0.00	1.00
NC	-do-	-do-		0.12	0.32	0.00	1.00
SC	-do-	-do-		0.07	0.26	0.00	1.00
TN	-do-	-do-		0.06	0.23	0.00	1.00
TX	-do-	-do-		0.03	0.18	0.00	1.00
VA	-do-	-do-		0.08	0.27	0.00	1.00
PINE3	Softwood type timberland	Percent	FIA	24.65	13.60	0.07	75.72
MIXED3	Mixed type timberland	-do-	FIA	11.10	6.57	0.11	56.80
HARD3	Hardwood type timberland	-do-	FIA	28.51	11.66	0.61	79.63
AGRI3	Cropland and pasture	-do-	NASS	20.60	12.21	0.46	82.95
OTHSH3	Urban and other land	-do-	Residual	15.15	9.57	0.21	69.69
PSTPR	Pine stumpage price (real $)[a]	$/MBF	TMS	158.55	60.49	44.79	284.26
OSTPR	Oak stumpage price (real $)[a]	$/MBF	TMS	109.61	61.25	38.54	286.74
NETAGRET	Net agricultural returns (real $)[a]	$/Acre	NASS	88.08	170.19	–383.55	2147.59
PD	Population density	Persons/ 1000 Ac.	Census Bureau	153.27	243.08	6.23	3200.32
PCINC	Per capita income (real $)[b]	$1000/ person	BEA	9.95	2.81	2.39	26.48
AVLCC	Weighted average land quality rating	1(high) to 8 (low)	USDA	4.03	0.86	2.36	6.68
MHVAL	Median house value (real $)[b]	$1000/unit	Census Bureau	38.03	13.66	11.01	126.06
CONTI	Urban–rural continuity code	1 (more) to 9 (less)	USDA	5.43	2.42	1.00	9.00
BEDU	Education: bachelor's degree or higher	Percent	USNCES	10.74	5.98	0.70	51.50

FIA = Forest Inventory and Analysis; NASS = National Agricultural Statistics Service; TMS = Timber Mart-South; BEA = Bureau of Economic Analysis; USDA = United States Department of Agriculture; USNCES = United States National Center for Educational Statistics.

[a]PPI (all commodities) adjusted (1982 = 100).

[b]CPI (urban) adjusted (1982–1984 = 100).

that since county-wise stumpage prices are not available, we were compelled to use stumpage prices for two regions only in each state. Although we use stumpage prices as proxies for returns for timberland use instead of the theoretically correct forest returns as annualized values of discounted future returns, the estimated results are reasonably good.

The marginal effects and elasticities for population density and per capita income show a significant negative impact on all types of land use, in line with our expectations. The average land quality variable is coded in such a way that as the code increases, the land is regarded as being of poorer quality. As expected, the effect of the average quality variable is positive towards all forest type land uses and negative towards agricultural land use.

The effect of the median housing value is significant and negative for softwood and mixed types of land use, and the marginal effects and elasticities are positive for agricultural land use, in line with our expectations. In another indicator of urban influence, as the CONTI code increases, the urban influence declines, showing

Table 27.3a. Estimation results according to land use by forest type (n = 2436)[a]. Softwood type, dependent variable: Ln(Soft/Urban and Other).

Variable	Coefficient (SE)	Marginal effect (SE)	Elasticity (SE)
Constant	1.0460 (0.174)[e]		
AR	0.1498 (0.092)	0.0080 (0.018)	0.0336 (0.075)
FL	0.3648 (0.094)[e]	0.1009 (0.018)[e]	0.4233 (0.077)[e]
GA	0.4973 (0.075)[e]	0.0786 (0.015)[e]	0.3298 (0.061)[e]
LA	0.2895 (0.092)[e]	0.0720 (0.018)[e]	0.3021 (0.076)[e]
MS	−0.4164 (0.081)[e]	−0.0280 (0.016)[c]	−0.1175 (0.066)[c]
NC	0.1654 (0.081)[d]	0.0423 (0.016)[e]	0.1776 (0.067)[e]
SC	0.3409 (0.085)[e]	0.0938 (0.017)[e]	0.3936 (0.07)[e]
TN	−0.8296 (0.108)[e]	−0.1620 (0.021)[e]	−0.6797 (0.089)[e]
TX	0.0245 (0.119)	−0.0556 (0.023)[d]	−0.2334 (0.098)[d]
VA	0.4168 (0.097)[e]	−0.0019 (0.019)	−0.0081 (0.08)
PSTPR	0.0030 (0.001)[e]	0.0004 (0.0001)[e]	0.2891 (0.08)[e]
OSTPR	0.0006 (0.001)	−0.0002 (0.0001)	−0.0802 (0.052)
NETAGRET	0.0001 (0.0001)	0.000004 (0.00002)	0.0016 (0.008)
PD	−0.0014 (0.0001)[e]	−0.00009 (0.00002)[e]	−0.0599 (0.014)[e]
PCINC	−0.0836 (0.014)[e]	−0.0004 (0.003)	−0.0162 (0.118)
AVLCC	0.0982 (0.024)[e]	0.0096 (0.005)[d]	0.1631 (0.08)[d]
MHVAL	−0.0261 (0.003)[e]	−0.0034 (0.001)[e]	−0.5422 (0.093)[e]
CONTI	0.0178 (0.01)[c]	−0.0016 (0.002)	−0.0359 (0.046)
BEDU	0.0253 (0.006)[e]	0.0026 (0.001)[d]	0.1160 (0.051)[d]
Adjusted R-Sq.[b]	0.40		
Predicted share	0.2384	Actual share	0.2465

[a]The marginal effects and elasticities are computed at the means of the variables, except for the dummy variables, which are based on value of dummy variable = 1.
[b]Conventional.
[c]$P < 0.10$, [d]$P < 0.05$, [e]$P < 0.01$.

Table 27.3b. Mixed type, dependent variable: Ln(Mixed/Urban and Other).

Variable	Coefficient (SE)	Marginal effect (SE)	Elasticity (SE)
Constant	0.1356 (0.166)		
AR	0.2159 (0.086)[c]	0.0107 (0.009)	0.0997 (0.085)
FL	−0.8305 (0.088)[d]	−0.0832 (0.009)[d]	−0.7720 (0.087)[d]
GA	−0.0585 (0.071)	−0.0243 (0.007)[d]	−0.2260 (0.069)[d]
LA	−0.0400 (0.089)	−0.0029 (0.009)	−0.0274 (0.087)
MS	−0.2886 (0.076)[d]	0.0011 (0.008)	0.0102 (0.074)
NC	−0.1555 (0.078)[c]	−0.0154 (0.008)[b]	−0.1433 (0.076)[b]
SC	−0.3925 (0.083)[d]	−0.0366 (0.009)[d]	−0.3398 (0.081)[d]
TN	−0.6385 (0.101)[d]	−0.0526 (0.011)[d]	−0.4887 (0.099)[d]
TX	0.1412 (0.115)	−0.0126 (0.012)	−0.1167 (0.112)
VA	0.0445 (0.094)	−0.0410 (0.01)[d]	−0.3804 (0.091)[d]
PSTPR	0.0011 (0.001)[b]	−0.00001 (0.0001)	−0.0154 (0.091)
OSTPR	0.0019 (0.001)[d]	0.00007 (0.0001)	0.0704 (0.06)
NETAGRET	0.0003 (0.0001)[c]	0.00002 (0.00001)[b]	0.0178 (0.01)[b]
PD	−0.0012 (0.0001)[d]	−0.00002 (0.00001)[b]	−0.0303 (0.016)[b]
PCINC	−0.0908 (0.014)[d]	−0.0009 (0.001)	−0.0877 (0.135)
AVLCC	0.1909 (0.022)[d]	0.0143 (0.002)[d]	0.5368 (0.089)[d]
MHVAL	−0.0188 (0.003)[d]	−0.0008 (0.0003)[c]	−0.2650 (0.108)[c]
CONTI	0.0168 (0.01)[b]	−0.0008 (0.001)	−0.0417 (0.052)
BEDU	0.0234 (0.006)[d]	0.0010 (0.001)	0.0956 (0.059)
Adjusted R-Sq.[a]	0.37		
Predicted share	0.1077	Actual share	0.1110

[a]Conventional.
[b]$P < 0.10$, [c]$P < 0.05$, [d]$P < 0.01$.

Table 27.3c. Hardwood type, dependent variable: Ln(Hard/Urban and Other).

Variable	Coefficient (SE)	Marginal effect (SE)	Elasticity (SE)
Constant	0.5007 (0.141)[d]		
AR	0.1469 (0.073)[c]	0.0094 (0.018)	0.0307 (0.058)
FL	−0.1366 (0.074)[b]	−0.0240 (0.018)	−0.0782 (0.059)
GA	0.1144 (0.059)[b]	−0.0163 (0.014)	−0.0531 (0.047)
LA	0.0284 (0.071)	0.0126 (0.018)	0.0411 (0.057)
MS	−0.3508 (0.063)[d]	−0.0159 (0.015)	−0.0519 (0.05)
NC	−0.1571 (0.064)[c]	−0.0445 (0.016)[d]	−0.1448 (0.051)[d]
SC	−0.0193 (0.067)	0.0103 (0.016)	0.0334 (0.054)
TN	0.3130 (0.085)[d]	0.1421 (0.021)[d]	0.4629 (0.068)[d]
TX	0.2480 (0.095)[d]	−0.0030 (0.023)	−0.0099 (0.075)
VA	0.8265 (0.077)[d]	0.1233 (0.019)[d]	0.4016 (0.061)[d]
PSTPR	0.0020 (0.0005)[d]	0.0002 (0.0001)[c]	0.1268 (0.061)[c]
OSTPR	0.0015 (0.0005)[d]	0.0001 (0.0001)	0.0256 (0.04)
NETAGRET	0.00001 (0.0001)	−0.00002 (0.00002)	−0.0055 (0.006)
PD	−0.0012 (0.0001)[d]	−0.00004 (0.00002)[b]	−0.0193 (0.011)[b]
PCINC	−0.1078 (0.011)[d]	−0.0079 (0.003)[d]	−0.2565 (0.09)[d]
AVLCC	0.2096 (0.02)[d]	0.0466 (0.005)[d]	0.6125 (0.064)[d]
MHVAL	−0.0087 (0.002)[d]	0.0010 (0.001)[b]	0.1197 (0.072)[b]
CONTI	0.0417 (0.008)[d]	0.0053 (0.002)[d]	0.0936 (0.035)[d]
BEDU	0.0176 (0.005)[d]	0.0010 (0.001)	0.0334 (0.039)
Adjusted R-sq.[a]	0.42		
Predicted share	0.3070	Actual share	0.2851

[a]Conventional.
[b]$P < 0.10$, [c]$P < 0.05$, [d]$P < 0.01$.

Table 27.3d. Agriculture, dependent variable: Ln(Agri/Urban and Other).

Variable	Coefficient (SE)	Marginal effect (SE)	Elasticity (SE)
Constant	2.3688 (0.148)[d]		
AR	0.0613 (0.078)	−0.0109 (0.014)	−0.0549 (0.07)
FL	−0.0709 (0.079)	−0.0025 (0.014)	−0.0124 (0.072)
GA	0.1016 (0.064)	−0.013 (0.011)	−0.0659 (0.058)
LA	−0.4345 (0.08)[d]	−0.0836 (0.014)[d]	−0.4218 (0.072)[d]
MS	−0.3072 (0.069)[d]	−0.0017 (0.012)	−0.0084 (0.062)
NC	0.0671 (0.069)	0.0157 (0.012)	0.0794 (0.063)
SC	−0.433 (0.074)[d]	−0.0753 (0.013)[d]	−0.3803 (0.067)[d]
TN	0.1039 (0.092)	0.0503 (0.017)[d]	0.2538 (0.083)[d]
TX	0.8115 (0.1)[d]	0.1096 (0.018)[d]	0.5536 (0.091)[d]
VA	0.3385 (0.083)[d]	−0.0171 (0.015)	−0.0864 (0.075)
PSTPR	−0.0014 (0.0005)[d]	−0.0005 (0.0001)[d]	−0.3983 (0.075)[d]
OSTPR	0.0024 (0.0005)[d]	0.0002 (0.0001)[c]	0.1249 (0.049)[c]
NETAGRET	0.0001 (0.0001)	0.000004 (0.00002)	0.0016 (0.008)
PD	−0.001 (0.0001)[d]	−0.000001 (0.00002)	−0.0004 (0.013)
PCINC	−0.0969 (0.012)[d]	−0.003 (0.002)	−0.1487 (0.11)
AVLCC	−0.2553 (0.021)[d]	−0.062 (0.004)[d]	−1.2624 (0.076)[d]
MHVAL	−0.0046 (0.003)[b]	0.0014 (0.0005)[d]	0.2738 (0.087)[d]
CONTI	0.0282 (0.009)[d]	0.0008 (0.002)	0.0206 (0.043)
BEDU	0.0028 (0.005)	−0.0023 (0.001)[d]	−0.1261 (0.048)[d]
Adjusted R-Sq.[a]	0.36		
Predicted share	0.1981	Actual share	0.2060

[a]Conventional.
[b]$P < 0.10$, [c]$P < 0.05$, [d]$P < 0.01$.

an expected positive impact on hardwood type land use and agricultural land use. The coefficient for educational attainment variable (BEDU) is positive and significant for all types of timberland use, and the marginal effects and elasticities are negative and significant in the case of agricultural land use, in accordance with our expectations.

A comparison of predicted and actual shares of various land uses from the balanced sample used in the analysis, shown in Table 27.3a–d, indicates that the analysis of timberland use by forest type results in predictions that are closer to actual land use shares. The analysis by forest types together predicted 65.31% in timberland use, 19.81% in agricultural use and 14.88% in urban and other use, as against 64.26%, 20.60% and 15.14%, respectively, in the sample used for our estimation.[8] With regard to forest types, our model predicted 23.84% in the softwood type, 10.77% in the oak–pine mixed type and 30.70% in hardwood type land uses, as against 24.65%, 11.10% and 28.51%, respectively, in our balanced sample.

Land Use Projections

Using our model estimates and mean values of variables in 2000 as a base scenario, we first derive the proportions of various land uses under the two scenarios by using the estimated coefficients and variable values under various assumptions. Applying these proportions to the total land area enables us to develop area projections by various land uses. Table 27.4 shows the assumptions of percentage increases in the values of different variables at 10-year intervals under two scenarios. While the model coefficients are obtained from a balanced sample of 2436 observations, we develop projections to the total land area of 204.973 million acres only related to the counties in the balanced sample for the year 2000 as the base scenario, extrapolating the projections on the basis of assumptions of the variables in different periods.[9]

In both scenarios, we assume the same rates of change in population (PD) and per capita income (PCINC), based on interim population projections for the 11 states developed by the US Census Bureau (2005) and per capita income projections for the 11 states developed by the US Department of Commerce (1995). We assume two sets of price increases for stumpage prices to reflect increasing real returns for timberland use and two rates of decrease in real net returns to agriculture land use. Scenario 1 assumes a 0.5% annual increase in real PSTPR, a 0.75% annual increase in real OSTPR and a −0.5% annual decrease in the real NETAGRET. A higher rate of increase for the oak sawtimber price is justifiable, since oak sawtimber prices grew historically at a higher rate than pine sawtimber.[10] A negative rate of change for NETAGRET is justifiable, as real farm product prices have declined 1.2% per annum historically during the last 50 years. Scenario 2 assumes a 1% annual increase in real PSTPR, a 1.5% increase in real OSTPR, and a −1% annual decrease in the real NETAGRET. All other variables are assumed to be constant at the year 2000 level.

Table 27.5a and b and Figs 27.1 and 27.2 provide the results of projections for different land uses under two scenarios. Under scenario 1, total timberland use is projected to decline by 14.0 million acres, agricultural land use declines by 8.1 million acres, whereas urban and other land use increases by 22.1 million acres by the year 2050 from the year 2000 level. Among the forest types, land use under softwood forest type

[8]These numbers do not tally with the numbers in Table 27.1, which are derived from a total sample of 3909 observations, whereas the estimation was carried out with a balanced sample of 2436 observations due to zero values for some land uses in certain years for some counties.

[9]The total land area of all counties in the 11 states of the southern USA (excluding Kentucky, Oklahoma and western counties of Texas) in the year 2000 is approximately 312.349 million acres, as shown in Table 27.1. However, we are not developing land use projections for this entire area, since it will not be reasonable to extend the land use projections based on the coefficients derived from a limited balanced sample that covers a total land area of 204.973 million acres.

[10]During the last 40 years, from 1964 to 2003, real stumpage prices have increased at an annual compound rate of 2.4% for pine sawtimber and 4.3% for oak sawtimber. However, we use conservative estimates in our projections.

Table 27.4. Assumptions in various scenarios (percentage changes per annum).

Scenarios	Variables	2010	2020	2030	2040	2050
Scenario 1	(a) Population density	1.37	1.28	1.29	1.28	1.27
	(b) Per capita income	1.10	1.00	0.90	0.80	0.70
	(c) Pine sawtimber price	0.50	0.50	0.50	0.50	0.50
	(d) Oak sawtimber price	0.75	0.75	0.75	0.75	0.75
	(e) Net agricultural returns	−0.50	−0.50	−0.50	−0.50	−0.50
Scenario 2	(a) Population density	1.37	1.28	1.29	1.28	1.27
	(b) Per capita income	1.10	1.00	0.90	0.80	0.70
	(c) Pine sawtimber price	1.00	1.00	1.00	1.00	1.00
	(d) Oak sawtimber price	1.50	1.50	1.50	1.50	1.50
	(e) Net agricultural returns	−1.00	−1.00	−1.00	−1.00	−1.00

declines the most (7.9 million acres), while land use under oak–pine mixed forest type declines the least (2.0 million acres) under this scenario by the year 2050. This scenario projects a loss of 4.2 million acres in hardwood forest type land use.

Under scenario 2, the total land under timberland use declines from 133.1 million acres in the year 2000 to 132.0 million acres in the year 2050, a modest decline of 1.18 million acres. Land under agricultural use declines by 9.2 million acres, whereas land under urban and other uses increases by 10.3 million acres. In this scenario also, land use under the softwood forest type (4.1 million acres) will be the most affected. Land use under the mixed type decreases slightly (0.02 million acres), while hardwood forest type land use increases by nearly 3 million acres over the year 2000 level. This scenario closely resembles the historical land use change in the region and may thus be more realistic.

Our projections are not strictly comparable with previous projections developed by other researchers (e.g. Alig *et al.*, 2003; Alig and Butler, 2004), since we base our projections on the results derived from a limited balanced sample. None

Table 27.5a. Land use projections under scenario 1 in the southern USA.

	1970 actual	1970 predicted[a]	2000 actual	2000 predicted[a]	2010	2020	2030	2040	2050
				Percentage					
Soft	25.46	24.31	25.22	22.95	22.81	22.59	22.28	21.87	21.39
Mixed	11.49	10.96	10.74	10.70	10.56	10.39	10.20	9.98	9.75
Hard	28.01	30.81	28.99	32.00	31.05	30.05	29.02	27.97	26.96
Timberland	64.95	66.08	64.95	65.65	64.42	63.03	61.49	59.83	58.10
Agriculture	22.83	21.63	18.44	17.91	17.26	16.59	15.90	15.20	14.50
Urban and other	12.22	12.29	16.61	16.44	18.32	20.38	22.61	24.98	27.40
Land area	100.00	100.00	100.00	100.00	100.00	100.00	100.00	100.00	100.00
				1,000 acres					
Soft	52,206	49,848	51,690	47,047	46,751	46,306	45,663	44,833	43,834
Mixed	23,566	22,473	22,016	21,931	21,639	21,297	20,899	20,457	19,985
Hard	57,435	63,195	59,425	65,592	63,646	61,599	59,476	57,341	55,264
Timberland	133,207	135,516	133,131	134,571	132,037	129,202	126,038	122,631	119,082
Agriculture	46,817	44,366	37,791	36,706	35,385	34,005	32,584	31,149	29,731
Urban and other	25,064	25,206	34,051	33,696	37,552	41,766	46,351	51,193	56,160
Land area	205,088	205,088	204,973	204,973	204,973	204,973	204,973	204,973	204,973

[a]These predicted values for the years 1970 and 2000 are generated by using the average variable values for the respective years in the equations.

Table 27.5b. Land use projections under scenario 2 in the southern USA.

	1970 actual	1970 predicted[a]	2000 actual	2000 predicted[a]	2010	2020	2030	2040	2050
	Percentage								
Soft	25.46	24.31	25.22	22.95	23.06	23.15	23.19	23.21	23.20
Mixed	11.49	10.96	10.74	10.70	10.66	10.63	10.62	10.65	10.73
Hard	28.01	30.81	28.99	32.00	31.51	31.07	30.70	30.48	30.44
Timberland	64.95	66.08	64.95	65.65	65.23	64.84	64.52	64.34	64.37
Agriculture	22.83	21.63	18.44	17.91	17.07	16.23	15.43	14.67	13.97
Urban and other	12.22	12.29	16.61	16.44	17.71	18.92	20.05	21.00	21.65
Land area	100.00	100.00	100.00	100.00	100.00	100.00	100.00	100.00	100.00
	1,000 acres								
Soft	52,206	49,848	51,690	47,047	47,274	47,451	47,543	47,571	47,551
Mixed	23,566	22,473	22,016	21,931	21,841	21,782	21,768	21,828	21,997
Hard	57,435	63,195	59,425	65,592	64,584	63,676	62,937	62,474	62,403
Timberland	133,207	135,516	133,131	134,571	133,699	132,910	132,248	131,872	131,951
Agriculture	46,817	44,366	37,791	36,706	34,982	33,274	31,619	30,060	28,636
Urban and other	25,064	25,206	34,051	33,696	36,293	38,790	41,106	43,041	44,387
Land area	205,088	205,088	204,973	204,973	204,973	204,973	204,973	204,973	204,973

[a]These predicted values for the years 1970 and 2000 are generated by using the average variable values for the respective years in the equations.

the less, our projections show relatively more loss of timberland due to urbanization pressures. If our limited projections are any indication of the future picture, there might be a larger loss of timberland use under the softwood forest type than under the hardwood forest type. A limitation of our projections is that we have not been able to incorporate the prominent tree-planting behaviour of private timberland owners in the region.

Conclusions

In this chapter, we conduct a modified multinomial logit analysis on land use change and timberland by forest type in 11 states of the southern USA between 1964 and 2003, estimated by a maximum likelihood method using a multiplicative heteroscedastic regression procedure. The results show that relative returns to respective land uses, land quality, economic, demographic

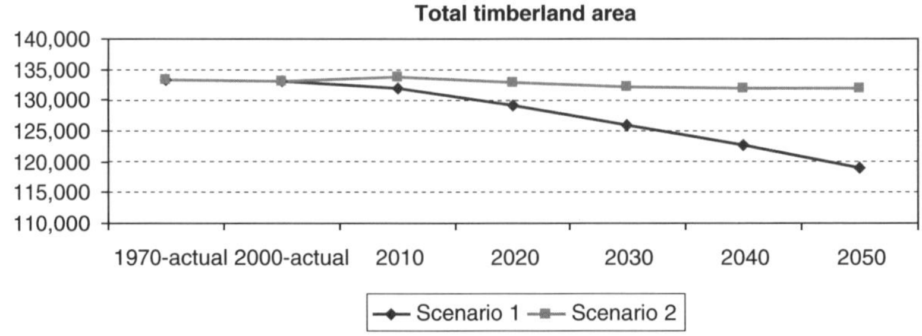

Fig. 27.1. Timberland projections in two scenarios in the southern USA.

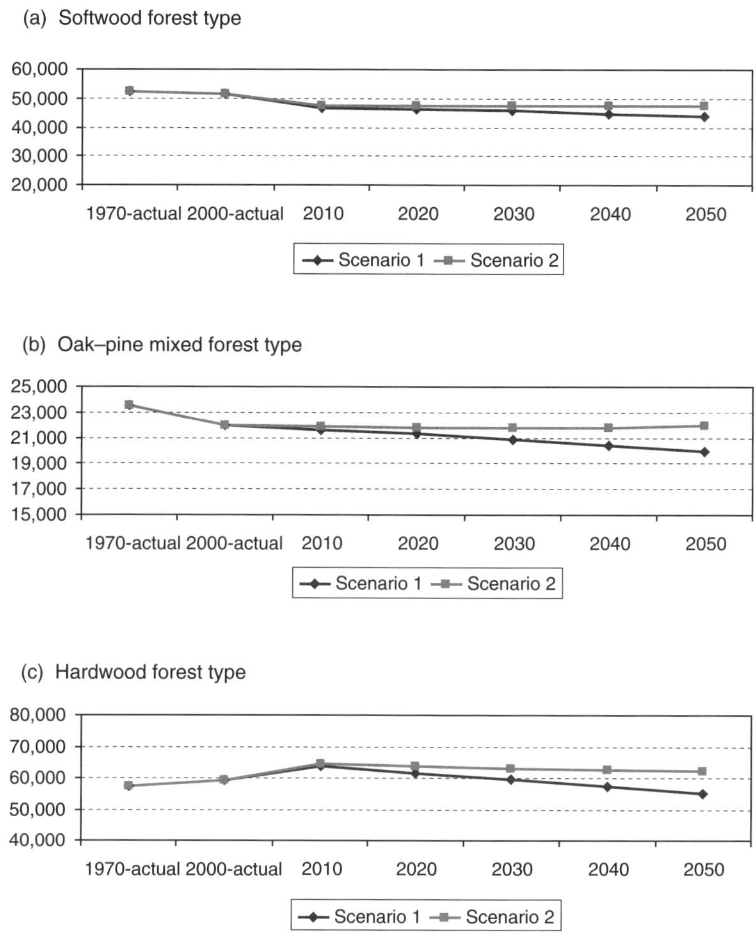

Fig. 27.2. Area projections by forest type in two scenarios in the southern USA. (a) Softwood forest type. (b) Oak–pine mixed forest type. (c) Hardwood forest type.

and urbanization variables are the key driving factors in the land use change.

Our results also show that urbanization variables have a negative impact on timberland use in general (i.e. reducing the probability that land would be devoted to timber production) and, specifically, have the largest negative impact on softwood timberland use and the least impact on hardwood timberland use. Thus, urbanization and production forestry (represented by softwood timberland and timberland in general) appear to compete for land. In the wake of declining real net returns to agricultural land use relative to other uses, the agricultural land closer to urban areas is likely to be converted to urban and developed uses, while agricultural land away from the urban areas is likely to be converted to timberland. Housing prices and urban influence (as indicated by the CONTI code) are also key factors in the decline of timberland use.

Counties with a higher proportion of good land quality and residents with higher incomes are likely to have less land used for timber production in general. On the other hand, the proportion of good land quality increases the probability of land being used for agriculture, while additional population pressures as indicated by population density and increased real incomes have the opposite effect. In summary, different factors affect timberland by forest type in different magnitudes.

Using our model estimates, we develop projections of land use in different uses under two scenarios for total land area based on our limited balanced sample. Our projections indicate that there will be a large loss of timberland use under timberland due to urbanization pressures in the southern USA by the year 2050. In addition, the loss in timberland use under the softwood forest type will be more than the loss in timberland use under the hardwood forest type. If the government implements a tree-planting programme, the loss of softwood forest type could be curtailed to some extent.

Our results and projections show that timberland in the region is influenced by and subject to cross-sectoral forces, such as urbanization (population, economic growth), returns to agricultural land and other socio-economic factors. Although changes in relative economic returns favour timberland use over agricultural land use, the greater impact of urbanization could lead to greater losses of land under both agriculture and forestry use. Should timberland continue to decline, production forestry in the southern USA – one of the largest timber baskets in the world – could decline or at least not grow as fast as in the past.

Furthermore, urbanization pressures have led to the creation of an urban–wildland interface, an area in which houses, structures and other forms of human development meet or intermingle with undeveloped wildland or vegetative fuels, creating risks of wildfire destroying forests, wildlife habitats and human property.

Finally, loss of timberland in the region portends negative consequences for various ecosystem services provided by the forested landscapes, such as the protection of water quality, preservation of species biodiversity, provision of open spaces for outdoor recreational benefits and mitigation of global climate change. If these ecosystem benefits based on forestry land use are going to be protected or enhanced, there may be a necessity to adopt policy measures to curtail the losses in timberland use and to promote extensive tree planting under cost-share programmes.

References

Ahn, S., Plantinga, A.J. and Alig, R.J. (2002) Determinants and projections of land use in the South Central United States. *Southern Journal of Applied Forestry* 26, 78–84.

Alig, R.J. (1986) Econometric analysis of factors influencing forest acreage trends in the Southeast. *Forest Science* 32, 119–134.

Alig, R.J. and Butler, B.J. (2004) *Area Changes for Forest Cover Type in the United States, 1952 to 1997, with Projections to 2050*. General Technical Report PNW-GTR-613. US Department of Agriculture, Pacific Northwest Research Station, Portland, Oregon.

Alig, R.J., Plantinga, A.J., Ahn, S. and Kline, J.D. (2003) *Land Use Changes Involving Forestry in the United States: 1952 to 1997, with Projections to 2050*. General Technical Report PNW-GTR-587. US Department of Agriculture, Pacific Northwest Research Station, Portland, Oregon.

Amemiya, T. and Nold, F. (1975) A modified logit model. *Review of Economics and Statistics* 55, 255–257.

Bureau of Economic Analysis (BEA) (2005) Regional Economic Information System (REIS). Available at: http://www.bea.gov/bea/regional/reis

Greene, W.H. (1993) *Econometric Analysis*. Macmillan, New York.

Greene, W.H. (1995) *LIMDEP Version 7.0. User's Manual*. Econometric Software, Bellport, New York.

Hardie, I.W. and Parks, P.J. (1997) Land use in a region with heterogeneous land quality: an application of an area base model. *American Journal of Agricultural Economics* 73, 300–310.

Hardie, I.W., Parks, P.J., Gottlieb, P. and Wear, D.N. (2000) Responsiveness of rural and urban land uses to land rent determinants in the US South. *Land Economics* 78, 659–673.

Harris, T.M., Siry, J. and Baldwin, S. (2003) How competitive is the Southern timber industry? An examination of Georgia's pulp and paper sector. In: *2003 SOFEW Proceedings*. Mississippi State University, Mississippi. Available at: http://sofew.cfr.msstate.edu/papers/0204harris.pdf

Harvey, A.C. (1976) Estimating regression models with multiplicative heteroskedasticity. *Econometrica* 44, 461–465.

Howard, J.L. (2003) *U.S. Timber Production, Trade, Consumption, and Price Statistics 1965–2002*. Research Paper FPL-RP-615. US Department of Agriculture, Forest Service, Forest Products Laboratory, Madison, Wisconsin.

Klingebiel, A.A. and Montgomery, P.H. (1961) *Land-Capability Classification*. USDA Agricultural Handbook, no. 210. US Department of Agriculture, Washington, DC.

Lubowski, R.N., Plantinga, A.J. and Stavins, R.N. (2003) *Determinants of Land-Use Change in the United States 1982–97*. Discussion Paper 03–47. Resources for the Future, Washington, DC.

Mauldin, T.E., Plantinga, A.J. and Alig, R.J. (1999) *Land Use in the Lake States Region: an Analysis of Past Trends and Projections of Future Changes*. (Res. Pap. PNW-RP-519.) US Department of Agriculture, Pacific Northwest Research Station, Portland, Oregon.

Miller, D.J. and Plantinga, A.J. (1999) Modeling land use decisions with aggregate data. *American Journal of Agricultural Economics* 81, 180–194.

Nagubadi, R.V. and Zhang, D. (2005) Determinants of timberland use by ownership and forest type in Alabama and Georgia. *Journal of Agricultural and Applied Economics* 37(1), 173–186.

Norris Foundation (various years) *Timber Mart-South Quarterly Report*. Norris Foundation, Daniel B. Warnell School of Forest Resources, University of Georgia, Athens, Georgia.

Parks, R.W. (1980) On the estimation of multinomial logit models from relative frequency data. *Journal of Economics* 13, 293–303.

Prestemon, J.P. and Pye, J.M. (2000) A technique for merging areas in Timber Mart-South data. *Southern Journal of Applied Forestry*, 24(4), 219–229.

Smith, W.B., Miles, P.D., Vissage, J.S. and Pugh, S.A. (2004) *Forest Resources of the United States, 2002*. General Technical Report NC-241. US Department of Agriculture, Forest Service, North Central Research Station, St. Paul, Minnesota.

US Census Bureau (2001) *Population Change and Distribution, 1990 to 2000: Census 2000 Brief*. US Department of Commerce, Economics and Statistics Administration, US Census Bureau, Washington, DC.

US Census Bureau (2005) *Table A1: Interim Projections of the Total Population for the United States and States: April 1, 2000 to July 1, 2030*. Population Division, Interim State Population Projections, US Census Bureau, Washington, DC.

US Department of Commerce (1995) *Bureau of Economic Analysis Projections to 2045: Volume 1, States*. US Department of Commerce, Washington, DC.

Wear, D.N. (2002) Land use. In: Wear, D.N. and Greis, J.G. (eds) *Southern Forest Resource Assessment*. (Gen. Tech. Rep. SRS-53.) US Department of Agriculture, Forest Service, Southern Research Station, Ashville, North Carolina, pp. 153–173.

Wu, J. and Segerson, K. (1995) The impact of policies and land characteristics on potential ground water pollution in Wisconsin. *American Journal of Agricultural Economics* 77, 1033–1047.

28 Towards Joint Transboundary Watershed Management in Patagonia: Lessons from Argentina and Chile

Samuel J. Francke-Campaña

Introduction

A watershed can be defined as an area in which all flowing streams are fed by a common source and form part of the same hydrological system. Humans interact with a variety of natural resources, such as water, soil, flora, fauna and mineral resources through different socio-economic and cultural processes. Some of these are directed towards immediate resource utilization, transformation and consumption, while others aim at the protection, restoration or preservation of resources.

In South America, 60% of the population live in watersheds with transboundary water resources shared by two or more countries, and these represent 60% of the total area (Falkenmark *et al.*, 1987). For example, the watersheds of the Amazon and Rio de la Plata rivers are shared by eight and five countries, respectively. This chapter discusses the ways in which the increasing scarcity of water and other natural resources in the transboundary region of Patagonia have been addressed in order to develop a more integrated and multisectoral management approach that is locally specific and operates at different time scales (Francke, 2005).

Helsinki Agreement on Transboundary Watershed Management

Water conservation and the reduction of pollution in watersheds that belong to two or more countries are regulated by the international Agreement on Protection and Utilization of Transboundary Watercourses and International Lakes (Anonymous, 1992). The agreement defines negative impacts that are to be avoided in using the resources of such areas in the following manner: 'The term transboundary impact shall be understood as any adverse significant effect produced by a modification in the state of the transboundary waters caused by a human activity, whose physical origin is totally or partially located in an area under jurisdiction of a party, which may be generated to the environment of an area located under the jurisdiction of another party.' Negative factors resulting from human interventions to be avoided in this context are in particular 'those affecting the health and security of humans, flora, fauna, soil, atmosphere, water, climate, landscape and historic monuments or other physic structures, or the interaction of the mentioned factors, and those having effects upon the cultural patrimony or the socioeconomic conditions derived from the alterations of the mentioned factors.'

The third article of the agreement requires that the parties shall elaborate, adopt, apply and, as far as possible, take legal, administrative, economic, financial and technical measures in order to guarantee the following objectives:

- The prevention, control and reduction of pollutant emissions through the use of

technologies for treating pollutants and waste without residues;
- The protection of transboundary waters against pollution originating from specific sources as a result of previous authorization by the competent national authorities for discharges of wastewater, and vigilance and control over these authorized discharges;
- Establishment of limits for wastewater discharges on the basis of the best available technology for the discharge of hazardous substances;
- Imposition of strict regulations, including total prohibition in specific cases, when the quality of the receiver waters or the ecosystem requires it;
- Application of at least one biological treatment or an equivalent process to urban wastewater, including gradual introduction of these processes when necessary;
- Adoption of appropriate measures, such as the best available technology, in order to reduce the contribution of nutrients to industrial and urban water sources;
- Elaboration and application of appropriate measures and the best environmental practices in order to reduce the contribution of hazardous nutrients to area sources, particularly when the main sources are based on agriculture;
- The application of environmental impact assessments and other means of evaluation;
- Requirement for the sustainable management of water resources, including the application of an environmental approach;
- Development of emergency plans;
- Adoption of specific additional measures to prevent the pollution of subsurface waters;
- Reducing the risks of accidental pollution to a minimum.

The above principles and regulations have to be taken into account when formulating agreements between countries that are signatories to the Helsinki Agreement. In developing processes for environmental cooperation on transboundary watersheds, the following problems have to be addressed:

- Lack of hydrological, environmental, socioeconomic and cultural information;
- Absence of management plans for international watersheds;
- Absence of institutions capable of appropriate transboundary management, at the political and administrative levels;
- Lack of operating mechanisms, financial instruments and general coordination;
- Poor participation by the stakeholders and economic sectors involved;
- Traditional approaches to management relating exclusively to water resources;
- Difficulties in measuring the benefits versus the costs of externalities;
- Absence of economic mechanisms or instruments to provide compensation for delivering or protecting environmental goods and services.

Argentina and Chile have a common border approximately 4500 km long. In the Patagonian region, the water resources shared between the two countries extend south of the 39th parallel, with permanent watersheds from old flows, representing superficial water flows and lacustrine water resources of the Andean–Patagonian zone (Table 28.1). The watersheds of Chilean Patagonia, involving water resources shared with Argentina, represent 40% of the total area of Chilean Patagonia, including 57% of the surface of Chilean native forests and 41% of the surface of wild protected areas, including national parks, natural monuments and national forest reserves, which contribute to maintaining the ecological balance of watersheds (Francke, 2002).

The Patagonian ecosystems involve vegetation resources, pristine landscapes and water resources of high quality with low levels of pollution, considering the modest population density and the limited industrial and production activities in this region. This makes it possible to maintain the fragile balance of the ecosystems in relation to the watersheds, and contributes to increasing ecotourism in the Patagonian region. Both countries use their water resources for human consumption, irrigation and also for recreational activities, fishing and ecotourism.

There are also areas – for instance, in Chilean Patagonia – that are suffering degradation of the forest cover due to forest fires and intensive overgrazing, and dispersion of the soil due to wind erosion in semiarid cold steppes. In areas with higher precipitation, the susceptibility to water erosion is increased and hydrological processes of erosion and blockage of the watersheds

Table 28.1. Watershed water resources shared by Chile and Argentina. (From Francke, 2002.)

Argentina	Chile	Flowing into
• Hua-Hum	Valdivia	Pacific
• Manso–Puelo	Puelo	Pacific
• Encuentro/Engaño/Pico River	Palena	Pacific
• Futaleufú, Puelo and parts of Manso rivers	Yelcho	Pacific
• Simpson River	Aysén	Pacific
• Buenos Aires Lake/Jeinimeni River/Pueyrredón and parts of Mayer rivers	Backer	Pacific
• Vizcachas, Zanja Honda and Guillermo	Serrano	Pacific
• Gallegos River	Gallegos	Atlantic
• Chico River	Chico	Atlantic
• Grande River, Tierra del Fuego	Grande	Atlantic
• Fagnano Lake	Azopardo	Atlantic
• Fagnano Lake	Lapataia	Atlantic

are becoming more intense. Similarly, processes of desertification are noticeably and intensely affecting vast areas of the Patagonian zone in Argentina, mainly due to overgrazing and wind erosion. In both countries, the degradation of natural resources is largely due to the application of unsustainable development models.

Policy Instruments for Managing Transboundary Watersheds in Patagonia

Environmental Agreement (1991)

The Environmental Agreement of 1991, signed by the republics of Argentina and Chile, forms the framework for further bilateral measures in accordance with the international commitments assumed by the two countries. In Part I of the Agreement, it is stated that the parties (Argentina and Chile) shall carry out coordinated or joint actions regarding the protection, preservation, conservation and restoration of the environment, taking into account existing links between the environment and development. Each party commits itself to avoid unilateral actions causing damage to the other party's environment. The parties agree to maintain their positions during the process of negotiation that may take place during multilateral forums on subjects relating to the present agreement. Among the policies and measures that are necessary, special attention is to be given to the requirements of native people.

Protocol on Shared Water Resources (1991)

The parties agreed on current and future actions and programmes to be carried out in relation to the use of shared water resources, in accordance with the concept of integrated watershed management. The use of water on the territory of one of the parties that originates from a transboundary watershed should not cause any damage to the shared water resources, other watersheds or the environment. Under this agreement, the signatories consider the term 'shared water resources' to mean 'those waters whose natural drainage crosses or coincides, in part or in whole, with the international terrestrial Chilean–Argentinean border'. Actions and programmes for the use of the shared water resources will be carried out through coordinated or joint action using general plans for use in the framework of an environmental subcommittee.

Protocol on Cooperation in Forest Matters (1997)

In order to establish applicable criteria for cooperation in forest matters affecting transboundary watersheds, the following areas of application of the protocol are listed:

- Forest promotion;
- Protection, conservation and management of wild fauna;
- Rational administration and management of protected areas;

- Prevention and suppression of wildfires;
- Evaluation of renewable natural resources, especially regarding the use of geographic information systems, pest and forest disease control, and use of wood-related products.

In order to deal with the topics mentioned above, the signatories may interchange experts or jointly call for specialists from third countries; organize seminars and workshops on matters of mutual interest; train technicians through courses at different universities or specialized institutions; establish permanent channels for passing on technical or academic information and use appropriate means to improve awareness among the public and in private institutions pertaining to each party. The parties are to establish a working group, under the environmental subcommittee, to include members of Chile's National Forestry Service and Argentina's Governmental Office of Natural Resources and Sustainable Development. This working group is to be in charge of implementing the measures established during bilateral cooperation.

In the framework of the environmental and integration agreement and the additional protocol on shared water resources, the Argentine and Chilean working parties have succeeded in the following:

- Selecting priority watersheds;
- Developing an information system on shared water resources and natural resources in the two countries;
- Defining a structure for general utilization plans in accordance with the concept of integrated watershed management.

The following watersheds have been identified as first priority for integrated management measures:

- Hua-Hum River/Valdivia;
- Puelo River (Chile)/Manso–Puelo River (Argentina);
- Futaleufú River/Yelcho Grande River in Tierra del Fuego.

A second series of watersheds for high-priority intervention includes:

- Baker and Jenimeni Rivers, Buenos Aires Lake;
- Palena River/Palena Lake and River/Encuentro River, Tigre and Figueroa Rivers;
- Gallegos River/Penitente and Rubens Rivers.

The exchange of legal, technological and scientific information through seminars and workshops has been promoted as a complementary action to the protocol on shared water resources. The bilateral meetings have gathered together considerable numbers of scientists, technicians, experts, government officials and representatives of local communities in the Patagonian watersheds. Regional and provincial seminars and workshops that have been of particular relevance include:

- Puerto Varas, Chile, 1997;
- San Carlos de Bariloche, Argentina, 1998;
- Coyhaique, Chile, 1999;
- Río Gallegos, Argentina, 2000;
- Punta Arenas, Chile, 2001;
- Ushuaia, Argentina, 2002;
- Valdivia, Chile, 2003;
- Trevelin, Argentina, 2005.

The following goals for expanding bilateral cooperation have been defined:

- Encouraging the exchange of experience;
- Assessing and carrying out continuous monitoring of the bilateral agreements;
- Developing a mechanism for coordinating the multidisciplinary institutions and professionals directly and indirectly involved in this subject;
- Supporting mechanisms for regional and provincial participation;
- Increasing awareness and sensitivity on the part of the political and technical authorities, communities and the general public;
- Promoting the integrated management of transboundary natural resources, with a view to achieving sustainable development;
- Improving awareness of the importance of the Patagonian region in the southern cone part of Latin America from an ecological, economical and sociocultural point of view.

Projects related to forestry, national parks and wildlife

Among the projects and agreements jointly operated together with international partners, with direct or indirect links to watershed matters, the following may be mentioned:

- Project for conservation and sustainable management of native forest.
- Project of the temperate forest fund for the Chiloé model forest.
- Project for the ecoregion of Valdivia; public and private mechanisms for the conservation of the biodiversity in the Tenth Region of Chile.
- Programme of the International Network of Model Forests in Chilean–Argentinean Patagonia.
- Epic–Force project on 'Policies for the Integrated Control of Forest Watersheds' for extreme events (European Union, World Bank, Costa Rica, Ecuador, Argentina and Chile).
- Creation of private national parks for the conservation of nature in Chilean–Argentinean Patagonia.
- Project for the creation of transboundary national parks in Chilean–Argentinean Patagonia, as a natural heritage for humankind.
- Specific and mutual agreements on matters involving fighting wildfires.
- Food and Agriculture Organization of the United Nations cooperation network on watershed management for Latin America and the Caribbean region.

It is expected that the exchange of experience through seminars and workshops, support for agreements on specific programmes for watershed management, the submission of specific projects to international organizations in areas of common interest and reactivation of the forest protocol, will all continue in the future.

Conclusions

The integrated management of transboundary watersheds should be recognized as an important task for riparian states, and it needs to be included in the national plans for economic and social development, and environmental protection. The guiding principles for international and bilateral cooperation are the protection, prevention and restoration of the environmental heritage, social equity in the redistribution of benefits and sharing of costs, and rational and balanced economic uses of the available natural resource potential. In implementing multisectoral policy measures, it is essential to regard transboundary watersheds as systemic units and to promote planning and management with an integrated development approach. A point of particular importance in transboundary watershed management is the participation by users and local communities, which has to grow over time and requires substantial support from national public institutions, as well as from the international community.

The creation of binational 'round tables for watershed affairs' at the levels of the central, regional and provincial governments, as in the case of the cooperation between Argentina and Chile, makes it possible to develop and reorient institutional activities based on agreements, projects, training courses and expert advice. In these two countries, which share a common history, it has been possible to initiate a legal and practical process and to generate change in many of those involved to promote a 'new culture of managing natural resources'. This may lead to a promising future for the Patagonian region.

References

Anonymous (1992) International Agreement on Protection and Utilization of Transboundary Watercourses and International Lakes. Helsinki, Finland.

Falkenmark, M., da Cuhna, L. and David, L. (1987) New water management strategies needed for the 21st century. *Water International* 12(3), 94–101.

Francke, S. (2002) *Gestión Integrada de Cuencas Transfronterizas Argentino-Chilenas*. Actas 6° Seminario Binacional. Ushuaia, Taller.

Francke, S. (2005) Cuencas hidrográficas transfronterizas y derecho ambiental internacional con énfasis en América del Sur. In: Gallardo, E. and Schmithüsen, F. (eds) *La Contribución del Derecho Forestal – Ambiental al Desarrollo Sustentable en América Latina*. IUFRO World Series, Vol. 16. Internacional Union of Forest Research Organizations, Vienna, pp. 162–173.

29 Environmental Policy and Law Influencing Forest Management Practices in Chile

Enrique Gallardo Gallardo

Introduction

Since 1972, the year of the first worldwide conference on human environment of the United Nations, environmental issues have become subject to a new and special branch of law. Environmental law today addresses the complex aspects that govern the utilization, development and preservation of natural and sociocultural resources associated with global, regional and local environmental systems. It increasingly regulates the management and conditioning of economic and social development, as well as maintaining the diversity of vegetation, wildlife and landscapes. This new branch of law forms the framework for improving sectoral policies and legislation and making them more effective. As a consequence, forest policies and law in Latin America are increasingly taking into account multiple interests in relation to the environmental role of forest resources and their sustainable use and management, and are putting greater emphasis on participation and commitment from a variety of public and private stakeholders (Gallardo, 2005; Schmithüsen, 2005).

Environmental Policy and Law Development

In Chile, regulations relating to the environment have existed for a long time, but they were limited only to matters affecting specific resource utilization. For example, forest regulations contained rules for maintaining the forest cover, protecting rare forest ecosystems and conserving the production potential of forest stands. This type of regulation provided protection mainly as a consequence of the interdependent rational use of a definite natural resource and the implicit positive implications for the environment as a whole.[1] A study undertaken by the National Commission on the Environment identified 718 bodies of laws and regulations, in force in Chile in 1985, as 'significant' for the environment. Subsequent analyses showed that approximately one-third of the texts qualified in this manner were part of the forest legislation.[2]

In 1990, the Chilean government formulated a new environmental policy aimed at protecting and preserving the environment, nature and the country's wildlife heritage. The main assumption was that this policy had to be based on progressive actions and realism. Progressive actions meant taking into account the fact that the environmental problems that the country had to face were the

[1] Chapter 11, on law, in the document *Environment in Chile: Investigation and Planning an Environment Centre,* Universidad Católica de Chile Editions, Santiago de Chile, 1985.

[2] *Repertory of Significant Environmental Legislation in Force in Chile,* National Commission of the Environment, Santiago de Chile, 1992.

result of decades during which a range of sector policies had been adopted without considering their local, national and global environmental impacts. Changing the course of environmental deterioration and finding a way to reconcile economic development with the preservation of the environmental heritage would require a phased approach based on a combination of measures, some of which could be implemented immediately, whereas others had to be adopted over a longer time perspective. Processes that had to be initiated urgently were the institutionalization of environmental regulations in the public sector, the revision of sectoral policies and the launching of educational measures aiming to achieve a change of attitudes towards the environment and nature protection. All societal groups would have to be included in formulating a comprehensive national environmental policy and would have to contribute their part in making it a reality.

Considerable experience in other Latin American countries on formulating environmental codes and supporting legislation already existed, but there was only little evidence on appropriate policy instruments to ensure the implementation and monitoring of results. Realism in formulating a new Chilean environmental law meant that the goals and objectives had to be defined in such a manner that there was a chance of achieving tangible results within a determined period of time. Consideration had to be given to the scale of the environmental problems to be solved, ways and means of solving them and the resources available in the country to put them into practice. This required defining political orientations that could serve for future public and private environmental management. The principles of prevention, participation, the 'polluter pays' principle, environmental responsibility and efficiency had to become the guidelines for future public policy interventions.

In 1994, just 2 years after the United Nations Conference on Environment and Development was held in Rio, the Chilean parliament approved a Law on the General Basis of the Environment.[3] As indicated in the presidential message to parliament, the law was intended, in accordance with the Constitution of the Chilean Republic, to contribute to sustainable development, to safeguard land, water and genetic resources, to avoid further degradation of the environment, and to be technically and economically viable and socially acceptable. The new environmental text has the character of a 'framework law', which makes it possible to intervene from a global and systemic perspective in public regulation of all environmentally relevant activities at the national level.

The law endeavours to put into reality the constitutional guarantees of living in an environment free of pollution and the obligation of the state to watch over the preservation of nature.[4] It serves as a point of reference for amending the current body of legislation, provides for new instruments for environmental management and establishes an institutional basis for coordinating sectoral competencies for environmental and nature protection. During the first 12 years in which it has been in force, a number of complementary regulations have been adopted that have strong influences on the forestry sector, including rules on maintaining environmental quality, controlling pollutant emissions and prevention and decontamination plans. Following the trend in other Latin American countries (Schmithüsen, 2005) Chile has incorporated significant social and environmental components into its public statute on forest management, particularly with regard to supporting owners of small forests, restoring degraded soils and promoting afforestation or reforestation activities.

The process of defining a new environmental policy led to the production of two other important political documents approved by the Board of Directors of the National Commission for Environment, which counts among its members 12 ministers with competence in environmental matters. The document *An Environmental Policy for Sustainable Development* of 1998 compiles the experience gained since the promulgation of the environmental basic law and defines new strategies with respect to two constitutional demands. The document addresses rational use, preservation and where necessary restitution of the national heritage formed by unique, limited and representative natural components, the permanence and regenerative capacity of which needs to be secured. In addition, it makes reference to

[3]Law 19300, published in the Official Gazette of 3 September 1994.

[4]Section 19 no. 8 of the Political Constitution of the Chilean Republic of 1980, published in the Official Gazette of 24 October 1980.

the preservation of nature, with the understanding that this objective has to be attained through a combination of policies, programmes and regulations safeguarding specifically the preservation and development of endangered species and ecosystems, including natural forest.

The second political document, adopted in 2000, *An Environmental Policy for Sustainable Use of the Renewable Natural Heritage*, concretizes the mid-term and long-term objectives and determines environmental measures to be initiated during the coming years. A national strategy for conservation and sustainable use of biodiversity followed in 2004, together with a national plan of action determining short-term, medium-term and long-term measures that provides for the approval of national policies focusing on protected areas, protection of endangered species and conservation and rational use of wetlands. In 2006, the National Strategy for Climate Change was approved. Chile is also developing a number of forest projects located between the Eighth and Eleventh Regions (southern part of the country) in order to implement the clean development mechanism of the United Nations Framework Convention on Climate Change or Kyoto Protocol.[5]

Environmental Impact Assessments of Industrial-sized Forestry Projects

The Assessment of Environmental Impact[6] constitutes an important instrument for preventing degradation of natural resources. Since 1997, development projects considered as having an industrial dimension relating to wood exploitation, pulp and paper industries, chip mills and sawmills and other wood-processing facilities are subject to this type of assessment. 'Industrial size' in this context refers to developmental activities that cover an aggregated area of more than 20 ha to be worked annually in the First to Fourth Regions; of 200 ha in the Fifth to Seventh Regions, including the Metropolitan Region; of 500 ha/year in the Eight to Eleventh Regions; and of 1100 ha/year in the Twelfth Region. For pulp and paper industries, the term 'industrial size' refers to industries with an annual wood consumption equal to or higher than 350,000 solid cubic meters. For chip mills and sawmills, wood consumption – as raw material supply – must be equal to or higher than 25 solid cubic metres per working hour. For wood-panel production plants, wood raw material supply must be equal to or higher than 10 solid cubic metres per working hour.[7] An impact assessment has to be undertaken as well in case of forestry activities on fragile soils susceptible to suffer severe erosion and on lands covered by native forest as defined in the pertaining regulation.

Up to the present, the only industrial undertaking that has been subject to an environmental impact assessment is the Rio Condor Project of the Trillium Forest Company. The project involves an approximate area of 270,000 ha and is targeted for wood extraction in an area of approximately 130,000 ha of primary and secondary native *Nothofagus* forests, with subsequent plantations of the same species on degraded soils. It is located on the island of Tierra del Fuego in the Twelfth Region (Magallanes y Antártica Chilena). The project represents an enormous challenge to forestry and an emblematic cause for forest conservation on the frontiers of a territory located at the southern limit of South America.

During the preparation of the assessment, the firm stated its agreement to practise sustainable forest management, understanding this to mean the use of forests and forest lands in such a manner as to avoid damage to other ecosystems and to safeguard biological diversity, productivity and the regenerative capacity of the forest. The firm was advised by an independent scientific commission on the preparation of baseline studies evaluating the feasibility of an ecologically sustainable project. More than 100 environmental clauses were identified in the Resolution qualifying the project.[8] Most of these referred to forest management practices. Among them were recom-

[5] P. Richard Torres, National Forestry Service, personal communication on 16 June 2006.

[6] Supreme Decree no. 30, 1997, of the Ministry of General Secretariat of The Presidency, published in the Official Gazette of 3 April 1997.

[7] Section 3, letter m, of the said Supreme Decree no. 30, 1997.

[8] Resolution no. 03 of 11 February 1998, of the Regional Commission of the Environment of the 12th Region of Magallanes y Antártica Chilena, modified by Resolution no. 032 of 3 June 1998.

mendations by the scientific commission requiring periodical performance assessments every 5 years, with publication of the results and with the obligation to modify harvesting practices if required as a result of such assessments.

During the first 3 years of the project, annual wood harvesting could not exceed 340,000 m^3/year and the total of the first 5-year harvesting period was limited to 1,970,000 m^3. Commercial cutting of the Coigüe of Magallanes forest type was prohibited as recommended by scientific investigations and feasibility studies. The cutting quota was to be subject to a detailed area planning of logging interventions as determined by the approved management plan balancing wood removals in relation to the estimated tree growth in each management compartment. Protected reserves had to be established in order to safeguard representative samples of biological diversity within the project area, including 10,000 ha of potentially marketable full-grown forests.

The project constituted a model demonstrating for the first time that new environmental policies and laws regarding sustainable forest management could be put into practice. During the environmental assessment process, there was significant participation by concerned citizens, using several administrative and judicial resources to confront the governmental authorities in charge of project preparation with their responsibilities as set by law. However, after all these actions, the firm finally decided to desist from carrying out the project. At present, the owners of the forest lands are keeping them for conservation as natural resources.

General Regulations Addressing the Development and Protection of Forest Resources

Forest management plans

The Chilean environmental law of 1994 requires that public institutions in charge of regulating the utilization of natural resources in a specified area must require the submission of forest management plans integrating environmental considerations referring to the protection of water flows, soil conservation, maintenance of landscape values and to the protection of endangered, vulnerable, rare or insufficiently known species.[9] Complementary to these requirements of the environmental law, the actual Chilean forest development law requires formal approval or registration of management plans by the National Forestry Service for tree cutting in native forests and forest plantations. It establishes an obligation to reforest, in accordance with approved standards, an area of equal size in relation to the area cut. Tree cutting on forest land without a previously approved or registered management plan is punishable with a fine equivalent to twice or three times the commercial value of the illegally cut wood. Lack of implementation of an approved or registered management plan is punishable with a fine equivalent to 5 or 15 monthly tax units.

Criteria and indicators

Important aspects in assessing the actual and future impact of the expanding environmental legislation on forest management are the definition of sustainability criteria and indicators for recognized forestry practices and the promotion of certification systems confirming that forestry practices are actually implemented in conformity with them. Certification is a typical market-oriented and consumer-oriented instrument that has a high level of acceptance among private and public forest owners, as well as in the wood-processing industry. At the moment, there are more than 1.5 million ha of certified forests in the country that qualify for the seal of the Programme for the Endorsement of Forest Certification Schemes. At present, a new standard for the sustainable forest management of native forests has been officially presented and is now subject to public consultation.[10] The Forest Stewardship Council for Chile already approved forest certification standards for native forests and forest plantations in the country in 2005. The area that has been certified in Chile using these standards amounts to 425,000 ha.[11]

[9]Sections 41 and 42 of Law 19300.

[10]Sistema de Certificación Forestal CERTFOR. Available at: http://www.CERTFOR.ORG

[11]Diario de la Sociedad Civil. Available at: http://www.sociedadcivl.cl/icefi

Protected wild areas system

The environmental base law establishes an obligation on the part of the state to place extensive wild areas, which include unique native forests, under a special protection regime. This requirement has led to the establishment of the national system of protected wild areas, which is making a significant contribution to environmental heritage conservation and to the preservation of nature and landscape.[12] It allows a regulatory distinction to be made between productive or economic outputs from forests on the one hand, and its social, cultural, recreational functions and environmental services on the other. Chile has thus adopted a balanced approach in conceiving forestry development as a multidimensional issue in terms of integrating protection and production demands in public and private forest management and in expanding the wood-processing sector. This approach has an important effect on land use planning and assessing forestry development projects. At present, 14 million ha – 19% of the country's surface – are under this system of legal protection.

Public awareness

The multiplying effect of preserving the natural heritage according to new demands of the national, as well as of the international community, is a cultural and educational element of enormous importance. It increases public awareness in relation to the variety of economic goods and social services provided by the forest. It makes the public better acquainted with the concepts of sustainable ecosystem management and promotes the understanding of the manifold social, cultural and spiritual values that are at stake in utilizing renewable natural resources. Environmental educational activities have been developed by the national forestry service and are at present being implemented in cooperation with the ministry of education through the 'Outdoor School Activity Programme' and the 'National System of Environmental Certification for Educational Institutions'.[13] Chile is thus formulating and implementing its national environmental policy in accordance with the 'Decade of Education for Sustainable Development' proclaimed by the United Nations for the years 2005–2014. Together with six other Latin American countries and cooperating with six international networks, the country is also participating in the 'Global Environmental Citizenship Activity', which is coordinated by the United Nations Environment Programme.[14]

International Environmental Cooperation

The global concept of sustainable forest management was agreed on at the summit in Rio de Janeiro in 1992 and has served as a foundation for subsequent international initiatives – addressing, for instance, the identification of specific criteria and indicators for the conservation and management of forests. Among such intergovernmental initiatives, Chile is actively participating in what is known as the 'Montreal Process', sharing an interest with other participant countries that have temperate and boreal forests in order to develop practical guidelines for sustainable forestry practices at different management scales. In the Santiago Declaration of 1995, seven criteria were agreed on, which contribute to a more consistent approach in protecting the environment, preserving pristine nature and landscape values and ensuring a sustainable economic base for forestry development. Among them, the establishment of a coordinated institutional and regulatory framework received special attention, in recognition of the growing impact of environmental laws and regulations on forest policy and forest management practices.[15]

---

[12] Sections 34, 35 and 36 of Law 19300.

[13] *ECD Environmental Performance Reviews – Chile,* Organization for Economic and Cooperation Development (OECD) (2005), p. 206.

[14] Global Environmental Citizenship Programme, 2005. United Nations Environment Program and International Union for the Conservation of Nature Regional Offices for Latin America and the Caribbean.

[15] *The Montreal Process in Chile,* Technical Paper 128 of *Chile Forestal Magazine*, National Forestry Service, August 1999.

Conclusions

Chile has demonstrated its commitment to sustainable forest management through the organization of the Ninth Meeting of the Technical Advisory Committee, held in November 2005 in Malalcahuello in the Ninth Forestry Region. At this meeting, important advances regarding the review, modification and development of new management indicators consistent with those in the Montreal Process were made.[16] During several years, the Chilean government has received valuable technical cooperation from countries such as Canada, France, Germany, Japan and the Netherlands. In turn, Chile has offered technical cooperation to a number of countries in Latin America and the Caribbean and has actively participated in international commissions, forums and conventions on forest and environmental issues.[17]

Dynamic developments in environmental policy and law, as well as the growing impact of environmental regulations on current forest practices, underline the need to implement differentiated but clearly targeted forest management strategies that are supported by the political will of a number of governmental authorities operating in a coordinated, process-oriented system of policy formulation and implementation. The incorporation of the environmental dimension into forest management also shows that there is an urgent need for a coordinated approach in land use planning, combined with forest management strategies that embrace production, protection and preservation objectives in a balanced and realistic manner.

This has been well expressed in a recent evaluation report on environmental performance in Chile.[18] The report stresses the urgent need to establish an agreement on national strategic orientations related to forest resources, addressing the protection and sustainable management of the remaining native forest area, as well as the establishment and management of forest plantations on appropriate sites. The preparation of an agreement of this type needs to involve all of the parties concerned. It should propose practical measures for securing multifunctional use and management, including incentives for providing environmental services, mechanisms of mutual fulfillment, association and cooperation among the parties concerned and reinforcement of the monitoring, supervisory and implementing capacity of the National Forest Service.

[16] P. Richard Torres, National Forestry Service, personal communication on 16 June 2006, when preparing the report for the said meeting.

[17] Corporación Nacional Forestal (CONAF): Foundations, Guidelines and Plans of Action of the Secretariat of International Affairs, 2006–2010.

[18] *OECD Environmental Performance Reviews – Chile*, OECD (2005), p. 160.

References

Gallardo, E. (2005) Efectos de la legislación ambiental en las prácticas de manejo forestal en Chile. In: Gallardo, E. and Schmithüsen, F. (eds) *La Contribución del Derecho Forestal – Ambiental al Desarrollo Sustentable en América Latina*. IUFRO World Series, Vol. 16. International Union of Forest Research Organizations, Vienna, Austria, pp. 113–120.

Schmithüsen, F. (2005) *El Papel de la Legislación Forestal y Ambiental en Países de América Latina para la Conservación y Gestión de los Recursos Naturales Renovables*. FAO, Rome (Estudio de la FAO en línea, no. 49).

30 Cross-sectoral Linkages in Mexican Forestry

Víctor Sosa Cedillo

Introduction

Mexico has one of the highest rates of deforestation in the world, with an annual loss of 631,000 ha of forest (FAO, 2003), the main cause being changes in land use to make way for agricultural and livestock activities. As in other countries, millions of hectares have been stripped of their forests for such purposes. The negative effects are very serious in terms of soil erosion, biodiversity loss, reduced water production, CO_2 generation, etc. Incentives to promote agricultural and livestock production (such as the programme of direct support to the sector) have not always considered sustainable practices, thus aggravating deforestation and forest degradation. It is estimated that most of the loss takes place in tropical zones (75%) due to the conversion of forests to livestock and agricultural activities, whereas the main cause of deforestation in temperate zones is changes in land use from forest to agricultural activities and with urbanization intentions. 'Open access' to forests and non-sustainable uses also cause deforestation. There is also a high rate of forest degradation in densely populated areas, with problems of water supplies, underdevelopment and poverty in forest zones. This means that government policies have so far been unable to establish any effective intra- and intersectoral coordination, so that mechanisms and instruments need to be improved to remedy the situation.

On the other hand, Mexico's indigenous groups – numbering more than 10 million people (Presidencia de la República, 2004) – live in forest zones. The agrarian, social and rural sectors need to be better coordinated in order to improve the level and quality of life of these groups and reduce pressure on forest resources. If deforestation and degradation are to be reduced, more appropriate public policies need to be designed and implemented, while institutional coordination among the various sectors involved needs to be improved.

Mexico has an annual water availability of 4986 m^3 per inhabitant, and there are problems of shortages, distribution and quality. Water production is closely linked to forest resources. In the future, most forests in mountain areas will be dedicated mainly to water production. Close coordination is thus needed among the forest sector, the national and state bodies responsible for water and agriculture and the country's main cities.

This chapter analyses the impact of other government policies on the forest sector and vice versa in the case of Mexico, particularly the state of Chihuahua. This issue is extremely important in Mexico because the effect on the forest sector of policies outside this sector is greater than that of policies specifically relating to it. The document has the following specific objectives:

- To identify the main impacts of external sectoral policies on the forest sector;
- To revise mechanisms in order to cope with these external influences;
- To recommend changes in policies, instruments or institutional mechanisms in order to maximize positive impacts and minimize negative ones.

Context of National and State Planning

The National Development Plan 2001–2006 (Presidencia de la República, 2001) is the main planning instrument in Mexico, establishing the priorities, objectives and strategies of the federal government for the period in question. The plan is the basis for sectoral, institutional and regional programmes, which constitute the mechanisms for implementation of the various government policies (Fig. 30.1). The main indicators for the country and Chihuahua state are shown in Table 30.1.

Inter- or *cross*-sectoral coordination means coordination among different secretariats and sectors or levels of government (federal, state, municipal), while *intra*-sectoral coordination refers to coordination within the forest sector. The main problems of coordination between national and state level are as follows:

- Forest policies and programmes drawn up with different methodologies, timeframes and perspectives;
- Different views on the role to be played by forest resources;
- Different views on central and state administrative responsibilities for the forest sector (decentralization);
- State governments' lack of resources to support the forest sector.

Both federal and state governments want greater decentralization with regard to the forest sector. However, more resources are needed for the sector, together with clear decentralization mechanisms and the creation of forest administration structures within state and municipal governments.

Mexico's Strategic Forest Programme 2025 (CONAFOR, 2001a). This programme was formulated with the assistance of the government of Finland and the Inter-American Development Bank with the aim of promoting and boosting the sustainable development of forest ecosystems

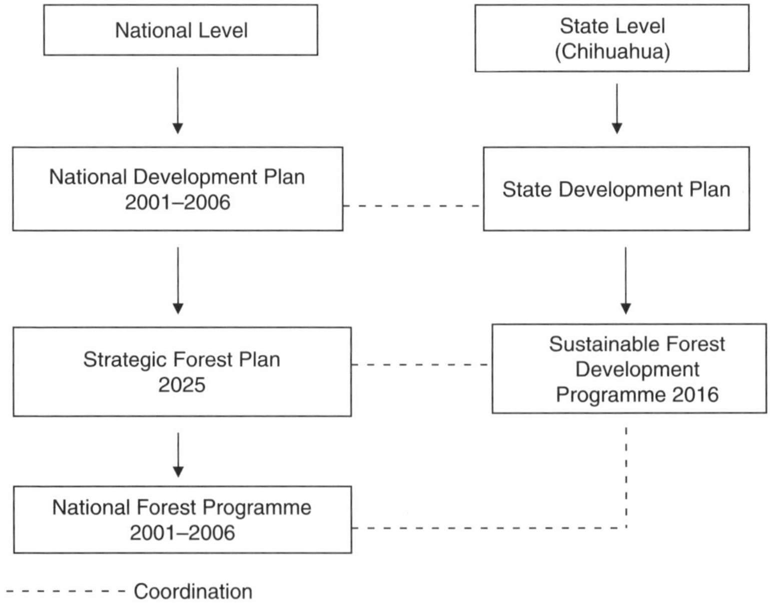

Fig. 30.1. National and state planning structure.

Table 30.1. Main socio-economic and forestry indicators. (From SEMARNAT, 2001c; INEGI, 2006.)

Indicator	National level	Chihuahua state level
Nominal gross domestic product (1998) (thousands of pesos at current rates)	3,517,781,860	152,384,281
Agricultural, forestry and fisheries product (1998) (thousands of pesos at current rates)	152,384,281	9,209,638
Total population (2000) (millions of inhabitants)	97	3
Employed population (1999) (millions of inhabitants)	39	0.065
Total area (millions of hectares)	196	25
Total forested area (1994) (millions of hectares)	57	7.6
Annual deforestation rate (1980–1990) (thousands of hectares)	615	4.4
Forest timber production (1999) (millions of m^3 of logs)	8.5	1.9

through conservation, protection, rehabilitation, promotion and production activities. Various consultative and participatory activities were carried out for its formulation, such as workshops and the gathering of opinions by Internet. The 2025 strategic programme encompasses the 2001–2012 programme for investment, which is divided into various lines of action, for a total of US$18 billion. The first evaluation of the 2025 programme was recently carried out with the support of the Food and Agriculture Organization of the United Nations (FAO).

National Forest Programme 2001–2006 (CONAFOR, 2001b). This programme focuses on objectives and actions for the period, based on the National Strategic Forest Programme and the National Development Plan 2001–2006. It includes a framework for the participation of the various institutions, sectors and bodies in its implementation, as based on the strategies it encompasses. Chihuahua state also has a Sustainable Forest Development Programme 2016 (Consejo Técnico Consultivo Forestal Estatal de Chihuahua, 1996), which was finalized in April 1996 and integrated into the State Forest Advisory Council. Participants are foresters, federal and state departments, forest owners, forest companies and others. The scope and areas of responsibility are forest ecosystems, population and culture, infrastructure and industry, as well as legislation and authority. Figure 30.2 shows the structure and main forest programmes in Chihuahua state.

Chihuahua's forest areas offer many excellent opportunities for developing ecotourism, and there are already various projects, such as the Barrancas Project, requiring special coordination between the forest and tourism sectors at both national and state levels. There are indigenous groups with about 106,000 people, 91% of whom are Tarahumaras living in the state's forest zones and having forest *ejidos* (communally held forest land). These groups have a higher level of poverty than the other inhabitants of the state. Strong coordination is needed with the rural, social and agrarian sectors, and with the state government and municipal governments in the forest zone.

Chihuahua has a relatively low rate of deforestation, because of the low population density and the fact that the climate and soil tend to be unsuitable for agriculture in upland forest zones (see Table 30.1). Forest degradation caused by inappropriate forestry practices, forest fires and illegal logging is more of a problem, so that institutional coordination between the state and national levels is important in order to ensure that more resources are allocated for promoting sustainable forest management through incentives, such as those allocated under the Forest Development Programme, the building of roads and rural and social development. In the mountain zones where Chihuahua's forests are located, runoff flows mainly towards the Pacific Ocean, and such water is extensively used in irrigated farming zones in the states of Sinaloa and Sonora. Chihuahua therefore has a good opportunity to obtain funds by producing water from its forests and setting up payment mechanisms for this service. Coordination with the National Water Commission and the governments of Chihuahua, Sonora and Sinaloa is needed.

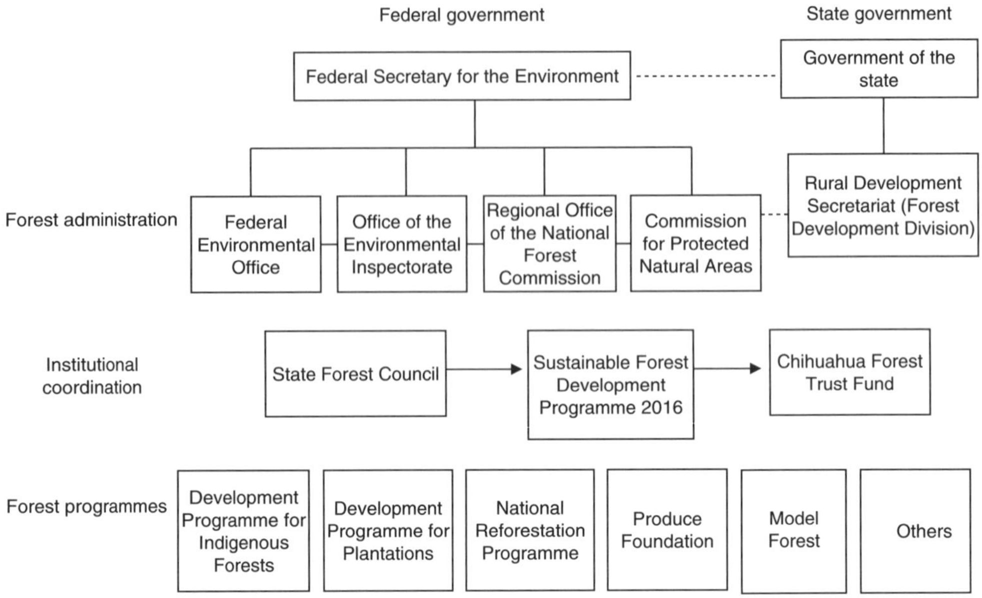

Fig. 30.2. Planning and forest administration structure in Chihuahua.

Legal and Administrative Framework

The regulatory framework for Mexico's forest sector is composed mainly of specific legislation for the sector and regulations concerning the environment, wildlife and agriculture. The General Law on Sustainable Forest Development (SEMARNAT, 2002) governs the obligatory principles of forest policy, the management, conservation and promotion of forest resources, grass-roots participation, surveillance measures and penalties. The General Law on Ecological Balance and Environmental Protection (SEMARNAP, 1997) legislates on the ecological aspects of land use planning, biodiversity, protected natural areas and environmental impact evaluation. The Law on Wildlife (SEMARNAP, 2000) governs the use and harvesting of plant and animal wildlife, and non-wood forest products. The Agrarian Law legislates on the joint property of *ejidos* and forest communities, who account for more than 80% of Mexico's land area, and the limits on small private holdings, including forest land (SARH, 1992). Table 30.2 shows the objectives of all these laws.

A number of state governments have forest offices or institutional mechanisms, and in line with the new Forest Law, these are starting to exercise authority. Federal forest administration is divided between various bodies, a situation that causes problems over intra- and intersectoral coordination. More functions and resources need to be decentralized to state and municipal levels. Table 30.3 shows the main functions and external structures of the various bodies of the Environmental Secretariat that share in forest administration, while Table 30.4 shows forest administration in Chihuahua. Taking account of the extent of the state's forest resources, a specific body responsible for the forest sector should be set up, with the highest possible rank and the necessary resources to operate.

Impacts of Government Policies at the National Level and in Chihuahua[1]

Policies establishing the institutional framework

Mexico's forest sector has traditionally been given a very low priority in relation to macroeconomic

[1]The positive and negative impacts of other sectoral policies on the forest sector are analysed below, as based on the grouping of policies suggested by Schmithüsen (2003).

Table 30.2. Main laws affecting Mexico's forest sector. (From SEMARNAP, 1997; 2000; SARH, 1992; SEMARNAT, 2002.)

Law	Objectives
General Law on Sustainable Forest Development (2002)	To govern and promote the conservation, protection, rehabilitation, harvesting, management, cultivation and production of ecosystems and their resources, and to divide responsibility for forest affairs among federal, state and municipal levels
General Law on Ecological Balance and Environmental Protection (1996)	To foster the sustainable development, preservation, rehabilitation and enhancement of the environment, together with the necessary mechanisms for coordination and consensus building
Law on Wildlife (2000)	To foster the conservation, promotion and sustainable harvesting of plant and animal wildlife
Agrarian Law (1992)	To prescribe regulations for Article 27 of the Constitution, concerning agrarian matters

policies, and this has been reflected in the lack of fiscal and direct incentives and a very low budget. The situation has been one of the main reasons for the lack of the conditions and interest needed to conserve and develop forest resources, which has, in turn, had a negative impact on forest owners, the forest industry and the general population, in terms of the quantity and quality of goods and services generated by these resources. The overvaluation of the Mexican peso at many

Table 30.3. Forest administration within the Secretariat for the Environment and Natural Resources. (From SEMARNAT, 2001a,b.)

Institutional mechanism	Main function	Regional structure
National Forest Commission (decentralized)	Development, fostering and promotion of forest production, conservation and rehabilitation activities, and application of the sustainable forest development policy	13 regional offices operating in coordination with the National Catchment Area Division
National Institute for Ecology (decentralized)	Research on the use and sustainable harvesting of natural resources	Regional offices of the Secretariat for the Environment
Management Subsecretariat for Environmental Protection (central government)	Exercise of authority in forest matters; evaluation of environmental and forest impacts	Regional offices of the Secretariat for the Environment
Federal Inspectorate for Environmental Protection (decentralized)	Monitoring of observance of environmental and forest legislation, and imposition of penalties	Offices in each state
National Commission for Protected Natural Areas (decentralized)	Federal-level administration of protected natural areas	Regional offices of the Secretariat for the Environment
Subsecretariat for Environmental Promotion and Legislation (central government)	Issuing of forest rules and regulations	Regional offices of the Secretariat for the Environment

Table 30.4. Forest administration in Chihuahua. (From SEMARNAT, 2001a,b.)

Level	Office	Function
Federal	Federal Division of the Secretariat for the Environment	Exercise of federal-level forest authority
Federal	State Office of the Federal Inspectorate for Environmental Protection	Monitoring of observance of environmental and forest legislation, and imposition of penalties
Federal	Regional Office of the National Forest Commission	Forest promotion and development
State	Secretariat for Rural Development (Forest Development Division)	Forest monitoring and silvicultural promotion

periods has also led to an increase in imports of forest products, which has a further negative effect on forest producers and the national industry. All these factors make it harder to achieve and maintain forest protection and sustainable management, inasmuch as forests are destroyed or degenerate when their owners lose interest because of the lack of income from them. A similar negative impact occurs with the elimination or reduction in import duties on forest products because of a free trade policy. The country's timber production falls, causing the closure of forest companies and the loss of jobs and income in certain – mainly rural – zones.

In 1997, Mexico started granting direct forest incentives through programmes for the development of indigenous forests and commercial forest plantations (SEMARNAT, 2003a,b). These incentives have had positive impacts, with an improvement in forest management, a rise in the legal production of timber (under the previous federal government), the training of forest producers and workers, capacity-building for foresters and the creation of jobs and income opportunities in forest zones. Since 1997, public expenditure on the forest sector has risen considerably, with very positive and ongoing impacts on the sector. Mexico's free trade policy has had both positive and negative impacts. The positive impacts are represented by development of a segment of the forest industry focusing on exports, for example, of furniture, and by the fact that Mexican consumers are being offered a sufficient quantity and variety of forest products at competitive prices. The need to be competitive is a major factor driving modernization of the Mexican forest industry and the large-scale expansion of commercial forest plantations, mainly of fast-growing species.

In many regions, the policy of privatizing and/or liberalizing many aspects of the forest production chain has hindered consolidation of more efficient 'management units' and increased illegal logging, a situation that has led to more deforestation and forest degradation. On the other hand, the policy of privatizing owners' rights over their forests has had positive impacts in certain zones where forest producers (mainly *ejidos* and communities) are well organized and trained to implement sustainable forest management practices (e.g. in San Juan Nuevo, Michoacán; El Salto, Durango; El Largo, Chihuahua and Sierra de Juárez, Oaxaca).

In Mexico, institutional coordination among the various secretariats involved in federal civil protection and state governments has had very positive impacts on the forest sector through actions to prevent and combat forest fires. In the case of Chihuahua, direct forest incentives under the programme for the development of indigenous forests have had positive impacts in terms of an increase in areas under authorized technical management, the training of forest producers and the use of silvicultural techniques, such as thinning. The federal and state governments have approved and released considerable – albeit still insufficient – resources for these purposes. On the other hand, recent years have seen an increase in illegal logging in Chihuahua, and this practice is causing large-scale forest degradation with a loss of biomass and a fall in forest quality. Government policies have not been sufficient to increase domestic and foreign investment in Chihuahua's forest sector, a situation that has left the state's timber production potential underexploited, which in turn affects forest producers, the forest industry and the state and national economies. Rural and social policies have not helped to raise the standard of living of indigenous groups and the rural population.

Ejidos and communities hold more than 90% of Mexico's forests. The largest smallholding allowed is 800 ha, and it is forbidden to subdivide or sell communal forest holdings. There are also conflicts because of overlapping landholdings, which again have negative effects on forest conservation and development. Rural and social policies have on the whole been unsuccessful in improving the standard of living in forest zones, with a further negative effect on forests and the forest sector in general because of pressure on resources and the difficulty of implementing sustainable management and involving the rural population in this activity. The policy of promoting grass-roots participation through Chihuahua's State Forest Council has had such positive results as the organization of forest producers, the consolidation of forest management units, the creation of the Chihuahua Forest Trust Fund and the Forest Programme 2016. All these instruments and actions promote sustainable forest management and associated benefits.

Policies of specific economic sectors

Recent years have seen an increase in the interest and concern of the Secretariats of Agriculture and Environment in developing and implementing farming and livestock practices that will help reduce deforestation, forest degradation and soil erosion. There have been some 'productive reconversion' programmes and research focusing on the non-use of fire in farming and livestock activities, the use of annual plants that improve soil fertility, reduced tillage and the intensification of livestock farming – all practices that help reduce pressure on forest resources. There have also been positive ecotourism initiatives in the forest zones of Chiapas (the Selva Lacandona Reserve), Michoacán (the Monarch Butterfly Reserve), Chihuahua (the Tarahumara Model Forest) and other places where forest owners are in charge and receive benefits from these activities.

More than two-thirds of the annual timber extraction in Mexico is carried out to obtain fuelwood and for rural uses – one of the main causes of forest degradation in temperate to cold climates. Fuelwood is not obtained in a technically appropriate or sustainable manner, and there is no policy that focuses sufficiently on fuelwood production in terms of forest management, incentives, research and study on this major activity. The impact is felt by the rural population, who do not use fuelwood to the best effect and suffer from health problems due to inadequate combustion, the exploitation of women and children, the low value of fuelwood, and the lack of the incentives and technical assistance that would convert the present problems into opportunities to improve forest protection and their own income.

In tropical zones, road building has caused major deforestation, whereas in temperate forests the lack of roads in some regions has led to overexploitation of forests and high costs of extraction and transportation. Inappropriate road building has also caused soil erosion, mainly in mountain zones. Transport costs are the main element in the cost of timber, and in the case of natural forests they make timber production uncompetitive on the international market. Moreover, Mexico's water resources have been harnessed only in the lower part of catchment areas, a policy that has had negative impacts on forests, inasmuch as until very recently resources have not been allocated for water production by forests.

In Chihuahua, there are some specific positive impacts in connection with forest research and training. The state has some good experience in managing natural forests, and professional forest education was established at Chihuahua University many years ago. The state has experience of ecotourism in forest zones, and has also seen the diversification of production activities, such as aquaculture – all of which has positive effects for the rural population. These effects are basically the same as at the national level, with the difference that Chihuahua has a lower deforestation rate than the rest of the country.

Sectoral policies promoting protection and development

At both the national and Chihuahua state levels, this group of policies generally has a positive impact on the forest sector, inasmuch as it promotes programmes and lines of action aiming at sustainability, biodiversity, conservation, appropriate soil use, water production, protection against fire and pests, and activities connected with research, training, education and forest and environmental awareness.

In terms of the design and definition of this type of policy, in theory there should be no possibility of its producing permanent negative effects, inasmuch as it is drawn up with the aim of sustainability. However, there *are* cases of negative effects, for example, when the authorities do not apply the rules and regulations correctly – as is seen when authorization for forest management programmes is restricted without taking the established legal framework into account, because conservationist-type concepts are being wrongly applied in the belief that it is better not to manage or harvest forests. Such an approach again means that their owners have little interest in caring for them, if they have no alternative sources of income. Other negative impacts occur when forest management programmes or other activities connected with production are carried out without complying with the corresponding technical norms.

Impacts of the forest sector on other sectors

For the future, all forest policy instruments are designed to have generally positive impacts on other sectors. The level of such positive impacts will depend on the resources available and the effectiveness of institutional coordination and policy implementation. Some of the specific positive impacts on other sectors are the generation of environmental services, increased employment in forest zones, an increased contribution to the gross domestic product, a longer useful life for dams and other waterworks, a reduction in areas under non-sustainable farming and stockraising practices and an improvement in the forest trade balance. In the past, policy instruments have had various negative impacts on other sectors. Deforestation and degradation affect the whole population through a reduced capacity to generate goods and services. The inability to meet domestic demand for forest products has a negative impact on employment, rural income and the forest trade balance.

Analysis of the Effectiveness of Government Policy Coordination

Instruments to coordinate government policies and build consensus regarding Mexico's forest sector have to date operated in more intra- than intersectoral terms. The Presidential Coordination Commissions for environmental matters have not worked under the present government. The Board of Directors of the National Forest Commission has acted to approve administrative actions but has played no major role in coordinating government policies regarding the forest sector. The national and state Forest Advisory Councils, including Chihuahua's, have worked to coordinate actions and build consensus within the forest sector, and in Chihuahua the council has played a major role in building consensus, formulating and implementing the long-term state forest programme, and creating and operating a trust fund to run various projects within the forest management units. Table 30.5 shows the main instruments for the coordination and implementation of government policies connected with the forest sector in Mexico and Chihuahua.

The National Committee for Protection against Forest Fires has been very effective in coordinating actions and building consensus in this sphere among the federal departments involved and state governments. The National Forest Service, which was created by the Forest Law to coordinate the actions of the central departments concerned and state governments, has not worked. The committees to grant various incentives to natural forests and plantations have worked well. A number of mechanisms – such as foresters' organizations, microregional councils and the regional forest studies of management units – are in the process of being established under the General Law for Sustainable Forest Development and will play a major role in improving the coordination of government policies and actions in the forest sector. In the case of Chihuahua, the instruments that have been particularly important, as mentioned above, are the State Forest Council, the Chihuahua Forest Trust Fund and the Tarahumara Model Forest, although it has proved impossible to consolidate the last-mentioned well enough to make it self-sufficient and extend it to other regions.

Table 30.5. Main instruments of government policies concerning the national and Chihuahua state forest sectors. (From Presidencia de la República, 2001; SEMARNAT, 2001a,b; 2002; 2003a,b.)

Instrument	Objective
Presidential Coordination Commissions	Avoidance of duplicated actions, more efficient use of resources and improved results
Board of Directors of the National Forest Commission	Administration and management of the National Forest Commission and intersectoral coordination
Forest Councils	Advice to the forest sector at national and state levels
National Committee for Protection against Forest Fires	Coordination of actions of federal divisions and state governments to prevent and combat forest fires
National Forest Service	Coordination of activities connected with the forest sector among all divisions and levels of government; established under the General Law for Sustainable Forest Development
National Forest Development Programme	Granting of direct incentives to owners and holders of natural forests for sustainable forest management; decisions are taken in state committees made up of forest producers and representatives of state governments
National Programme for the Development of Commercial Forest Plantations	Granting of incentives for the establishment and maintenance of commercial forest plantations; operates through a national committee made up of representatives of the departments involved and planters
Grass-roots participation under the General Law on Sustainable Forest Development	Promotion of grass-roots participation in the forest sector through such mechanisms as state forest councils, forest development campaigns, microregional forest councils, foresters' organizations within forest management units throughout the country
Community Forest Development Programme	Support to *ejidos* and forest communities for sustainable forest management through community forestry schemes
Programme for Development of the Environmental Services Market	Promotion of access to national and international environmental services markets (CO_2 fixation, biodiversity and water production)
Programme for Regulation and Boosting of Forest Self-management	Empowerment of foresters' organizations with a view to regional planning, efficient resource management and management by forest producers
Chihuahua Forest Trust Fund	Financing of sustainable forest management activities in Chihuahua's various regions and forest management units
Chihuahua Model Forest	Coordination of actions and consensus building with a view to sustainable forest management in a 110,000 ha area in Chihuahua's Tarahumara forest zone, with a view to extension to other regions
State Produce Foundation	Raising of funds to carry out agricultural, livestock and forest research projects
Programme of Incentives to the Field	Granting of subsidies to farmers

Conclusions and Recommendations

Various forest programmes have recently been consolidated, and there has been other major progress, such as adjustment and expansion of the first direct incentive programmes for indigenous forests and plantations, devolution of authority to state and municipal level with the allocation of responsibility for forest matters by the new General Law of Sustainable Forest Development, creation of a National Forest Commission, establishment of the obligatory principles of forest policy, launching of new programmes, such as those for incentives to the environmental services of forests and support to producers' organizations, and adoption of a Strategic Forest Programme 2025 and the promotion of similar programmes in all the country's states. However, there are still a number of problems, such as the following:

- The perspective and appraisal are focused more on the forest sector as such than on the relationship between this sector and other factors in the social, economic and political environment and the policies of other related sectors.
- Solutions to the problems of the forest sector are therefore still to a large extent focused on effects rather than causes; for example, forest surveillance by police or military units, the application of penalties, restrictions on forest management because in certain government departments it is thought that the main cause of deforestation and degradation is logging, rather than underlying factors, such as poverty, the lack of clear property rights over forest land, the low levels of benefits received by forest owners and holders and the lack of productive alternatives that would reduce pressure on forests.
- Similarly, the existing coordination and consensus-building mechanisms still focus their actions mainly within the forest sector (where there are some successful examples, as described above) and to a much lesser degree on coordinating actions and building consensus with other sectors or cross-sectoral policies.
- Mexico's forest policy is not yet set within a more general perspective in the national and international contexts; little account has to date been taken of international trends in the timber trade, certification, tariffs and international conventions and initiatives, and consideration is given mainly to the interests of national forest producers and industries, and not always to those of the users of forest goods and services.

The General Law on Sustainable Forest Development states that the country is to be divided into forest zones in order to group and reorganize forest and primarily forest land, foster better administration and contribute to sustainable forest development. The specific recommendation is that, apart from other types of zoning based on natural criteria, a zoning should be carried out with criteria based on Hyde's categories (Hyde, 2003) for the following stages in forest development: (i) areas of expansion of the agricultural frontier (deforestation); (ii) 'open access' farmed areas (degradation); and (iii) managed and remnant or inaccessible forests.

Zoning according to development stages would enable appropriate policies and programmes to be formulated and implemented for each case and region in order to maximize positive impacts and minimize negative ones. It would also be a practical mechanism to coordinate the actions of those taking part in the operation and to build consensus. To this end, long-term state strategic forest programmes (to 2025) should be based on this type of zoning, as should the regional forest studies of each management unit. The next step in order to bring actions still further down to real situations on the ground would be to take advantage of the effort being made by the National Forest Commission to organize the country into 240 forest management units and microregional forest councils, whose lead programmes should also encompass zoning according to forest development stages with a view to joint formulation and agreement of programmes and actions by all participants, including particularly implementation of the policies of other government departments.

So far as the administrative structure of the forest sector is concerned, it is clear that a single department should be in charge. At present, promotion, management, legislation, research, surveillance and protected natural areas, to mention only the main spheres, are divided between different departments, causing a continuing lack of coordination within the forest sector, and to an even greater degree with other sectors of the public administration. The National Forest Service should at least work as laid down by the General Law on Sustainable Forest Development. In the case of Chihuahua and in view of the extent and potential of the state's forest resources, the forest administration should be upgraded to a secretariat or autonomous body with the authority and resources needed to develop this important sector to its potential level.

The objectives and mechanisms of the various federal forest programmes need to be revised in order to avoid duplication and actual or potential overlapping, as is seen, for example, with the Mountains Programme and the Forest

Management Units Programme. Planning should also be carried out with clearly defined and coordinated aims in each region in order to optimize available resources. A new appraisal of Mexico's forest sector should be carried out, with a different perspective from the traditional or primarily forest-focused one, in order to identify the true relationship between society and forest resources, taking account of poverty level, income, survival alternatives, population density in forest areas, etc. The impact of other government policies on the forest sector should also be identified and measured, particularly those concerning the macroeconomic, agrarian, agricultural, livestock, social development, infrastructure, education and tourism spheres. Such an appraisal would give a clearer idea of the real underlying causes of the problem, so that appropriate policies can be implemented. It would also enable the population to be better informed and allow it to participate more effectively. Lastly, with the support of international organizations, public and private research institutions should undertake systematic study of the influence of other government policies on the forest sector and vice versa, inasmuch as this would gradually give an idea of how these complex cross-linkages work.

References

Comisión Nacional Forestal (CONAFOR) (2001a) Programa Estratégico Forestal para México 2025, Comisión Nacional Forestal, México.

Comisión Nacional Forestal (CONAFOR) (2001b) Programa Nacional Forestal 2001–2006, México.

Consejo Técnico Consultivo Forestal Estatal de Chihuahua (1996) Programa Estatal de Desarrollo Sustentable Forestal de Chihuahua, Chihuahua, México.

FAO (Food and Agriculture Organization of the United Nations) (2003) State of the World's Forests. FAO, Rome, Italy.

Hyde, W. (2003) Economic considerations on instruments and institutions. In: *Cross-sectoral Policy Impacts between Forestry and other Sectors*. FAO Forestry Paper 142: Chapter 3. Rome, Italy.

Instituto Nacional de Estadística, Geografía e Informática (INEGI) (2006) Sistemas Nacionales Estadístico y de Información Geográfica, Inegi,gob,mx

Presidencia de la República (2001) Plan Nacional de Desarrollo 2001–2006, México.

Presidencia de la República (2004) Anexo del Cuarto Informe de Gobierno 2004, México.

Schmithüsen, F. (2003) Understanding cross-sectoral policy impacts – policy and legal aspects. In: *Cross-sectoral Policy Impacts between Forestry and other Sectors*. FAO Forestry Paper 142: Chapter 1. FAO, Rome.

Secretaría de Agricultura y Recursos Hidráulicos (SARH) (1992) Artículo 27 Constitucional. Ley Agraria, SARH, México.

Secretaría de Medio Ambiente, Recursos Naturales y Pesca (SEMARNAP) (1997) Ley General del Equilibrio Ecológico y la Protección al Ambiente, SEMARNAP, México.

Secretaría de Medio Ambiente, Recursos Naturales y Pesca (SEMARNAP) (2000) Ley General de Vida Silvestre, SEMARNAP, México.

Secretaría de Medio Ambiente y Recursos Naturales (SEMARNAT) (2001a) Reglamento Interior, México.

Secretaría de Medio Ambiente y Recursos Naturales (SEMARNAT) (2001b) Estatuto Orgánico de la Comisión Nacional Forestal, SEMARNAT, México.

Secretaría de Medio Ambiente y Recursos Naturales (SEMARNAT) (2001c). Anuario Estadístico de la Producción Forestal 1999, México.

Secretaría de Medio Ambiente y Recursos Naturales (SEMARNAT) (2002) Ley General de Desarrollo Forestal Sustentable, México.

Secretaría de Medio Ambiente y Recursos Naturales (SEMARNAT) (2003a) Reglas de Operación del PRODEFOR. Diario Oficial de la Federación del 26 de marzo de 2003, México.

Secretaría de Medio Ambiente y Recursos Naturales (SEMARNAT) (2003b) Reglas de Operación del PRODEPLAN. Diario Oficial de la Federación del 26 de marzo de 2003, México.

31 Major Policy Changes in National Forest Management in the USA, 1975–2005

Dennis C. Le Master and James B. Snow

The National Forests comprise 60.1 million ha of forest land – 19.8% of the forest land in the USA. They are administered by the Forest Service of the US Department of Agriculture in 155 separate units. Management of these lands is generally governed by federal laws, based on the principles of multiple use and sustained yield.

Over many decades, policy changes in the management of the National Forests have been both reactive and evolutionary, and reflect many political and social influences. Generally, policies have evolved from broad discretion given to federal land managers during the early part of the 20th century to more prescriptive substantive and procedural requirements since the early 1970s, resulting from legislation, administrative rule making and court decisions. The origin of the changes is also of interest. As elsewhere in the world, changes frequently come from outside the forestry sector rather than within, including the agriculture, economic, energy or environmental sectors. These are often referred to as cross-sectoral impacts. This chapter provides an overview of some of the changes in policy in National Forest management over the last three decades, and the forces and policies that have influenced them.

Historical Context

The US Constitution grants the Congress plenary authority over the management of the public lands, and the Congress has generally exercised that power through enactment of statutes that grant land management authority to various federal agencies. Beginning with the enactment of the Forest Reserve Act in 1891 and the Organic Administration Act of 1897, and for most of the first half of the 20th century, the Congress exercised its authorities with a loose hand, giving considerable latitude and discretion to professional land managers. To manage the National Forests, land managers had broad authority to 'regulate the occupancy and use and to preserve the forests ... from destruction'.[1] Broad authority was given to federal land managers to harvest timber, acquire and exchange lands, grant use permits and rights of way, lease minerals, fight fires and provide public recreation – all under the principles of multiple use. That these laws were broad and discretionary, with little or no public input or procedural constraints, reflected general societal norms, which did not view federal land management or environmental quality as political or legal issues.

The need for lumber in the Second World War and the post-war housing boom dramatically affected the demand for higher timber harvest levels from the National Forests. At the same time, the public's increasing environmental awareness led to the enactment of laws profoundly affecting National

[1] 16 United States Code (U.S.C.) §551.

 Cross-sectoral Policy Developments in Forestry (Dubé and Schmithüsen)

Forest management. In the decade of the 1960s, the Congress codified the prevailing forest management principles in the Multiple Use–Sustained Yield Act. The Congress also afforded special protections to portions of the federal lands through enactment of laws, such as the Wilderness Act of 1964 and the Wild and Scenic Rivers Act. Although these and other laws enacted during this period have substantive and procedural differences, they all reflect a public policy shift away from broad grants of administrative discretion to federal land managers to more prescriptive management with numerous procedural requirements. Many of these policy shifts were the result of environmental advocacy from the private sector, which increasingly turned to the courts to enforce the substantive and procedural requirements of law.

The National Forest Management Act and Land Management Planning

One of the most contentious environmental issues of the latter half of the 20th century was the timber harvesting technique known as clear-cutting, in which all merchantable timber is removed from a stand. The legality of this practice in the National Forests, as well as the scope of the Forest Service's discretion, was struck down in 1975 in a landmark lawsuit arising in West Virginia. In that case, the courts found that clear-cutting violated requirements of the Organic Administration Act, because individual trees were not marked and designated for harvest.[2]

The judicial clear-cutting ban set the stage for the enactment of the National Forest Management Act (NFMA) and the Federal Land Policy and Management Act (FLPMA), the most dramatic restructuring of public land management policy and practice in the USA in over 75 years. As a result, most policy analysts and decision-makers look upon the late 1970s as a watershed period for public policy pertaining to the National Forests.

The NFMA amended the earlier Forest and Rangeland Renewable Resources Planning Act of 1974 (RPA) by setting standards and guidelines for the management of the National Forests, especially for National Forest land management planning. The Secretary of Agriculture was to promulgate regulations with the advice and counsel of a committee of scientists for the development and revision of land management plans for each National Forest. Such plans were to be prepared with public participation and in accordance with the Multiple Use–Sustained Yield Act. Among other things, plans were to 'provide for diversity of plant and animal communities'.[3] Restrictions were placed on the use of clear-cutting and other even-age silvicultural systems, as well as the quantity of National Forest timber that could be sold annually, the so-called non-declining even-flow provision.

As noted above, NFMA amended the RPA. The premise of the RPA was that conflicts over the use of federally owned natural resources could be resolved through a process of comprehensive long-term planning. To this end, RPA provided for preparation of a 5-year programme for management of the National Forests, Forest Service cooperative forestry assistance to states and private landowners, and Forest Service research to be based on a comprehensive 10-year assessment of the long-term demand and supply for forest and rangeland renewable resources in the USA. RPA established, in contemporary parlance, a *strategic planning* process for Forest Service programmes. The Congress was of the view that the application of this process would minimize conflicts over the use of natural resources under the administration of the Forest Service. Strategic planning has also come to be recognized as an effective method of programme implementation both within and outside government.

Relationship of RPA and NFMA to Selected Environmental Laws

The National Environmental Policy Act of 1969 (NEPA) is the most comprehensive statement of national policy toward the environment. Section 101 commits the federal government to use 'all practicable means ... to create and maintain conditions under which man and nature can exist in productive harmony, and fulfill the social, economic, and other requirements of present and future generations of Americans'.[4] Section 102

[2] *West Virginia Division of the Izaak Walton League of America, Inc. v. Butz*, 522 F.2d 945 (4th Cir., 1975).

[3] 16 U.S.C. §1604(g)(3)(B).

[4] 42 U.S.C. §4331.

directs that the policy contained in section 101 be implemented 'to the fullest extent possible' by all federal agencies.[5] Section 102(2)(C) requires 'a detailed statement' – what has come to be known as the environmental impact statement (EIS) – 'on proposals for legislation and other major federal actions significantly affecting the quality of the human environment'.[6] Even though the foregoing requirements are clear, NFMA reinforces them by stating that the regulations that set out the process for development and revision of land management plans shall specify 'procedures to insure that land management plans are prepared in accordance with the National Environmental Policy Act of 1969, including, but not limited to, direction on when and for what plans an environmental impact statement ... shall be prepared'.[7] Hence, the connection between NFMA and NEPA is tight.

A similar connection exists between the NFMA and the Endangered Species Act of 1973 (ESA). The first two stated purposes of ESA are 'to provide a means whereby the ecosystems upon which endangered species and threatened species may be conserved' and 'to provide a program for the conservation of such endangered species and threatened species ...'.[8] Endangered and threatened species are protected through their formal listing (based on the best scientific and commercial data available), identification of critical habitat and development and implementation of a recovery plan. All persons under the jurisdiction of the USA are prohibited from importing, exporting, taking, possessing, selling, delivering, receiving or transporting threatened or endangered species. ESA declared as policy that 'all Federal departments and agencies shall seek to conserve endangered and threatened species and shall use their authorities in furtherance of the purpose of this Act'.[9] Section 7(a)(2) of ESA further requires federal agencies to consult with the US Fish and Wildlife Service (for land and freshwater species) and the National Marine Fisheries Service (for marine species, including anadromous fish) to 'insure that any action authorized, funded or carried out by such agency is not likely to jeopardize the continued existence of any endangered species or threatened species or result in the destruction or adverse modification of habitat of such species which is determined by the secretary to be critical'.[10]

NFMA requires that regulations be developed to 'provide for diversity of plant and animal communities based on the suitability and capability of the specific land area in order to meet overall multiple-use objectives ...'[11] Hence, the Forest Service must not only consult with other agencies to ensure that any action it undertakes does not jeopardize the existence of a threatened or endangered species, it must also provide for biological diversity on the lands it administers.

Regulations for National Forest Planning

NFMA provided guidelines for the Secretary of Agriculture to promulgate regulations for National Forest planning. The Secretary was to be assisted by a Committee of Scientists to 'provide scientific and technical advice and counsel on proposed guidelines and procedures to assure that an effective interdisciplinary approach is proposed and adopted'.[12] Seven scientists were appointed, and the Committee began its work in May 1977. It met 18 times, and the final planning regulations were issued by the Department of Agriculture on 17 September 1979.

The final regulation set forth a technical 10-step planning process and also required an EIS for development of plans, significant amendments and revisions. The reasoning at the time was that a forest plan EIS would be sufficient for all decisions on projects during the life span of the plan, which was anticipated to be about 15 years.

This reasoning proved to be erroneous, for subsequent court decisions established that projects viewed as significantly affecting the quality of the human environment must be accompanied by their own site-specific NEPA analysis. A single EIS for a forest plan would not work. This precedent substantially expanded the work involved in forest plan implementation.

A summary of National Forest policy as provided for in RPA/NFMA and its initial (1979) implementing regulations are given in the text box below.

---

[5]42 U.S.C. §4332.
[6]42 U.S.C. §4332(2)(C).
[7]16 U.S.C. §1604(g)(2).
[8]16 U.S.C. §1531(b).
[9]16 U.S.C. §1531(c).
[10]16 U.S.C. §1536(a)(2).
[11]16 U.S.C. §1604(g)(3)(B).
[12]16 U.S.C. §1604(h)(1).

Policy Summary of RPA/NFMA, *c.*1979

RPA/NFMA and the initial implementing regulations set the following policies for National Forest planning and management in 1979.

- A hierarchical, integrated, iterative strategic planning process will be used by the Forest Service in administering the National Forests.
- National Forests are to be managed according to plans developed and revised in accordance with regulations promulgated by the Secretary of Agriculture – with the advice and counsel of a Committee of Scientists – and procedures established under the National Environmental Policy Act of 1969.
- The public shall have the opportunity to participate in the development, review and revision of land management plans.
- Such plans will provide for:
  - Multiple use and a sustained yield of renewable surface resources, including wilderness, in accordance with the Multiple Use–Sustained Yield Act of 1960;
  - Fish and wildlife habitat managed in each National Forest so as to maintain viable populations of existing native and desired non-native vertebrate species;
  - Certain species identified and selected as management indicator species for assessing the effects of management alternatives on fish and wildlife populations.
- Plans will ensure that:
  - Timber will be harvested generally only where timber production is viable on a sustainable basis;
  - Protection is provided for watercourses, lakes and wetlands;
  - Even-age management regimes are used only where it is appropriate, and clear-cutting is used only when it is optimal;
  - Maximum size limits are established and applied for one-harvest operations.
- With a few possible exceptions, timber sale levels from each National Forests shall be limited to a quantity equal to or less than a quantity that can be removed from such forest annually in perpetuity on a sustained-yield basis.

Authority and Discretion for National Forest Land Managers Under RPA and NFMA

As noted above, National Forest land managers historically had broad authority and discretion in the conduct of their management programmes and activities. That land managers had such authority and discretion was believed necessary because of the circumstances in which they operated. First, National Forest land managers dealt with one or more forest ecosystems, communities of plants and animals that are tremendously complex, where site-specific knowledge is essential. Second, land managers operated under enormously varying physical, social and economic conditions. Third, the forests were scattered in location, often occupying rough terrain, making close supervision of activities impossible. Hence, decentralized decision making from Washington, DC to the National Forests to the ranger districts was a principal organizational feature of the Forest Service, something about which the agency was proud.

The authority and discretion of National Forest land managers was eroded by NEPA with the requirement that environmental analyses be conducted for any action on the forest significantly affecting the quality of the human environment. The effect of such analyses was not so much that the decisions would have been different. Instead, it made the decision-making process open to public scrutiny with the requirements of detailed documentation of analyses, involvement of the public and public availability of all the underlying documentation.

ESA also eroded the authority and discretion of National Forest land managers. The consultation requirement with the US Fish and Wildlife Service or the National Marine Fisheries Service in the event of an effect on a threatened or endangered species was designed to ensure that any action carried out on the National Forest would not likely jeopardize the continued existence of a threatened or endangered species or result in the destruction or adverse modification of critical habitat of such species. It was plain, that the presence of threatened or endangered species in a National Forest could constrain a variety of activities that might be conducted.

RPA/NFMA eroded the authority and discretion of land managers still further by the procedural requirements that management plans be developed and revised with public participation and in accordance with NEPA procedures. The authority and discretion of land managers to determine where timber harvesting could occur was constrained by the RPA/NFMA requirements that provision be made for diversity of plant and animal communities, on the use of even-age silvicultural systems and on the quantity of National Forest timber that could be sold annually.

Cross-sectoral Impacts of Endangered Species Protection on National Forest Management

No issue has been more problematic in National Forest planning and management than protection of threatened and endangered species under ESA and implementation of the diversity provision of NFMA, including the viability clause in the former implementing regulations.[13] As noted earlier, the agency must consult with the US Fish and Wildlife Service (for land and freshwater species) and the National Marine Fisheries Service (for marine species, including anadromous fish) before undertaking any activity likely to affect a threatened or endangered species, and it must act affirmatively to rehabilitate these species when they are present on National Forest land. At the same time, the Forest Service must manage fish and wildlife habitats sufficiently 'to maintain viable populations of existing native and desired non-native vertebrate species in the planning area'.[14] While the purposes of the ESA and NFMA provisions are similar, they also have the potential for conflict in their implementation, for indeed the orientation of ESA is on *species diversity*, while the orientation of NFMA is on *ecosystem diversity*. The measures to protect a listed species might necessitate habitat modification, which might be at the expense of other species of the same ecosystem.

Nowhere has this tension between endangered species and forest management been more pronounced than in the Pacific Northwest, where litigation over the northern spotted owl severely impacted logging in old-growth forests. The Northwest Forest Plan (NWFP) was developed to address this problem. It was based on the report of the Forest Ecosystem Management Assessment Team, established at the direction of President William Clinton as a result of the Forest Summit held in Portland, Oregon on 2 April 1993 (Forest Ecosystem Management Assessment Team, 1993). The team delivered to the President 10 management options, with full assessments of each. The President selected option 9, and the report was issued in July 1993, together with the Draft Supplemental EIS. A Final Supplemental EIS was issued in February 1994, and the Record of Decision was jointly signed by the acting Secretary of Agriculture and the Secretary of the Interior on 13 April 1994 (US Department of Agriculture Forest Service and US Department of the Interior Bureau of Land Management, 1994a,b).

The NWFP was challenged in federal court, and Judge William Dwyer of the Western District of Washington upheld the plan against all claims on 21 December 1994.[15] The case was appealed to the Ninth Circuit Court of Appeals, which upheld the lower court on 10 April 1996.[16] The decision made clear that the Forest Service must comply with the biodiversity requirements of both ESA and NFMA *simultaneously*. The decision also had important implications for the regulatory provisions concerning species diversity and viability.

The litigation was unambiguous in finding that the viability regulation is but one part of the guidelines for multiple-use planning and management, not a separate, unique requirement.

[13] 36 Code of Federal Regulations (C.F.R.) 219.19 (9 November 2000).

[14] Loc. cit.

[15] *Seattle Audubon Society v. Lyons*, 871 F. Supp. 1291 (W.D. Wash., 1994).

[16] *Seattle Audubon Society v. Moseley*, 80 F.3d 1401 (9th Cir., 1996).

Another result was that the viability regulation is not a concrete standard, but a requirement to analyse and disclose the projected effects of planning alternatives on the future viability of species and then to use that information during selection of a multiple-use alternative. The litigation also suggested that species viability is a primary requirement of planning, and as such it is important, closely scrutinized and enforced by the courts. No further example is needed beyond the painful, agonizing history of northern spotted owl litigation. Finally, the litigation demonstrated that the techniques for projections of species viability are evolving and not static. For example, the use of minimum viable populations and management indicator species changed since the 1979 planning regulations were promulgated, and the NWFP added advances in ecosystem viability and risk assessment and management. Future applications of the species viability concept are likely to be different as a result of new scientific knowledge and analytical capabilities.

Indeed, the 2005 amendments to the NFMA planning regulations do not include a requirement to provide for viable populations of plant and animal species. Rather, they require 'ecological sustainability', which is intended to provide 'a framework to contribute to sustaining native ecological systems by providing ecological conditions to support diversity of native plant and animal communities'.[17] The regulation reads further: 'This will satisfy the statutory requirement to provide for diversity of plant and animal communities based on the suitability and capability of the specific land area in order to meet over multiple-use objectives.'[18] It also states that National Forest planning procedures must be consistent with both ecosystem diversity and species diversity, and it is under the latter that additional provisions are to be made in the plan for threatened and endangered species, species-of-concern and species-of-interest.

Although it is still too soon to tell, the decision of the Ninth Circuit Court of Appeals may have gone a long way to reducing the uncertainty associated with the cross-sectoral impacts of protection of threatened and endangered species in management of National Forests.

[17]36 C.F.R. 219.10 (2005).

[18]Loc. cit.

New Planning Regulations: a Fourth Attempt to Better Relate NFMA with NEPA

National Forest planning regulations and amendments were promulgated in 1979, 1982, 2000, and most recently, in 2005. The reason for this repeated effort is that virtually no one has been satisfied with National Forest planning. The planning process has been generally criticized as being too complex, too opaque, too lengthy and too costly – needing to be simplified, clarified, shortened and made less costly.

As indicated above, NFMA is clear that regulations for development and revision of land management plans shall specify 'procedures to insure that land management plans are prepared in accordance with the National Environmental Policy Act of 1969, including, but not limited to, direction on when and for what plans an environmental impact statement ... shall be prepared'. However, the perennial issue since 1979 has been that the procedures set forth in the planning regulations have been sufficiently burdensome that efficient, effective management of the National Forests is impeded – a situation that has been recognized not only within the Forest Service, but both inside and outside of government (Shands *et al.*, 1990; US Congress Office of Technology Assessment, 1992; US Department of Agriculture Forest Service, 1995, 2002; Floyd, 1999).

In response, the 2005 planning regulations promised a *paradigm shift* in National Forest land management planning. They did this first by proposing a categorical exemption from NEPA for plans, plan amendments and plan revisions on the basis that National Forest plans 'are strategic and aspirational in nature and generally will not include decisions with on-the-ground effects that can be meaningfully evaluated and that may be major'.[19] In other words, if National Forest plans, plan amendments and revisions do not have any major environmental effects, they should be and are excluded from NEPA documentation. Of course, projects and activities implementing forest plans and viewed as 'significantly affecting the quality of the human environment' must

[19]*National Forest System Land Management Planning*, 70 Fed. Reg. 1023 (5 January 2005).

continue to be accompanied by their own site-specific NEPA analysis.

Second, if land management decisions are categorically excluded from NEPA documentation, it is possible to react to changes much more swiftly and efficiently because a National Forest plan can be amended or revised at any time by the 'responsible official', defined as the official with the authority and responsibility to oversee the planning process, or the National Forest supervisor. Hence, 'adaptive management', the *nom du jour* for 'being responsive', is a major theme of the 2005 planning regulations. Plans will be changed as new information becomes available, and a systematic monitoring programme is to be established and maintained with regard to 'key social, economic and ecological performance measures relevant to the plan area'.

Third, to make forest planning more 'strategic and adaptive', procedural and technical details in the planning regulation are moved from the planning documents to the *Forest Service Manual* and the *Forest Service Handbooks*. The *Manual* and *Handbooks* are internal management guides for the Forest Service and do not have the legal importance of administrative regulations.

Fourth, the 2005 planning regulations require each planning unit to develop and implement an environmental management system (EMS) corresponding to the international consensus standard published by the International Standards Organization (ISO) as *ISO 14001: International Management Systems – Specification with Guidance for Use*.[20] Executive Order 13148, issued in April 2000, requires federal agencies to use an EMS approach for improving environmental performance. Earlier, the National Technology and Advancement Act of 1995 provided that federal agencies adopt, when possible, technical standards developed by consensus organizations such as ISO.

An EMS is a set of procedures and policies directed at *continual improvement* of environmental performance and compliance with environmental laws by an agency through measurement, evaluation, feedback and change. In contrast, the NEPA process is designed to apply environmental analysis and documentation in accordance with law to *individual decision points* with respect to a given agency programme. In other words, NEPA is applied to specific points over the duration of an agency action, while application of an EMS is designed to be continuous.

The addition of EMSs to forest planning is a major change in the 2005 National Forest planning regulations, as is the elimination of the requirement for NEPA documentation for plans, plan amendments and revisions – part of a concerted effort to streamline and improve National Forest planning.

Wilderness and Roadless Areas: the Lasting Struggle Between Development (Economic) and Preservation (Environmental) Interests in National Forest Management

There was a perception in the 1970s that planning would allow National Forest land use decisions to be fully aired and decided, and the Congress would not have to be involved. This perception has proven illusory, particularly with respect to management of roadless areas and wilderness designations.

For decades, the Forest Service attempted to get a handle on the 'roadless' issue, referring to lands without roads, which are therefore eligible for wilderness designation. Wilderness has been one of the major issues of National Forest management, because wilderness is the only land use designation that effectively blocks all development. Within wilderness areas, among other things, there can be no commercial enterprise, no permanent roads, structures or installations, no logging and no motorized uses.

Since only the Congress can declare an area of federal land as wilderness, public advocates for and against wilderness designations have not allowed forest land management planning to arbitrate the debate. Rather, wilderness designation has played out in the political arena. Often, the best that plans can accomplish is to highlight the wilderness issue for congressional resolution by identifying roadless lands eligible for designation.

The Forest Service twice attempted in the 1970s to inventory all roadless lands for their wilderness suitability, in ill-fated efforts referred to as the Roadless Area Review and Evaluation

[20] 36 C.F.R. 219.5 (2005).

(RARE). The ostensible purpose of the reviews (known as RARE I and RARE II) was to inventory all roadless lands, make judgements as to their suitability for wilderness, and recommend wilderness designation by the Congress of those suitable lands. Inherent in the process was the decision that lands not deemed suitable for wilderness would not have to be reconsidered later and would thus be available for multiple-use management. Lands considered and not deemed suitable for wilderness were deemed 'released' from further consideration. The RARE process was ultimately set aside by the courts due to insufficient site-specific analyses. This effectively denied the Forest Service a mechanism for one nationwide inventory of lands suitable for wilderness. Thereafter, every forest plan had to consider wilderness anew.

When the land management planning process could not solve a wilderness issue, the Congress often intervened. Usually, this intervention took the form of a state wilderness bill offered by a state's congressional delegation. Such bills usually designated one or more wilderness areas along with 'release' language, which would prevent reconsideration of non-designated roadless areas for wilderness for a prescribed period of time. State wilderness Acts were enacted for several states where a political consensus could be reached.

The roadless management debate shifted back to the administrative arena in rule-making conducted during the Clinton and Bush administrations. During the second term of the Clinton administration, the Forest Service undertook three regulatory initiatives designed to address roadless area management. One administrative rule prohibited new road construction in roadless areas.[21] The Forest Service also issued rules concerning road management, which would have reduced new road construction and decommissioned other roads deemed unnecessary.[22] Finally, the Forest Service amended its planning rules to provide enhanced protective management for roadless areas.[23]

In 2001, the incoming Bush administration began to modify the previous Clinton administration's roadless initiatives. The roadless area rule was proposed to be modified by the Bush administration to allow road-building pursuant to a process in which a state Governor could file a petition with the Secretary of Agriculture to establish management requirements for roadless areas in that state.[24] Similarly, the restrictions on road construction were largely lifted.[25] Finally, the Bush administration proposed modifying land management planning procedures with respect to roadless areas to emphasize local decision making in lieu of national standards.

It should not be surprising that the legality of the roadless rule was challenged in several lawsuits. It is equally unsurprising that the rule was upheld by one court[26] and struck down by another.[27] This apparent impasse among the various branches of government on the roadless issue illustrates the level of controversy and lack of consensus that marks many contemporary forest management issues.

Often, congressional intervention by enactment of a statute is the only way to cut the Gordian knot of conflicting administrative and judicial responses to land management conflicts. The best example is arguably the Alaska National Interest Lands Conservation Act of 1980 (ANILCA), which culminated years of debate over the future management of federal lands in Alaska. ANILCA designated millions of hectares in Alaska as new National Parks, National Wildlife Refuges and National Monuments, as well as established special management prescriptions for the National Forests.

It is probably inevitable that forest management policies in the USA will be increasingly dictated by the Congress. In the last three decades, most legislation affecting the public lands has been to make land allocation decisions that cannot be

[21]*Special Areas; Roadless Area Conservation*, Proposed at 65 Fed. Reg. 30,276 (10 May 2000); Final at 66 Fed. Reg. 3244 (12 January 2001).

[22]66 Fed. Reg. 3206 (12 January 2001).

[23]*National Forest System Land and Resource Management Planning*, 65 Fed. Reg. 67,514 (9 November 2000).

[24]*Special Areas; State Petitions for Inventoried Roadless Area Management*, Proposed 69 Fed. Reg. 42,636 (16 July 2004).

[25]*Forest Transportation System Analysis; Revisions to Road Management Policy*, 68 Fed. Reg. 69,986 (16 December 2003).

[26]*Kootenai Tribe v. Veneman*, 313 F.3d 1094 (9th Cir., 2002).

[27]*Wyoming v. U.S.D.A.*, 277 F.Supp.2d 1197 (D. Wyo. 2003).

effectively made administratively. The Congress is essentially zoning areas for development and preservation. This trend is likely to continue as land managers operate in an increasingly narrow range of discretion bounded by substantive and procedural requirements of law enforced by interest groups and advocates who have access to the courts.

The Threat of Catastrophic Wildfires to Human Health and Safety and Private Property

Perhaps the greatest forest management change of the last three decades has been in the area of fire management, in the broader context of what is generally referred to as 'forest health'. Forest health encompasses many concepts, including reducing susceptibility to fire, insects and disease, and maintaining a balanced ecological system. For nearly a century, the policy of the Forest Service was to eliminate fire from the forest ecosystems through aggressive suppression. In large measure, the policy was successful at dramatically reducing over decades the amount of hectares burned, but also with unintended consequences. Subsequent science has shown that periodic fires helped maintain a healthy forest environment by eliminating underbrush and adding nutrients to the soil. Vigorous fire suppression caused forest fuel loads to build gradually over time, ultimately resulting in more severe and catastrophic fires. At the same time, more lands are at risk of wildfires as population pressures have resulted in increasing numbers of homes and communities being built within and adjacent to forest areas, the 'wildland–urban interface'. Thus, the challenge is how to reduce the threat of wildfire to human health and safety in the wildland–urban interface, as well as the risk of economic loss from the destruction of private property, while at the same time maintaining healthy, sustainable forests.

A simple answer to promoting forest health and reducing catastrophic fires is to reduce fuel loads, either through prescribed burning or mechanical thinning. However, accomplishing fuel load reduction on a national scale is anything but simple and is highly controversial among environmental advocates, homeowners, communities, the forest products industry and the federal land managers.

In December 2003, the Congress enacted the Healthy Forests Restoration Act, which among other things authorized federal land managers to protect wildland–urban interface areas through hazardous fuel reduction projects. The Congress mandated a streamlined environmental analysis process to identify affected areas, mainly wildland–urban interface areas, municipal watershed lands and endangered species habitats. A truncated environmental review process is authorized with limited provisions for public comment and administrative appeal. Critics point out that inadequate funds have been appropriated to do projects requiring reliance on 'stewardship contracts', in which the federal land manager contracts with logging companies to conduct thinning. Under such contracts, removal of some merchantable timber is authorized in order to compensate the logging company. Time will tell whether the Healthy Forests Restoration Act has the effect of diminishing catastrophic fires and the threats they cause to human health and safety and private property, or whether it is primarily a mechanism for harvesting trees for economic gain, as some critics charge.

Summary of 2005 National Forest Policy

A summary of National Forest policy as provided for in RPA/NFMA and its 2005 implementing regulations is given in the text box below.

Trends and Prognostications

The last three decades have witnessed significant changes in policies governing National Forest management driven by a variety of forces, such as judicial decisions, administrative rule changes and catastrophic wildfires. Clearly, management of National Forests is no longer a job left to the discretion of professionals, but is a matter of increasing political and social consequence. Numerous laws have been enacted to attempt to reconcile the conflicting demands for forest resources with environmental protection, resulting in more mandated procedures and substantive legal requirements.

There is an unintended consequence of the legal evolution from discretion to prescription in

Policy Summary of RPA/NFMA, *c.*2005

RPA/NFMA and the 2005 implementing regulations set the following policies for National Forest planning and management. Changes from 1979 are shown in italics.

- A hierarchical, integrated, iterative strategic planning process will be used by the Forest Service in administering the National Forests.
- National Forests are to be managed according to plans developed and revised in accordance with regulations promulgated by the Secretary of Agriculture – with the advice and counsel of a Committee of Scientists – and procedures established under the National Environmental Policy Act of 1969, *except that a categorical exclusion from NEPA documentation is proposed for National Forest plans, plan amendments and revisions.*
- The public shall have the opportunity to participate in the development, review and revision of land management plans.
- *Systematic monitoring programmes shall be established for each planning unit with regard to 'key social, economic, and ecological performance measures' and the information they provide will be implemented through adaptive management.*
- *Each planning unit shall develop and implement an environmental management system (EMS).*
- Such plans will provide for:
  - Multiple use and a sustained yield of renewable surface resources, including wilderness, in accordance with the Multiple Use–Sustained Yield Act of 1960;
  - Fish and wildlife habitat managed in each National Forest *on the basis of the suitability and capability of the specific land area (ecosystem diversity) in order to meet overall multiple-use objectives and, simultaneously, species diversity, to provide appropriate conditions for specific threatened and endangered species, species-of-concern and species-of-interest;*
  - *Protection from threats of wildfire through reductions in hazardous fuels in wildland–urban interfaces, lands that are part of municipal watersheds, and habitats of threatened and endangered species, and from invasive species, including foreign diseases and their causative agents.*
  - *Management of roadless areas, where present, but their management will likely vary as various interests attempt to manipulate the substantive and procedural requirements of applicable law.*
- Plans will ensure that:
  - Timber will be harvested generally only where timber production is viable on a sustainable basis;
  - Protection is provided for watercourses, lakes and wetlands;
  - Even-age management regimes are used only where it is appropriate, and clear-cutting is used only when it is optimal;
  - Maximum size limits are established and applied for one-harvest operations.
- With a few possible exceptions, timber sale levels from each National Forest shall be limited to a quantity equal to or less than a quantity that can be removed from such forest annually in perpetuity on a sustained-yield basis.

public land management. The more rules and procedures that apply to any given management decision, the more exposure there is to appeals and litigation. Given the relative ease by which any disgruntled person or organization can get standing to challenge a management decision in court, the traditional management regimes are likely to get bogged down. Resort is then often made to the political process for resolution. Given the foibles of politics, it is no wonder the federal land manager is often thwarted and frustrated in attempting to apply basic science and professional judgement to basic land management decisions.

Within these legal and political constraints, the Forest Service continues to identify and address management issues. The current Chief of the Forest Service, Dale Bosworth, recently identified 'four threats' to the National Forests: unmanaged recreation, loss of open space, invasive species, wildfire and hazardous fuel reduction.

Unmanaged recreation particularly involves off-highway vehicles which, if unregulated, can cause tremendous damage to soils and vegetation. In November 2005, the Forest Service issued final rules regulating off-highway uses on the National Forests by requiring that roads, trails and areas be designated for motor vehicle use and prohibiting such uses elsewhere.[28]

Concerning open space, the Forest Service estimates that *c*.283,000ha of forests were annually lost to development between 1982 and 1997, and the trend continues at a rate of *c*.1600ha lost per day (Stein *et al.*, 2005). The Forest Service is promoting programmes to acquire conservation easements on land threatened with conversion to non-forest uses, but such programmes are largely dependent on the availability of appropriated funds.

As noted above, wildfire and hazardous fuel reduction is a major initiative of the Bush administration using authorities of the Healthy Forest Restoration Act. The concept of healthy forests also encompasses threats from insects and diseases. Invasive species pose a particularly insidious threat and include the southern pine beetle, western bark beetles, gypsy moth, emerald ash borer and others. In Fiscal Year 2005, over 352,000ha were treated to control such pests. As part of an overall healthy forest initiative, the Forest Service proposed a collaborative approach to invasive species including prevention, response, control and restoration effects.

In some ways, the policy changes in National Forest management over the last three decades are appropriate precursors to the Forest Service addressing these four threats and others. By identifying threats to the forests and marshalling resources to meet those threats, the agency is positioning itself away from traditional multiple uses and toward resource protection. While the debate will continue over issues, such as the appropriate methods of fuel reduction to meet healthy forest objectives, it is apparent that policy changes in the next decades will focus on maintaining diverse and healthy forest environments threatened by cross-sectoral impacts, such as economic development, global warming, fire and invasive species. The cynic may well predict that such policy changes and initiatives will face the usual gauntlet of environmental analyses, administrative appeals, litigation and congressional oversight. The optimist may hope that focusing on threats may be something everyone can agree upon.

[28]70 Fed. Reg. 68,264 (9 November 2005).

References

Floyd, D.W. (ed.) (1999) *Forest of Discord: Options for Governing Our National Forests and Federal Public Lands*. Society of American Foresters, Bethesda, Maryland.

Forest Ecosystem Management Assessment Team (FEMAT) (1993) *Forest Ecosystem Management: An Ecological, Economic, and Social Assessment*. US Department of Agriculture, US Department of the Interior (and other agencies), Portland, Oregon.

Shands, W.E., Sample, V.A. and Le Master, D.C. (1990) National forest planning: searching for a common vision. In: US Department of Agriculture Forest Service (ed.) *Critique of Land Management Planning*, Vol. 2. USDA Forest Service, 1990, Washington, DC, pp. 1–91.

Stein, S.M., McRoberts, R.E., Alig, R.J., Nelson, M.D., Theobald, D.M., Eley, M., Dechter, M. and Carr, M. (2005) *Forests on the Edge: Housing Development on America's Private Forests*. Gen. Tech. Rep. PNW-GTR-636. US Department of Agriculture, Forest Service, Pacific Northwest Research Station, Portland, Oregon.

US Congress Office of Technology Assessment (1992) *Forest Service Planning: Accommodating Uses, Producing Outputs, and Sustaining Ecosystems*. OTA-F-505. Government Printing Office, Washington, DC.

US Department of Agriculture Forest Service (1995) *Review of the Forest Service Legal and Regulatory Framework. Draft Thomas Task Force Report*. USDA Forest Service, Washington, DC.

US Department of Agriculture Forest Service (2002) *The Process Predicament: How Statutory, Regulatory, and Administrative Factors Affect National Forest Management*. USDA Forest Service, Washington, DC.

US Department of Agriculture Forest Service and US Department of the Interior Bureau of Land Management (1994a) *Final Supplemental Environmental Impact Statement on Management of Habitat for Late-Successional and Old-Growth Forest Related Species within the Range of the Northern Spotted Owl*. USDA Forest Service and US Department of the Interior Bureau of Land Management, Portland, Oregon.

US Department of Agriculture Forest Service and US Department of the Interior Bureau of Land Management (1994b) *Record of Decision for Amendments to Forest Service and Bureau of Land Management Planning Documents within the Range of the Northern Spotted Owl.* USDA Forest Service and US Department of the Interior Bureau of Land Management, Portland, Oregon.

Laws Referenced

Alaska National Interest Lands Conservation Act (Act of 2 December 1980; 94 Stat. 2371).
Endangered Species Act (Act of 28 December 1973; 87 Stat. 884).
Federal Land Policy and Management Act (Act of 21 October 1976; 90 Stat. 2743).
Forest and Rangeland Renewable Resources Planning Act of 1974 (Act of 17 August 1974; 88 Stat. 476).
Forest Reserve Act of 1891 (Act of 3 March 1891; 26 State. 1103), repealed in 1976.
Healthy Forests Restoration Act (Act of 3 December 2003; 117 Stat. 1887).
Multiple Use–Sustained Yield Act (Act of 12 June 1960; 74 Stat. 215).
National Environmental Policy Act (Act of 1 January 1970; 83 Stat. 852).
National Forest Management Act (Act of 22 October 1976; 90 Stat. 2949).
National Technology and Advancement Act of 1995 (Act of 7 March 1996; 110 Stat. 775).
Organic Administration Act (Act of 4 June 1897; (30 Stat. 11).
Wild and Scenic Rivers Act (Act of 2 October 1968; 82 Stat. 906).
Wilderness Act of 1964 (Act of 3 September 1964; 78 Stat. 890).

Index

Africa 82–83, 84, 85–87
Agriculture
cash crops 8, 97, 140, 142, 156
deforestation 1, 2–4, 8–9, 50, 110, 140, 231
land acquisition 142
tenure security 9–10
India 134–139
intensification 11, 93
pesticides 157
policy 46–47, 154, 156
slash and burn 6, 142, 153
Agroforestry 82–83, 84, 85
Argentina
deforestation 228
forest management 230
government policy 228
watershed management 227–229
Australia 38–39
Austria
area 175, 176
forest management
policy 175–176
stakeholders 175–176, 180
wood-product 177

Biodiversity 178–179, 204–205, 207, 237, 250, 251–253
Borlaug hypothesis 9

Cameroon
deforestation 5–6, 77
agriculture 8
forest management 41–42
model forests 77–78
Canada
deforestation 204
forest management
biodiversity 204, 205, 207
model forests 71–72
stakeholders 205, 206, 208
sustainable 71
government
land management 206–209
policy 205–206, 209
grizzly bear 72, 77
resource extraction 204, 205, 206–208
wood-product 24–25, 204
Central America 40–41
Chile
deforestation 227–228
forest 230
model forests 78
protection 235
sustainable management 233–234, 235
government
environmental law 231–233, 235
policy 228, 236
watershed 227–229, 230
China
environmental accounts 126, 127, 132
components 128–129, 130
timber production 127–128
forest
area 126, 127
sustainable management 131–132

government finance 131–132
population livelihood 129–131
wood-product
secondary products 27
trade 24–26, 27, 28–29, 31
Climate change 37, 52, 168
deforestation 54
forest management 52, 55–56
carbon offset 52, 53, 119–120
carbon sinks 53, 168, 169
Congo 41–42
Convention on Biological Diversity (CBD) 37–38
Cross-sectoral 180
agroforestry 82–83
ecological demands 16–17, 121
modelling 196
policy 14, 47–48, 110, 174–175, 183–184
decentralization 16, 18–19

Deforestation
Argentina 228
Cameroon 6, 77
Canada 204
causes
agriculture 1, 2–4, 8–9, 50, 117, 140, 142, 153, 227–228, 231
policy 1, 45
resource exploitation 4, 6–7, 21, 117, 204
road building 2, 6–7, 9, 11–12, 141, 142–143, 145, 243
timber production 4, 8–9, 68, 148
Chile 227–228
clear-cutting 249
climate change 52
Ecuador 4, 6–7
Ethiopia 89–91
Gabon 4
income 10
Indonesia 140, 141–143, 145
Mexico 237, 243
Nepal 148
Papua New Guinea 7
population 10, 89, 117, 212
rate 2–7, 89, 108, 154, 204, 220, 222, 231, 258
Thailand 153, 154
Venezuela 4
Desertification *see* Land degradation
Dutch disease 2, 4, 7

Ecuador 4, 6–7
Energy 31–32
Environmental accounts 64–67, 69, 126, 127, 132, 200–201
carbon accounts 67, 201
components 66, 128–129, 130
timber production 67, 127–128
Environmental law 231–232, 235, 253–254
Equality
policy 123–124
women's role 95–99, 100, 101–102
Ethiopia
deforestation 89, 91
forest management 90–91, 92
population growth 89
Europe
forest management 166
policy 163–164, 165, 166–167, 170–171
stakeholders 165, 167–168, 173
sustainability 165, 169
government
finance 167–168
policy 167–168, 172–173
trends 163, 171
wood-product
trade 27, 30, 31, 163, 171–172

Forest
area 4, 73, 89, 110, 117, 126, 147, 166–167, 175, 184, 190, 215, 248
growth 5
management 37, 47, 59
Argentina 230
Australia 38–39
Austria 175–177, 180
Cameroon 41–42, 77–78
Canada 71–72, 204, 208
Central America 40–41
Chile 78, 233–235
China 126, 127, 131–132
community forests 149–150
Congo 41–42
decentralization 15–16, 17, 18, 19–20, 37, 91–93, 105, 107, 111, 120, 123–124, 119, 238–239
ecosystem approach 36–38, 39, 41–42
Ethiopia 82, 90–91, 92
Europe 163–164, 166–167, 170–171
Ghana 95, 99–100, 101–102
India 17–20, 40
Mexico 237, 238–239, 240, 243
model forests *see* Model forests
Mozambique 104–105, 106–107
Netherlands 175–176, 178–179
Nepal 147–151
Nigeria 111, 114–115
Philippines 78–79
Romania 185–186, 187
Russia 42–43, 79–80
sustainable 32–33, 36–38, 61, 64, 122, 131, 165, 169, 233–234, 235

Forest (*continued*)
Sweden 80
Tanzania 119, 120, 123–124
Thailand 158–159
training 104–105, 176, 188, 242
USA 40, 195, 199, 251, 256–258
Rio summit 36, 58
subsidies 31, 49

Gabon 4–5
Ghana
forest management 95, 99–100
policy 95
women's role 95–99, 100, 101–102
landownership 98
tree management 99
Government
finance 61, 68, 92, 112, 124, 167–168, 178, 185, 186
China 131–132
Mexico 238, 240–242
industrial development 30
land management 206–209
policy 45–47, 48–50, 60, 85–87, 153, 242
Argentina 228
agroforestry 83, 84, 85–87
Chile 228, 236
decentralization 19–21, 91–93, 105, 107, 120
Europe 167–168, 172–173
environmental law 231–233, 235, 236
Mexico 237–238, 240, 243, 245, 246–247
Mozambique 103, 105, 107, 108
Netherlands 178, 179–180
Nepal 147, 148–149
Nigeria 111, 112, 113–114, 115–116
Romania 183, 184, 188
sustainable forest management 32–33, 89, 95, 122–123, 131, 240, 243, 244, 249–250
Tanzania 117, 118, 121, 122, 123
Thailand 154, 155, 159, 160
Turkey 193
USA 196, 197, 198, 199–200, 201, 212, 213, 248–250, 251–253

Illegal logging 101, 242
India
management 17, 18, 19, 20, 40
population 133–134, 138–139
social accounting matrix (SAM) 133–138
trade 28–29
Indonesia
deforestation
rubber plantations 140, 142
road building 141–143, 145
management 16, 17–18
population 142, 144, 145–146
Invasive species 85, 258

Kyoto
carbon trading 53–54
clean development mechanism 54, 233
energy production 31–32
forest management 53

Land degradation 58
global mechanism 60, 61, 62
landscape restoration 59, 60, 61
policy 59, 112–113

Mexico
deforestation 237, 243
forest management
biodiversity 237
decentralization 238–239
policy 239, 240, 243
stakeholders 239, 241–242, 244, 246
training 242
government
finance 238, 240–241, 242
policy 237–238, 240, 243, 245, 246–247
Model forests 71–72, 77–80
attributes 72, 73–76
Mozambique
community forests 103–104
livelihoods 106
management 104–105, 106–107
stakeholders 104, 106, 108
decentralization 105, 107
state management 103, 105
policy 105, 107, 108

Netherlands
area 175, 177–178
reforestation 178
forest management
biodiversity 178–179
policy 175–176
stakeholders 175–176, 178, 180
government
finance 178
policy 178, 179–180
wood-product 178
Nepal
agriculture 147
forest area 147
government policy 147, 148–149
management
community forests 149–151

stakeholders 148–149
state ownership 147–148
population 147, 150
Nigeria
forest area 110
government
finance 112
policy 111, 112, 113–114, 115–116
management 114–115
decentralization 111
stakeholders 111, 113, 115
population livelihood 110

Papua New Guinea 7
Philippines 78–79
Phytosanitary measures 33
Pollution 48
Population
education 235
growth
deforestation 10, 89, 117
rate 212, 215
livelihoods 14, 53, 58, 90, 97–99, 133, 138, 144, 147, 243, 246
medicinal plants 79, 84

Resource exploitation 2–7, 21, 117, 204
Romania
forest
area 184
privatization 185–186
stakeholders 185, 188–189
sustainable management 187
timber production 185, 186
government
finance 185, 186
infrastructure 187
policy 183, 184, 188
wood-product 184
Russia 42–43, 79–80

Slash and burn 6, 142, 153
Social accounting matrix (SAM)
agriculture 134–138
components 133–134
Sweden 80
Swidden production system 153, 155
deforestation 1, 142

Tanzania
forest 117, 118
government 124
policy 117, 118, 121, 122, 123
management
decentralization 119, 120, 123–124
stakeholders 120
population 122, 124
Thailand
agriculture 156–157, 158–159, 160
pesticides 157
conflict 157
forest
area 154
communal ownership 155–156, 159–160
deforestation 153, 154
management 158–159
government policy 154, 155, 159, 160
opium 154, 157
population 153, 156, 158
tourism 155
Timber production 4, 14–15, 17–18, 148, 185–186, 233–234, 248
Trade 24–26, 27–30, 31, 111, 163–164, 171–172, 191–192, 212
control measures 31
price trends 28–30, 215, 216–217, 220
Turkey
forest 190
government 190–191, 193
wood-products 191–192

UN
conventions 59, 233
FAO 103, 113
forum on forests (UNFF) 36–37
Urbanization 2, 39–40, 216, 258
deforestation 4, 5, 212, 222, 223–224
USA
environmental accounts 200–201
carbon credits 201
forest management 40, 195, 199, 251, 256–258
policy 248, 249, 255, 256
stakeholders 200
government
biodiversity 250, 251–253
infrastructure 254–255
policy 196, 197, 198, 199–200, 201, 212, 213, 248–250, 251–253
climate change 197
land use 201, 220–222, 248, 253–254, 255–256
modelling 195–198, 213–214, 216, 220–212, 224
trends 198
population
change 212
urbanization 212, 216, 222, 223–224, 258
wood-product
secondary products 27
trade 24–26, 27, 30, 212

Venezuela 4, 5

Watershed
- government
  - policy 228
- management
  - South America 226–227, 228–229, 230
  - training 229
- pollution 226–227

Wood products 177, 178, 184, 204
- industrial roundwood 24–25, 163, 212
- pulp and paper 25–26, 233
- sawnwood 24–25, 233
- secondary products 27–28
  - processing industries 30
- trade 24–26, 27, 31, 111, 163–164, 171–172, 191–192, 212
  - price trends 28–30, 215, 216–217, 220
- wood-based panels 25